Of Prairie, Woods, & Water

Of Prairie, Woods, & Water

TWO CENTURIES OF CHICAGO NATURE WRITING

EDITED BY *Joel Greenberg*

The University of Chicago Press ⁂ CHICAGO AND LONDON

Joel Greenberg is a naturalist, writer, and environmental consultant. He is the author of *A Natural History of the Chicago Region*, published by the Press, and coauthor of *A Birder's Guide to the Chicago Region*.

This book is supported in part by a generous grant from the Max McGraw Wildlife Foundation.

The University of Chicago Press, Chicago 60637
The University of Chicago Press, Ltd., London
© 2008 by The University of Chicago
All rights reserved. Published 2008
Printed in the United States of America

17 16 15 14 13 12 11 10 09 08   1 2 3 4 5

ISBN-13: 978-0-226-30660-5 (cloth)
ISBN-10: 0-226-30660-7 (cloth)
ISBN-13: 978-0-226-30661-2 (paper)
ISBN-10: 0-226-30661-5 (paper)

Library of Congress Cataloging-in-Publication Data
Of prairie, woods, and water: two centuries of Chicago nature writing / edited by Joel Greenberg.
    p. cm.
Includes bibliographical references and index.
ISBN-13: 978-0-226-30660-5 (cloth: alk. paper)
ISBN-10: 0-226-30660-7 (cloth: alk. paper)
ISBN-13: 978-0-226-30661-2 (pbk.: alk. paper)
ISBN-10: 0-226-30661-5 (pbk.: alk. paper)
1. Natural history—Illinois—Chicago Region. 2. Chicago Region (Ill.)—Description and travel. I. Greenberg, Joel (Joel R.)
QH105.I3044   2008
508.773' 11—dc22   2007033451

To David Beilenberg (1937–2007),
WHOSE GOOD WORK AS A SCHOLAR, TEACHER, ARCHITECT, AND PLANNER DESERVES
TO BE REMEMBERED

and

To Cynthia S. Kerchmar,
WHOSE HARD WORK MAKES IT MUCH EASIER FOR ME TO PURSUE MY NON-LUCRATIVE
PASSIONS

Contents

Preface

〜〜〜〜〜〜〜〜〜〜〜〜〜〜〜〜〜〜〜〜〜〜〜〜〜〜〜〜〜〜〜〜〜〜〜〜

᭗ ᭗ ᭗ One of the great joys in researching my last book, A *Natural History of the Chicago Region*, was coming upon early descriptions of this region that strain the imagination of today's resident. In those earlier times, it was possible to see clouds of passenger pigeons dense enough to blot out the sun for hours, paddlefish thrashing in the shallow waters of the Des Plaines River and its tributaries, and mountain lions stalking careless elk calves as they browsed on the edge of prairie groves. Some of the writing distinguished itself by its literary merit or by its glimpses into the personality of the author. And the more hidden the source, the more exciting the discovery.

Reviewers of early drafts of the book admonished me to quote less and paraphrase more. But there seemed to be something almost sacrilegious in "mutilating" the old texts. Still, my book was not meant to read like a legal brief, dominated by long quotations from established authority. Hopefully, in the end, the right balance was struck, but there remained so much wonderful material deserving to be shared with modern readers that I thought an anthology would be a worthwhile project. My model was Paul Angle's *Prairie State: Impressions of Illinois, 1673–1967* (University of Chicago Press, 1968). While some of Angle's early selections were of landscape descriptions, most dealt with other aspects of life in Illinois. My selections, on the other hand, would be restricted to natural history subjects, and I would emphasize the Chicago region, although not stay exclusively within the nineteen counties.

〜〜〜

Both Christie Henry and Jen Howard (my principal editors at the University of Chicago Press, who are among the most wonderful people I know) were enthusiastic about the idea and encouraged me to proceed. I dawdled longer than I care to think about, but an invitation by Craig Sautter to read some primary sources as a program for the Society of Midland Authors spurred me into action. Going through the boxes that make up my personal "archives" reintroduced me to many writings that were never used in my previous book and that I had forgotten.

I then compiled twenty-one pieces and submitted them to two friends who are unusually qualified to evaluate the merit of this kind of material. Ben Goluboff is both an associate professor of English at Lake Forest College (his doctoral work was on nineteenth-century travel writing) and an avid birder and botanist. Barbara Plampin also holds a doctorate in English and taught writing at the Illinois Institute of Technology until her retirement. She is now one of the area's premier field botanists, devoting most of her attention to the Indiana Dunes. I asked them to rate on a one-to-three scale whether a given article warranted reprinting (one being a definite yes and a three being a definite no). They agreed that eight were ones and four were threes. (To my chagrin, they both panned two articles that were among my favorites—these have been omitted.) They disagreed on the remainder, five of which I thought should be added to the "keep" list. The assistance these two generous friends rendered at this critical step in the project is appreciated beyond what I can express. Without their help, this book might well have died abornin'.

Next, I sent the thirteen plus four others to Christie for her assessment. She liked the sample but emphasized that I needed to enlarge the assortment. She thought twenty-five might be a minimum. Well, now more than a year later, this collection of writings has grown to almost a hundred. It is difficult even now to quit collecting, for I know that undetected gems still exist. Yet, at the end of the day, my hardest task was cutting the material that I had assembled to a size the Press quite reasonably thought would be affordable to potential readers. It pained me to cut Stephen Forbes's "The Lake as a Microcosm," Willard Clute's *Swamp and Dune*, Increase Lapham's "The Forest Trees of Wisconsin," and Jean Shepherd's hilarious "Harry Gertz and the 47 Crappies."

As the project matured, I decided that the material must either have been written before 1960 or, if not, be a reminiscence of events earlier than 1960. (My decision on this was based at least in part by my desire for the writings to predate my own memories.) I also wanted the pieces to be concentrated within the nineteen-county Chicago region (either through the author or subject), although I have not been as scrupulous in this as I was in my 2002 book. But the selections that do fall beyond those formal boundaries all possess both of these characteristics: their setting is within a few counties of the "border," and they deal with processes or organisms that also occurred within the nineteen counties.

Beyond the time and place thresholds, the criterion for inclusion really comes down to whether the piece is still of interest to a modern reader. This endurance usually stems from one or more of these elements: (1) quality of the writing; (2) ability to re-create a specific time and place; (3) nature of the story; (4) the personality of the writer is reflected in the words; and (5) the writers themselves deserve to be remembered due to their deeds. Much material was rejected because it was just too straightforward, generic, and/or lacked the sparkle I was looking for.

Because anthologies need some structure, I have organized the material into five sections: (1) "Landscape," emphasizing descriptions of scenery, sometimes including animals (at least four of these are by scientific-minded writers); (2) "Botany," descriptions of plants that were written by scientific-minded authors; (3) "Land Animals"; (4) "Water World," emphasizing organisms and/or activities related to water; and (5) "Mindscapes," emphasizing how people interacted with nature and/or the attitudes they held about nature.

I readily admit that these categories—to quote Thomas J. Lyon's disclaimer in his book on nature writing *This Incomparable Land* (Milkweed Editions, 2001)—"tend to intergrade" and are not "neat and orderly." I recognize that someone else might well have arranged things differently, and I ask readers to be forgiving if they think a particular piece appears in the wrong section.

The authors whose works make up this anthology range from those who are well-known (Jens Jensen, Nathan Leopold, and Theodore Dreiser, for example) to those who are anonymous, being identified by a pseudonym or mere initial. Since one motivation of this project is to keep alive the memory of local naturalists who toiled at a time when the guild was small, I have made an attempt to include biographical information where I have been able to find it. The length of each profile reflects in large measure the quality and quantity of the accounts that were available to me. I ask the reader for help: if you have knowledge about any of the writers included here whose biographies are missing or skimpy, please let me know.

As most nonfiction authors do, I received a great deal of aid and cooperation from many people. To all of them, I am greatly appreciative. These include the experts who are mentioned in the text, as well as those individuals and organizations identified elsewhere who granted permission to reprint the selected authors.

Several people and institutions deserve special thanks here for all the help they provided me:

Charles Potter and the Max McGraw Wildlife Foundation furnished financial assistance that enabled this project to be completed. Both their extreme generosity and belief in the value of my writings have touched me deeply, and I will be forever grateful.

Michael Steiber, Rita Hassert, and Nancy Faller facilitated my use of the Morton Arboretum's excellent library. I benefited greatly from their time, skill, diligence, and good cheer.

Professor Barbara Stedman of Ball State University and her students developed and maintain the excellent website *Our Land, Our Literature*, which introduced me to a number of Indiana writers whose work I have included. Subsequent communications with Barb have also proved very helpful and enjoyable.

Don Gorney, a superb birder and all-around naturalist whose ability to imitate the voices of other Indiana birders makes him the perfect companion on a slow birding day, offered to be my "Indianapolis agent." He spent many hours perusing local libraries for material unavailable here.

Jonathan Wuepper, an independent scholar whose work on the historical natural history of Berrien County has been an inspiration to me, very generously provided all the accounts of Michigan that appear in this anthology with one exception.

I also want to thank the following for their various contributions to this project: Dick Ammann, Doug Anderson, Jim Ballou, Jim and Pat Bland, Ken Brock, Janet Burden, Lee Casabeare, the Chicago History Museum, Larry Clark, Paul Clyne, Ed Collins, Erin DeWitt, John Elliott, Sandy Fabyan, the Field Museum Library, William Fitzhugh, Franklin College Library, Gary (Indiana) Public Library, Libby Hill, R. David Johnson, Steve Johnson, Ted Karamansky, Ken Klick, Dan McGrath, Steve McShane, Betsy Mendelsohn, Pat Patton, Steve Pescetelli, Margery Peters (and other librarians at College of DuPage), Mary Beth Prondzinski, Alan Resetar, Robert Rung, Ken Schoon, Steve Swanson, Janice and Paul Sweet, Martha Twigg, Chuck Westcott, Phil Willink, and the Wisconsin Historical Society.

For moral and other support my deepest appreciation goes to Renee and David Baade; Miriam, Robin, and Travis Greenberg; Lynn Hepler; David B. Jonson; Carol McArdle; Andy Sigler; Mike, Sam, and Ben Solomon; Celeste A. Thaine; and Tim Wallace.

And finally, to Cindy Kerchmar, who even more than before provides support on so many levels that I will be eternally in her debt.

The Chicago Region—*Introduction*

~~~~~~

⌒⌒ ⌒⌒ ⌒⌒ In understanding the rich diversity of life that inhabited the Chicago region, one must look at a number of factors. Location is a critical one. In general terms, being in the middle of the continent means the ranges of many organisms overlap here, including those whose presence are as outliers from larger populations situated elsewhere. And there are even a few species that occur here and nowhere else in the world, such as the plants *Iliamna remota* (Kankakee mallow) and *Thismia americana* (now almost certainly extinct), and the leafhopper *Paraphlepsius lupalus*.

The Chicago region is far enough north to have received a massive flow of ice that was part of the Wisconsinan Glacial Episode. The ice remained until around 13,800 years ago, and over the course of its stay, punctuated by expansions and contractions, it created the physiography that marks the region. There are plains where the ice scoured, and there are uplands where the glacier halted, depositing tremendous quantities of earth. Three of these morainal uplands parallel Lake Michigan in concentric U's: the Lake Border Upland, Valparaiso Upland, and Outer Upland. Examples of these include the north-shore bluffs between Wilmette and Waukegan and the rugged Palos area.

As the glacier receded, it left in its wake water in an array of configurations. Lake Michigan came into being as meltwater filled the giant trough gouged out by the ice. As the ice retreated, the lake also shrank to expose a low plain on its southern end that featured six shallow lakes; from west to east they were Lake Calumet, Hyde Lake, Wolf Lake (bisected by the Illinois-Indiana line), Lake George, Bear or Berry Lake (now gone), and Long Lake (Lake County, Indiana). Hundreds of other lakes formed among the ridges of the Valparaiso Moraine, most particularly in the region's northwestern counties. These include the lakes Fox, Geneva (the largest), Long (Lake County,

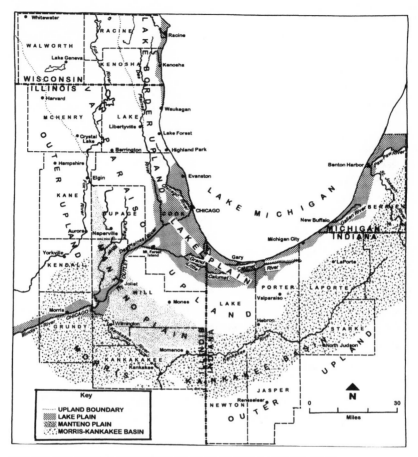

NATURAL DIVISIONS OF THE REGION OF CHICAGO BY F. N. FRYXELL (1927)

Illinois—the subject of Bill Beecher's article on egrets), and both Cedars (one in Lake County, Illinois, and one in Lake County, Indiana).

The Valparaiso Moraine was the spine that defined one of the midcontinent's principal divides. The waters that fell on the Lake Michigan side fed that lake through the Pike, Root, Chicago, Calumet, and St. Joseph rivers. Most of the water on the other flank would enter the Fox, Des Plaines, and Kankakee rivers. Where the last two merged, the Illinois River arose to become the third greatest tributary of the Mississippi. And where conditions allowed—on the low plains or along some rivers, for example—extensive wetlands were common.

Location, of course, also determines climate, which refers to meteorologic phenomena enacted over long periods. (In the short term, climate expresses

itself in events known as weather.) So with the ice age, a drop in temperatures and increased precipitation triggered it and increasing temperatures
ended it.

As the earth was once again free of its icy veil, forests of pine and spruce
became the dominant vegetation. But as the warming trend continued, deciduous trees took hold until they in turn were replaced by grasses from the
west. This became known as the prairie peninsula, a wedge of grassland
penetrating into the eastern forest. But a reversal occurred, with cooler and
moister weather that promoted the growth of trees. These ongoing climatic
shifts have spawned a contest between grass and trees that produced a third
kind of landscape. In this zone of overlap, savannas and shrublands represent
a transitional landscape where scattered trees and shrubs thrive with prairie
vegetation.

How the plant life distributed itself under a climate that was hospitable to
both trees and grass was not an accident. The critical factor was fire, set both
by lightning and the indigenous peoples who lived here. Winds mostly from
the west carried the flames unabated across the plains and low moraines,
thereby nurturing prairie. (Grasses and other prairie plants grow from the
roots and thus are not harmed by fire; woody plants grow from the tips and
are therefore more vulnerable to fire.) Where the land was broken by higher
moraines or water barriers, some woody vegetation survived: less frequent
burning would produce the shrublands; more burning, the savannas with
their fire-tolerant (but not totally resistant) oaks and hickories. In ravines and
along the eastern banks of rivers and lakes, fire rarely entered and forests
could develop. Berrien County is the region's exception, for it was predomi-
nantly forest as a consequence of, again, location: Lake Michigan on the west
provided more precipitation and acted as a firebreak, as did several rivers.

The interplay of topography, climate, and fire created the broad terrestrial
categories of prairie, wetlands, shrublands, savannas, and forests. But scientists recognize finer distinctions that more fully reflect the actual diversity of
the region. The Chicago Wilderness Biodiversity Recovery Plan, for example,
lists 11 types of prairies, 10 forests, 6 savannas, 4 shrublands, and 13 wetlands.
(Floyd Swink and Gerould Wilhelm record 2,530 species of vascular plants
that have been found in this area.*) Soil, moisture level, acidity, and amount
of light are among the numerous elements that affect which plants grow
in a particular location. Aquatic environments differ, too, in substrate, size,

---

*Floyd Swink and Gerould Wilhelm, *Plants of the Chicago Region*, 4th ed. (Indianapolis: Indiana
Academy of Sciences, 1994).

current, temperature, and many other variables. One more feature that adds to regional diversity is the weather spawned by Lake Michigan. The milder conditions created by this huge water body enable disparate floral elements to thrive in close proximity. Thus in the dunal regions, the prickly pear cactus of southwestern affinities is neighbor to the boreal bearberry.

With the great array of plant communities present and its midcontinental location, the region hosted a remarkable variety of animals: thousands of insects (14,000 estimated for Illinois), 400 or so birds, 60-plus reptiles and amphibians, 60-some mammals, and about 170 fish. In the early days, thick forest grew along much of the lakefront. This allowed northern mammals such as martins and fishers to include this area as the southernmost part of their range. Sharp-tailed grouse inhabited local savanna-prairie areas but today are found no farther south than northern Wisconsin. Grassland species such as bison, long-billed curlew, and whooping cranes followed the prairie peninsula eastward, although they were early victims of hunting and agriculture.

Some of the most spectacular manifestations of biological richness that grace this region involve bird migration. The immense flocks of passenger pigeons that nested in the forests of Wisconsin and Michigan came through here in numbers that would strain credulity if the accounts were not so many and consistent. Since time immemorial, a majority of eastern North America's sandhill crane population gathered in the Kankakee marshes as they made their way to and from breeding grounds to the north. And finally, what makes the Chicago region one of the premier inland birding locations according to some is that it is situated on the southern end of Lake Michigan, the only Great Lake on a north/south axis. Thus any bird moving south along the lakeshore or over the choppy surface will eventually enter this territory.

That the region straddles a major divide means that fish of both the Great Lakes and Mississippi basins course local waters. Restricted to the latter, such species as skipjack herring, red shiner, and slender madtom inhabited hospitable stretches of the area's southern counties. Within Lake Michigan, seven kinds of deepwater ciscoes foraged on the plankton that floated in the open waters of the lake. They, in turn, became prey for burbot and lake trout. And the biggest of all inland fish, the regal lake sturgeon, spawned in the shallows.

As I collected accounts of the region, it became clear that two specific sites have garnered more attention than any other. There are good reasons behind this, for they both were spectacular and of national significance. The first of these was the Kankakee Marsh, a mixture of water, grass, and woods that fringed in varying widths the Kankakee River from its origins near South Bend, Indiana, to Momence, Kankakee County, Illinois. Some authors estimate this vast marsh to have encompassed one million acres, making it one of the largest inland wetlands of the lower forty-eight states. The marsh, though, was a serious impediment to the development of northwest Indiana, and so state officials decided it needed to be drained, a fact that was largely accomplished by the early 1900s.

But it is the Indiana Dunes area that really eclipses the rest of the region. People spent careers studying, protecting, and celebrating this marvelously scenic and biologically rich province in the shadow of Chicago. The magic of the dunes themselves provided most of this inspiration, but the ongoing struggles to protect this sand-based wilderness explain at least some of the artistic productivity.

The stories of the Kankakee Marsh and the Indiana Dunes could provide metaphors for a larger trend that was going on in the country during the late nineteenth and early twentieth centuries. It was during those years that the nation's first environmental movement arose. At least two important changes, one related directly to natural resources and the other more social in character, generated increased concern for the loss of wild things and places.

The first was the decimation of wildlife due to unrestricted hunting and better weaponry. You did not have to be especially gifted to notice that during the previous decades of your life huge flocks of passenger pigeons passed overhead in spring and fall, but that in recent years few if any of the birds appeared. Nor did April hunting trips reap the hundreds of plover and curlews that they used to. Both Indiana and Illinois had lost their last deer by the onset of the new century.

Of course, even with the evidence as obvious as empty skies and gameless hours, mass action did not occur spontaneously, but rather through the concerted efforts of enlightened activists. Among just a very few of the early prophets were organizations like the state Audubon societies, National Audubon Society, and the American Ornithologists' Union; journals such as *Forest and Stream*, *American Field*, and *Outing*; scientists Spencer Fullerton Baird, Frank Chapman, and Edward Nelson; naturalists and authors John Muir, John Burroughs, William Hornaday, Florence Merriam, and Gene Stratton-Porter; and a most remarkable president the likes of whom we will probably never elect again, Theodore Roosevelt, the only modern naturalist to ever hold the office. It is also important to note that many of the foot soldiers who wrote letters and collected money were, in fact, unheralded women and children.

The second change was the increasingly high percentage of the American population who now lived in cities. This motivated some to regain the rural or even wilderness past that they believed had helped temper the nation's character. Others looked more practically and realized that these burgeoning urban populations needed places to recreate—that humans are better off if there are accessible areas offering open space and greenery.

If the wave of concern breaks upon the political landscape with enough force, a law is left as it ebbs. There were numerous legislative enactments that served to rectify the perceived problems. The American Ornithologists' Union drafted a model bird protection law that became the basis for statutes adopted by states across the country.

One of the earliest steps on the national level to address natural resource depletion was the establishment of the United States Commission of Fish and Fisheries in 1871. Congress passed the Lacey Act in 1900, thereby banning the interstate trafficking in birds taken in violation of state law. The first federal wildlife refuge became a reality in 1903 when President Roosevelt approved the designation for Florida's Pelican Island. In 1913 the Weeks-McLean Act codified the federal government's authority to regulate migratory birds under the interstate commerce clause of the Constitution. Three years after that, a treaty was signed with Great Britain to protect the many migrant birds that breed in Canada. To ratify that agreement, Congress passed the Migratory Bird Act Treaty in 1918, but the state of Missouri challenged both the treaty and act before the U.S. Supreme Court in *Missouri v. Holland*, 252 U.S. 416 (1920). The Court, however, upheld the federal position. Justice Holmes wrote: "To put the claim of the State upon title [to birds] is to lean upon a slender reed. Wild birds are not in the possession of anyone.... But for the treaty and the statute there soon might be no birds for any powers to deal with."

On the local level, government acted in a number of significant ways to provide places for citizens to play and relax. Indiana established the nation's second state park system shortly after celebrating its centennial in 1916. Urban parks had been around for a long time, but what Dwight Perkins proposed in the way of forest preserve districts was something different. As principal author of the enabling legislation, Perkins stated his intent when he assigned to the districts the power to hold lands "for the purpose of protecting and preserving the flora, fauna, and scenic beauties within such district … for the purpose of the education, pleasure, and recreation of the public" (70 ILCS 810/17).

Cook County created its forest preserve district in 1913, and DuPage County followed in 1915. Today all of the nine Illinois counties within the region have either forest preserve or conservation districts that hold well over 150,000 acres of protected land. (Cook County alone has around 70,000 acres.) Tens of thousands more acres are protected regionwide by park districts; municipalities; counties in Wisconsin, Indiana, and Michigan; states; the federal government; and private organizations. This is the legacy of those early conservationists, whose progressive views were forged in the era of profligate exploitation of our biological resources. Fortunately, they were able to save enough to enthrall and inspire the subsequent generations into whose hands the torches (and binoculars) have been passed.

Spelling and nomenclature in this volume have not generally been updated or corrected from the original version of the essays.

Landscape

Letters to the Dutchess of Lesdiguieres

Pierre-François-Xavier de Charlevoix

LETTER XXV

Sept. 17

Madam,

I did not expect to take up my Pen to write to you so soon; but my Conductors have just now broke their Canoe, and here I am detained the whole Day in a place where I can find nothing that can excite the Curiosity of a Traveller; therefore I can do nothing better than employ my Time in entertaining you.

I think I informed you in my last, that I had the Choice of two Ways to go to the *Illinois*: The first was to return to Lake *Michigan,* to coast all the South Shore, and to enter into the little River *Chicagou.* After going up it five or six Leagues, they pass into that of the *Illinois,* by Means of two Portages, the closest of which is but a League and a Quarter. But as this River is but a brook in this Place, I was informed that at that Time of the Year I should not find Water enough for my Canoe; therefore I took the other Route, which has also its Inconveniences, and is not near so pleasant, but it is the surest.

I departed Yesterday from the Fort of the River *St. Joseph,* and I went up that River about six Leagues. I landed on the Right, and I walked a League and a Quarter; at first by the Bank of the River, then cross the Country in a vast Meadow, interspersed all over with little Clusters of Trees, that have a very fine Effect. They call it the Meadow of . . . *the Buffalo's Head* because they found here a Buffalo's Head of a monstrous Size. Why should there not be Giants among these Animals?—I encamped in a very fine Place, which they call the Fort *des Renards (of the Foxes),* because the *Renards,* that is to say, the *Outagamis,* had here, and not long since, a Village fortified after their Manner.

This Morning I walked a League further in the Meadow, having almost all the Way my Feet in Water. Then I met with a little Pool, which communicates

Pierre-François-Xavier de Charlevoix, *Letters to the Dutchess of Lesdiguieres* (London, 1763), 272–81.

with several others of different Bigness, the largest of which is not one hundred Paces in Compass. These are the Sources of a River called *Theakiki*, and which our *Canadians* by Corruption call *Kihakiki*. *Theak* signifies a Wolf, I forget in what Language; but this River is so call'd because the *Mahingans*, which are also called the *Wolves*, formerly took Refuge here.

We put our Canoe, which was brought hither by two Men, into the second of these Springs, or Pools, and we embarked; but we found scarce Water enough to keep it afloat: Ten Men, in two Days, might make a straight and navigable Canal, which would save much Trouble, and ten or twelve Leagues Way; for the River, at the first coming out from its Spring, is so narrow, and we are continually obliged to turn so short, that at every Moment one is in Danger of breaking the Canoe, as it has just now happened to us.

LETTER XXVI

October 5

Madam,

The Night of the 17th of this Month, the Frost, which for eight Days past was perceivable every Morning, increased considerably. This was early for this Climate.... The following Days we went forward from Morning to Night, favored by the Current, which is pretty strong, and sometimes the Wind: In Fact, we made a great deal of Way, but we advanced very little on our Journey: After having gone 10 or 12 Leagues, we found ourselves so near our last Encampment, that Persons in both Places might have seen each other, and even have talked together, at least with a Speaking-Trumpet. But it was some Consolation to us, that the River and its Borders were covered with Wild-Fowl, fattened with wild Oats, which were then ripe. I also gathered some ripe Grapes, which were of the Shape and Bigness of a Musket-Ball, and soft enough, but of a bad Taste. This is probably the same that they call in LOUISIANA *the Plumb Grape*. The River by Degrees grows less winding; but its Borders are not pleasant till we are fifty Leagues from its Source. It is also for all this Space very narrow, and as it is bordered with Trees, whole Roots are in the Water, when one falls it bars up the whole River, and it takes a great deal of Time to clear a Passage for a Canoe.

Having got over these Difficulties, the River, about fifty Leagues from its Source, forms a small Lake, and afterwards grows considerably wider. The Country begins to be fine: The Meadows here extend beyond the Sight, in which the Buffaloes go in Herds of 2 or 3 hundred: But one must keep a good Lookout, not to be surprised by the Parties of *Sioux* and *Outagamis*, which are drawn hither by the Neighborhood of the *Illinois*, their mortal Enemies, and who give no Quarter to the *French* they meet on their Route. The Misfortune is, that the *Theakiki* loses its Depth as it grows wider, so that we are often obliged to unload the Canoes and walk, which is always attended with some Danger, and I should have been greatly perplexed, if they had not given me an Escort at the River *St. Joseph*.

What surprised me at seeing so little Water in the *Theakiki* was, that from Time to Time it receives some pretty Rivers. I saw one among the rest, above sixty Yards wide at its Mouth, which they have named the *Iroquois River*, because these gallant Men suffered themselves to be surprised here by the *Illinois*, who killed a great Number of them. This Blow humbled them the more, as they greatly despised the *Illinois*, who for the most Part can never face them.

The 27th of *September* we arrived *at the Fork*; this is the Name the *Canadians* give the Place where the *Theakiki* and the River of the *Illinois* join. The last, after a Course of sixty Leagues, is still so shallow, that I saw a Buffalo cross it, and the Water did not come above the Middle of his Legs. On the contrary, the *Theakiki*, besides bringing its Waters a hundred Leagues, is a fine River. Nevertheless it loses its Name here, without doubt because the *Illinois* being settled in many Places of the other have given it their Name. Being enriched all at once by this Junction, it yields to none that we have in *France*; and I dare assure you, Madam, that it is not possible to see a better nor a finer Country than that it waters; at least up to this Place, from whence I write. But it is fifteen Leagues below the *Fork* before it acquires a Depth answerable to its Breadth, although in this Interval it receives many other Rivers.

In this Route we see only vast Meadows, with little clusters of Trees here and there, which seem to have been planted by the Hand; the Grass grows so high in them, that one might lose one's self amongst it; but every where we meet with Paths that are as beaten as they can be in the most populous Countries; yet nothing passes through them but Buffaloes, and from Time to Time some Herds of Deer, and some Roe-Bucks.

〰〰〰 THE HONOR OF BEING THE FIRST EUROPEANS TO VISIT THE Chicago region generally goes to Louis Joliet, the Jesuit priest Jacques Marquette, and their subordinates who arrived in the summer of 1673. But because their descriptions have been so widely quoted,* I have chosen to include

---

*For example, see Paul Angle, *Prairie State: Impressions of Illinois, 1637–1867* (Chicago: University of Chicago Press, 1968); and Joel Greenberg, *A Natural History of the Chicago Region* (Chicago: University of Chicago Press, 2002).

instead the writings of Father Pierre-François-Xavier de Charlevoix (1682–1761), also a Jesuit, who made his trip in 1721. The land he saw was virtually, if not exactly, the same as it had been during the days of his better-known predecessors.

Father Charlevoix was born in Saint-Quentin, France, to a prominent family. As a youngster, he manifested both an interest in religion and a first-rate intellect. After completing his training for the priesthood at the College of Louis-le-Grand in 1705, he embarked on a voyage to Quebec, where he was to teach rhetoric at the Jesuit college there. Also on the ship was the king's principal envoy to Canada, the Marquis de Vaudreuil. The two men became friends, and Charlevoix's four-year stay in Quebec cemented his relationship with the colony's elite. He returned to France, where he taught at his alma mater. (Among his students was François-Marie Arouet de Voltaire.)

Ten years later Charlevoix was tapped to conduct a secret exploration of North America that included the task of finding another way to the Orient. Under the ruse of inspecting Jesuit facilities, he worked his way from the St. Lawrence to the southern end of Lake Michigan, where he descended the Kankakee River into the Illinois and Mississippi rivers. (A reading of this selection makes it clear that he was unaware of the Des Plaines River, believing that the Kankakee ran directly into the Illinois River rather than converging with the Des Plaines to form the Illinois.) Arriving in New Orleans, he attempted to return along the East Coast, but his ship was destroyed while crossing the Gulf of Mexico. Eventually, however, he did succeed in getting back to Quebec, two years after embarking on the trip.

Although his diary appears to be a collection of letters written to a patron, this is merely a literary device, for the documents were never mailed. Apparently, some were based on notes and not even fully composed until his return to France. His wanderlust sated, Charlevoix spent the rest of his life immersed in scholarship and canonical duties. His history of the French in Canada was considered the authoritative reference for fifty years.

Source: Colton Storm, "They Called It the West," Bulletin of the Historical and Philosophical Society of Ohio 14, no. 1 (January 1956): 5–7.

William Johnston's Tour from Fort Wayne to Chicago, 1809

William Johnston

ᔓ ᔓ ᔓ The general course of the road is something North of West. For about three miles the land is thin oak land to Spy river, when immediately on crossing Spy river, a fine bottom commencing which continues for some distance. The timber is generally sugar tree hickory & buckeye [horse chestnut] all of very large growth.

Twelve miles further is Eel river, a branch of the Wabash. This little stream is very deep, and at the distance of ten miles on a straight line from its source and about seventeen by the meanders of the Stream, it is not more than five yards wide and is generally three feet deep with a very slow current. The land on this river is very rich, and appears to be well adapted to the culture of wheat or hemp. There are but few mill seats in this country, on account of the land being very level. All the rivers in this country have their sources in Swamps & ponds as there are but few Springs in the upper country—that is, on the high land that divides the waters of the lakes from the waters of the Ohio.

Passing on westward you travel through a fine rich level country; tho' it appears as if it had been under water at some former period.

Fifteen miles from Eel river you come to the little lakes. Here is one of the most enchanting prospects my eyes ever beheld. The traveller, after passing through a country somewhat broken for a few miles is immediately struck with the sight of two beautiful sheets of water, as clear & as pure as any spring water. They are about one fourth of a mile apart. I encamped on the border of the most westerly one all night. The border is so low that a large wave might roll out on the bushes. I perceived a number of fishes playing in the water near the shore where it was not a foot deep. I took my pistols and went in ten or fifteen yards and shot several ringeyed perch. Indeed they were so tame that they came close to me, as if wondering at the new monster that had got amongst them. This lake covers about 100 acres and has an outlet at the east end of it. From thence it runs about half a mile along the side of a small ridge that divides

"William Johnston's Notes of a Tour from Fort Wayne to Chicago, 1809," in *The Development of Chicago, 1674–1914*, ed. Milo Quaife (Chicago: Caxton Club, 1916), 55–60.

it from the other lake. It then turns suddenly round to N E and passes through a break in the ridge and empties itself into the other lake. There is a good mill seat here with three or four feet fall, and water sufficient for ten mills abreast.

The soil in the neighbourhood of these lakes is well calculated for wheat or any kind of small grain. The timber is chiefly white oak, Spanish Oak and some chestnut oak. The land is a mixture of sand & clay, and in some places a deep black soil, something like river bottom.

Eighteen miles further you come to the Elkshart river, a branch of the St Josephs of Michigan. For eight miles before you come to this river you come to a thicket of young hickories and oaks, about as thick as a mans thumb, and growing so close together that it is impossible to penetrate it at any other place than by the road. This land is as rich as any in Kentucky, and there is no doubt but it would be as fruitful if cultivated.

Immediately on crossing the river, which is here about fifty yards wide, a most delightful prospect is presented to view. There is scarcely a tree in an acre of ground for three miles. Here is an Indian village of about twenty houses. One of the principal chiefs resides here—his name is "Five Medals." The village is beautifully situated on the edge of a fine prairie containing about three thousand acres. About a mile west of this prairie the road comes to the bank of the river, at a good spring of water (a thing that is very scarce in this country). Here the timber is tall and thick on the ground principally white oak—the soil is a white clay.

Fourteen miles further is the junction of the Elkshart & the St Josephs. Here is a place formed by nature for a town. One half mile from the forks the rivers come within sixty yards of each other. They then seperate and form something like an oval piece of ground of about one hundred acres until it comes to low bottom that appears to be made ground. This bottom con- tains about fifty acres, and I suppose is overflowed at high water. Both these streams are navigable, without any falls or other obstruction, almost to their sources. From the forks down to the lake, about Sixty miles by water, may be navigated by any kind of small craft at any season. The channel is deep & the current gentle.

There is no Situation in this country better calculated for trade than at the forks of St Josephs & Elkshart. These two branches flow through the richest and dryest part of this country; and I think it would be an object with our government to make a settlement at this place.

Ten miles down this river from the forks is a portage of three miles west to the Theakiki, a branch of the Illinois river. Fifteen miles below the forks is a French trading house.[1] There are about twenty persons kept here for the purpose of trading with the Indians. These men in the winter take each a load of goods and ascend one of the branches of the St Josephs; thence across the portages into the other river, and to the Indian villages, where they continue until spring, from whence they return with their peltry. They all collect in May and make up their packs when they proceed down the river into Lake Michigan & round to Michillimakinac. It will scarce be believed that these men perform a voyage of upwards of a thousand miles in a bark canoe heavily laden with packs, the greater part of the voyage in a boisterous lake.

The soil on this river varies, but none but what is equal to our third rate in Kentucky, and has the advantage of being level. There are several fine Springs in this part of the country.

At the factory I was told that there was the remains of a British fort[2] three miles below, where was a fine orchard of apple trees. Twenty miles from the fort to the mouth of the river. —Here I was informed that a trader had raised several crops of wheat, and that it was as good as could be raised any place. I crossed the St Josephs at the French factory. Twelve miles further is an Indian village called Turcope [Terre Coupe]. The town Stands on an eminence, and may be seen about seven miles. There is not a tree to interrupt the view for about nine miles. This prairie I was told *extends to the Mississippi, a distance of four hundred miles!* From this village to Lake Michigan, a distance of about forty miles, the land is about one half timbered, and the other half prairie, but all of a good quality, except about four miles adjoining the lake, where it is very Sandy. Here are some of the finest white pine trees I have ever seen. The road strikes the lake at the most Southerly end, at the mouth of the river *du Sma* [Du Chemin].

The country from Fort Wayne to Lake Michigan is I think of the greatest importance to the United States to have it settled. It may be said that about one half of the unceded lands of the Indians lying north of the present boundary in Illinois & Indiana West of the Wabash & Miami of the lake & East of Illinois & Chicago rivers, and including Michigan Territory, is rich and level. The other half may be divided between the Swamps ponds or lakes and prairies, the latter of which are by far the most extensive and would support

---

1. This was at Parc Vache (The Cowpens), where the Chicago-Detroit and Chicago–Fort Wayne trails forked.

2. Fort St. Joseph, destroyed by the Indians in Pontiac's War of 1763, and captured, having been restablished, by the Spaniards in 1781.

immense herds of cattle at very little cost; for it is a fact that salt can be got cheaper at Detroit than it can in Lexington in Kentucky.

The East* end of Lake Michigan is bounded by a mountain [ridge] of sand about 100 feet high. This hill has been accumulating since the formation of the lake. The lake is about three hundred miles long from north to south, and about sixty miles wide from East to west. The North west wind prevails here the greater part of the year. This wind blowing over an extensive level country, acquires such force when it arrives at lake Michigan, that in a dry day it will raise as much sand as darkens the air, which in time raised up a bank, and every storm adding more to it till the present. This ridge is covered with Stunted cedars and junipers. I think there could be as much junipers gathered here as would supply the United States with that article. The wind is always changing the position of this ridge. A strong wind will make a breach in the top; the wind will then deepen the breach, and the sand is carried back and deposited in the valley [*plain*] behind the hill which in time becomes the foundation of another mountain. The traveller may see hundreds of these hills behind the mountain. The whole body of the mountain is carried away in some places for several hundred yards, and deposited behind. The wind is constantly in motion, taking more & more until the whole hill will change its place. The lake of course will advance as the bank recedes; and there is not the least doubt that the water has [thus] gained considerably on the land.

This mountain [ridge] is not to be found on the west side of the lake....

The road still keeps the shore of the lake. Twelve miles further is the mouth of the Great Calumet. Here the sand mountain ends. Twenty miles further is the mouth of the little Calumet. These two rivers are of the greatest consequence to the traders on the lake. They are both about twenty yards wide at their mouth, but very deep. One of them is considerably longer than the other; & there is a communication between them, which in case of storm on the lake the trader can go up one several miles, then across into the other, and down it into the lake. It is twelve miles from the mouth of the little Calumet to the mouth of Chicago river. Here the United States have erected a garrison [Fort Dearborn] for the protection of the trade in this quarter of the country. This garrison does great honor to Capt. John Whistler who planned & built it. It is the neatest and best wooden garrison in the United States. This place guards the entrance of Chicago river.

---

* Mr. Johnston probably means the South end of the lake. [Copyist's note.]

Between the Chicago and the Illinois rivers, there is a direct water communication. The river Plein, which is one of the main forks of the Illinois, has its source near the bank of the lake, and nine miles from fort Dearborn it turns West. At this bend there is a long pond communicates with it, which runs Eastwardly towards the lake and terminates in a small creek which runs into the Chicago river. This creek is about two miles long; *and in the Spring of the year any kind of Craft may sail out of the lake to the Mississippi without being unladen.* The U. S. factor at fort Dearborn measured the elevation of land between the lake and the river Plein, and found it to be four feet on the side of the lake and five on the side of the Illinois. Thus by digging a canal of *nine** feet deep, a passage could be got at any season in the year from the Falls of Niagara to the mouth of the Mississippi without a Single foot of land carriage. The Canal would be about six miles long, through a beautiful prairie; and there is a quarry of limestone near this place which would make excellent casing for the Canal.

While at Fort Dearborn I was informed that there were some boats at the portage which would cross the next day. I accordingly went...to the portage. The water was low, it being about the 28th of June. The boats could not pass loaded; but I saw them sail out of the river Plein into the pond, & through it into the creek before mentioned, and down it into the Chicago river. The loads were brought over the portage in waggons; & they were re-loaded [into the boats] at the head of the Chicago.

...There are about sixty soldiers in garrison at Fort Dearborn; and so healthy has the place been that Capt. Whistler informed me he had lost but six men in nearly eight years, and he has the same men now that he had when he built the garrison; and although their term of enlistment expired yet they all enlisted again—a sure sign that he is a good officer.

Fort Dearborn is beautifully situated on the bank of the lake. It is bounded on the land side by an extensive prairie, interspersed with groves of trees, which gives it a beautiful appearance.

Lake Michigan abounds with fish of an excellent quality. The white fish is caught here in great plenty. This is probably the best fresh water fish in the waters of the U. S. The surge of the lake beating always against the Shore, frequently throws out large fish on the Land. I took up several perch & pickeral, that would weigh ten pounds, some of them alive. The shore is frequented by flocks of crows, buzzards, gulls &c which soon destroy the fish that is thrown out on the shore.

---

*I do not understand this—why an elevation of 4 feet on one side & 5 on the other should be *added together.* Does he not include in this, 4 feet for necessary depth of channel. [Copyist's note.]

〜〜〜〜 BEGINNING IN THE EARLY 1700S, HOSTILE INDIANS PUT A brake on travel through the Chicago region and much of the rest of the upper Midwest. Memories of the Fort Dearborn Massacre of 1812, when a force of up to five hundred Potawatomi slaughtered fifty-two soldiers and civilians before burning the fort, and the short-lived Black Hawk Rebellion of 1832 kept most settlers at bay. Not until the 1830s did local tribes agree to cede their lands and move west, thereby allowing the new immigrants to pour in unmolested. Visitor accounts before then are therefore of especial value.

William Johnston's document provides excellent descriptions of the landscape during his 1809 trip. As interesting as its contents, however, is the mystery of its origins, about which evidently nothing is known. Milo Quaife summarizes the facts in *The Development of Chicago*. The journal was donated to the Chicago Historical Society in 1894 by J. Fletcher Williams of St. Paul, Minnesota. Williams, however, knew nothing about it or Mr. Johnston, except that he had found the notes among his father's papers many years earlier. Quaife goes on to say that "although yellowed as if from age it is obvious that the manuscript is itself a copy of the original journal. Of its history, as of its author, nothing has been learned." A check with the staff of the Chicago History Museum confirms that in the ninety-plus years that have elapsed since Quaife wrote his comments, no new information has come to light.

Sources: Ulrich Danckers and Jane Meredith, A *Compendium of the Early History of Chicago* (River Forest, IL: Early Chicago, Inc., 1999), 102–3; Milo Quaife, ed., *The Development of Chicago, 1674–1914* (Chicago, Caxton Club, 1916).

A Visitor to Chicago in Indian Days: "Journal to the 'Far-Off West'"

Colbee Benton

ↁↁↁ FRIDAY AUG 23—We arose about daybreak, caught our horses, and commenced our journey without even thanking the Indians for our entertainment or even bidding them goodbye. We seemed to go off as if we

Colbee Benton, A *Visitor to Chicago in Indian Days: "Journal to the 'Far-Off West'"* (1833), ed. Paul M. Angle and James R. Getz (Chicago: Caxton Club, 1957), 83–96. Courtesy of the Caxton Club.

were mad, but I did not feel right to do so. Yet I was obliged to be a sullen Indian. They will resent the offer of money. After having stuffed myself with so much corn, beans, fried pork and cakes, I have considered myself very fortunate in not finding another village until night.

We passed through a little grove onto a prairie and travelled north about two miles; then northwesterly about seven miles to a beautiful spring on the prairie, where we rested and refreshed ourselves, but I could not eat any breakfast. About one mile from this we came to a little brook on which was some very heavy timber, and near it was a little lake.* After travelling about eight miles farther, across small prairies and through oak openings and the land not so good, we rested again at a beautiful little lake.† The Indians had had an encampment here but they had moved to some other place and we were almost at loss what course to pursue. We took the saddles from our horses and turned them out to feed, and Louis laid down and went to sleep. I employed myself about changing my stockings, which had become wet when riding through the high prairie grass in the morning dew. It was like wading through water, for my pantaloons and feet were completely soaked and I could not contrive any way to prevent it. I had another opportunity to get wet here. The horses being rather dry and the flies plagueing them induced them to wade into the lake and I could not get them out by any persuasion. Therefore I was obliged to go in after them.

I took the gun and walked round a little point of the lake in quest of game, and when I returned the horses had disappeared. I waked up Louis and after hunting some time we found them about half of a mile from the lake, and we determined not to turn them out again without hobbling them as it is rather dangerous for they might wander away but a short distance and we should never be able to find them again. And then to be in a wilderness so far from home would make us feel and look rather sorry if we had lost our horses.

I saw a number of loons in the lake but could not get a shot at them.

There is a small island in the lake which the Indians said was covered with small pines, but I was not able to distinguish them, it was so far from the shore. The lake was surrounded by oak openings, had a stony bottom and the water looked very clear indeed.‡

---

*Either Gray's Lake or Round Lake and its companion, Highland Lake. The "little brook" was undoubtedly Squaw Creek.

† Cedar Lake. In this eight-mile leg of their journey the travelers had proceeded due north.

‡ The "small island," "stony bottom," and clear water make the identification of this lake (Cedar Lake) certain, since no other body of water in the vicinity has these characteristics.

From this lake we travelled about one mile to another and a larger one.* Here was the remains of an Indian encampment recently deserted, and we were as much at a loss which course to pursue as before, and finally more so, on account of the number of trails which forked off to the right and left, but fortunately took the direct one. This lake was connected by a little stream, through a marsh, with another lake, and it was also connected with some other large marshes and small ponds or lakes which are great places for game.† We saw a large flock of wild geese, but we could not get a shot at them. Also saw some ducks and loons, and when we crossed the little stream between the lakes we saw some monstrous pickerel and other large fish. These lakes are surrounded by groves of oak and it is very pleasant about them. The land is not so good as the large prairies.

After crossing the little stream, which was somewhat difficult, we had a low marsh of mire and water to go through for about thirty rods.‡ Louis went forward and very soon his horse was flouncing and down he went, and Louis was mid-leg deep in the mire. I followed and soon found myself in a worse predicament. The horse flounced, his forefeet sunk in the mud, the saddle partly turned and off I went, with one foot hanging in the stirrup without the power to extricate myself immediately, and my back, and finally my body nearly covered with mud. I then thought "*attitude* was everything." I succeeded in getting up and getting my pony onto a little more solid foundation, and not liking the idea of wading through such a deep mire, I got onto my pony again; but I had not proceeded one rod before I was off again. I almost gave up ever getting across, it grew so much worse. Louis kept moving; sometimes his horse would sink into the mud and it was with the greatest exertion that he could extricate himself. After waiting until I saw Louis safe over, I took my pony by the bridle and led the way, and was soon safe and sound on the other side; and such dirty, muddy-looking fellows were never seen before, I presume. I could not have been hired to have gone back for fifty dollars, and I rejoiced to find ourselves safe, but my rejoicing was of short duration, for we found this dry spot but an island surrounded by tremendous marshes. While shooting some pigeons, which were very plenty on this little island, we accidentally perceived an Indian wigwam across the marsh from near the place we had just left, but we determined not to go back at any rate. We found that

---

*Fox Lake.

† The two men evidently followed the shore of Fox Lake north and west until they came to the narrow channel—the "little stream"—leading to Petite Lake a short distance to the north.

‡ The marshy area between Petite Lake and Grass Lake to the west.

the Indians had had an encampment on this ground and it looked as if they had but just left it. We followed the trail to the opposite side where we had a view of some wigwams on the mainland, but there was an impenetrable barrier which we could not surmount without assistance from other sources besides our own.

On our right and left were two extensive marshes or lakes grown up with high grass, and through the centre was a river which I learned from the Indians to be the head of Fox River, which runs into the Illinois. In front was a little neck of water about sixty rods across, which connected the two marshes.* On the further side was high grass and a muddy bottom and the water ten feet deep most of the way and some places more. After shooting a large sand crane Louis climbed a large tree and yelled "Indian" and was answered very soon, but it was nearly an hour before an Indian appeared with his canoe. He got almost to us before we saw him; the grass was so high, the way so crooked, and he came so still. He was a savage-looking Indian. His face painted black, and almost a naked body. We took off our saddles and put them into the canoe, got in ourselves, and led our horses at the side of it. We paddled out into the stream, and we were obliged then to turn to the left and follow down some distance before we could strike for the shore on account of the high grass, and before we reached it Louis's horse crowded onto my pony and he sank; but we held his head out of water to keep him from drowning. He proved a good anchor for we could not move the canoe at all, and I began to think we should be obliged to let him go and save ourselves. However, I was disappointed, for very soon he gave a sudden spring and we were soon ashore.

The swimming of our horses was quite a curiosity to me, although I have seen horses swim before; but I never noticed them so particularly, and their snorting was very singular. It sounded so wild and loud & frightful. We met two Indians on the shore horseback, who led the way to the village which is situated in the timberland near and in sight of the great marsh.† Here we took up our quarters for the night. Hobbled our horses and turned them into the woods. The Indian horses pitched a number of battles with my pony, but he would drive them all.

After driving our horses some distance into the woods we returned to the first wigwam. A squaw spread a mat for us which we improved. The

---

* The main channel of the Fox River between Grass Lake and Fox Lake.

† The village was located on the high ground a short distance west of Grass Lake, "the great marsh."

Indian men were playing cards and the squaws were husking corn. Presently the chief of this village, whose name is "Warp-sa," in English "White Skin," appeared and shook hands with us. He is a tall, good looking Indian about forty-five years of age, and is a *notable* drunkard, whenever he can get whiskey enough—so says Louis, who is well acquainted with him.*

A very pretty little squaw roasted us some corn which I was obliged to refuse on my part, as I have not entirely recovered from my last night's meal of nut cakes, corn, and pork. But I did not go without a supper. We had killed six pigeons and a prairie hen in the course of the day which were finely cooked by the chief's squaw, and I can say that I never tasted anything of the kind so sweet, so tender, and so nice and so beautifully cooked. She cut a small piece out of each side of the breast, which made it as tender as any other part of the body, and which I thought a great improvement, and I thought it might be of some advantage to the *ignorant whites*. The same squaw cooked some corn soup on purpose for me, as Louis said, and out of politeness to her I eat as much of it as I could. It was very good indeed, but I felt very awkward eating it out of such a monstrous spoon.

The little squaw who roasted the corn for us I liked very well. She was the prettiest that I had seen, and she appeared different from most of them, and she was very *coy*, for I could not catch her eye without being very sudden in my movements. If I turned she would be looking serenely towards the sky, and I could not help thinking that she was some pure and sinless being whose noble spirit held converse with angels in a brighter world, far above the mortal things of earth. Her tawny complexion only made her more interesting, and there in such a wild place among such a rude class of beings; altogether, she seemed quite a Pockahontas.

After taking off my boots and stockings and hanging them on the wooden crane to dry, and putting on dry stockings and pumps, I felt like taking a little rest. An Indian gave us a mat to spread upon the ground for our bed, and with our saddles and portmanteau for our pillows, we laid ourselves down under a large tree a little distance from the wigwam. The mosquitoes troubled me so much that I was obliged to wear gloves, and tie a handkerchief over my head and face to keep them off, but that did not wholly protect me from their voracious appetite. And what was quite as troublesome and aggravating, it did not affect in the least their *eternal singing*.

---

*Probably Wapsë, "White Deerskin," who is said to have been responsible, in 1833, for the sale of the Potawatomi lands in Illinois and the removal of that tribe to Kansas.

It was a beautiful evening. The moon was shining bright, and very soon all was "*solemn stillness*" save the hooting of the owl and the occasional whistling of the raccoon. The Indians had retired into their wigwams leaving us masters of the field, and we—or I—laid down to dream of angels and pretty squaws. I had been asleep but a short time before I was awaked by the loud laughing of some squaws who stood so near my head that I could touch them. I waked up Louis and found that the Indians had been disturbed by the horses getting into the corn. I supposed them to be ours and went down to the field, and after wandering about some time I saw an Indian but no horses, and when returning I met some more, but I could not learn whether they were our horses or theirs. I went back and told Louis and he went down and found that they were Indian horses. While he was gone I armed myself against the mosquitoes in presence of the squaws, and laid down again. And they laughed outrageously, which almost provoked me. I began to think them worse than the mosquitoes. Louis told me when he returned that they were laughing at my appearance, and the pains that I had taken to protect myself from the mosquitoes. They soon left us and we were soon sleeping again.

SATURDAY AUG 24—I was awaked very early by the squawking of the ducks and other fowl, and could not sleep any more. It is the greatest place for game here that I ever saw. The ducks and wild geese and loons, sand cranes, and other fowls are continually passing this place. It seems to be their only thoroughfare. And these great marshes and the little streams and lakes must contain a great quantity of that cash article, *Fur*. It must be quite a home for the hunter—such a home as I should like, were it not so remote from friends and the comforts of civilized life. But I begin to think it rather hard work to be an Indian. To be a hunter is the same, and I think I shall give up the idea of being a hunter.

As soon as daylight I went in pursuit of the horses and found them about one quarter of a mile from the village. They had lived well, having found plenty of the wild bean, which the horses in this country are very fond of, and which will fat them very quick indeed. We rigged our horses and left without seeing anyone but an old Indian and the *pretty squaw* which I expected to dream about, but unfortunately did not.

Soon after leaving the village we came onto a beautiful and very rich prairie, covered with grass and flowers, and surrounded by oak openings.* Passed across the prairie which was about five miles and then travelled about seven

---

* This area, in McHenry County, is known as English Prairie. A group of Englishmen settled there in 1836.

or eight miles through oak openings of not quite so good land to two lakes, each about two miles square, divided only by a strip of land about four rods wide, and connected by a little stream.* The trail we were following passed across this strip of land between the lakes and it seemed to me like passing through the Red Sea on dry land. On each side was a dam or embankment formed by nature or art, which prevented the water from rushing in upon us, and I thought it quite a curiosity. It is covered with oak, maple, butternut, and basswood; saw a great many of the little frost grapes. We made a stop for dinner; eat some crackers and drinked some sweetened water. Saw some Indians in a canoe crossing the lake. Louis yelled at them and they yelled at him. Found the bones of a large buck; took one of his horns and intended to carry it with me but found it too troublesome and throwed it away. Saw a great quantity of small shells washed ashore, which lay in little windrows like sand.

In about half a mile we passed a little bay in the lake, where we saw a large flock of ducks, but they were so shy that we could not shoot any of them. Soon came out on a rich and beautiful prairie which was about three miles across. After travelling a short distance in the oak openings, which were very good land, we came to the most splendid lake, and the largest that I have seen since I left Chicago.[†] It has a smooth, stony bottom, clear, pure water, and it is surrounded by high banks covered with the heavy prairie grass and the great variety of flowers which are everywhere to be found in this country. And the tall, scattering oaks with their rich, beautiful, and comfortable shades, added the finish to this charming and lovely spot. I became so deeply in love with it, and so much interested with its romantic and delightful appearance, that I felt quite unwilling to leave; and I did not, until I had selected a location for my wigwam, which was a gentle elevation in the beautiful grove and overlooked the calm and pure waters of the lake. It seemed the sweetest, the most calm, the most peaceful and retired spot that I ever saw.

Having admired this Elyseum for some time without feeling any more contented with my situation, I concluded to break the enchantment which chained me to the spot, and proceed on my journey. I succeeded, and soon after leaving the lake passed two very extensive marshes on our right, and in the center of each was a cluster of tamaracks which looked like the spruce swamps in N. England.[‡] They were much more delightful and beautiful on

---

*The Twin Lakes, named Mary and Elizabeth, in Randall Township, Kenosha County, Wisconsin.
†Lake Geneva.
‡The two travelers had come onto Lake Geneva at its eastern end, and then proceeded northward. These marshes border the White River north and east of Lake Geneva.

account of their rarity and situation, and I could but think what a view for a painter of landscapes, the marsh extending as far as the eye could see, covered with the thick high grass, and the little cluster of tall and bright green tamaracks in the centre, and the marsh surrounded by groves of oak extending on here and there. Altogether the view was most splendid and far beyond the powers of a painter to describe.

We passed some other small marshes and two small lakes in the course of three or four miles, and then travelled over heavy swells of land covered with the oak timber. Crossed a little stream of water and found the swells much more abrupt.* It looked as if there had been an earthquake which had thrown the earth up into high knolls, one of which I ascended, and from the top of it I beheld the most splendid view of land that can be imagined. In front was a low prairie or valley which extended beyond the power of the eye, and on each side the elevated land was covered with the green grass and the scattered oaks with here and there a little prairie to make out a grand variety.

The spot seemed to me as if it had *once* been a vast and extensive lake, and had been destroyed by some natural cause which left only considerable of a stream that passed through the centre of the valley; which stream I afterwards learned from the Indians to be a branch of the Rock River.† We crossed part of the prairie and forded the stream, and after rising a little hill we had a glimpse of two little Indian boys who we overtook, and who guided us to the Indian village which was but a short distance. It is the largest village I have seen.‡ Situated on three little hills which centre into one hollow, where their corn fields are. The land is immensely rich and their corn looks well....

After resting a short time we proceeded, although it was nearly sunset. Here we changed our direction and travelled nearly south. Soon came onto a beautiful and rich prairie. It was rather small, in the form of a square, and surrounded by the handsomest oaks that I have yet seen. Also passed through some splendid groves, the trees very tall and slim and without a limb for a great many feet.

Arrived at a little prairie containing about twenty acres and camped on the border of it, near a little stream of water that ran through it.§ Hobbled

---

* The small creek was Ore Creek, and the mounds are located northwest of the town of Springfield, Walworth County. They are kames, or small hills of glacial drift.

† Turtle Creek. The route had been west from the vicinity of the present Elkhorn.

‡ The topographical features locate this village in Section 21, Richmond Township, Walworth County.

§ This camp site appears to have been on Turtle Creek in the northeast corner of Darien Township, Walworth County.

our horses and turned them out; then made a fire with the limbs from a tree which had been shivered by lightning and prepared for the night. We eat some crackers & drank some sweetened water for our supper. Roasted a prairie hen which we had just killed, for the dog, and were about to lay down for the night when we heard Indians talking. Louis looked a little wild and said that we must catch our horses or perhaps we should never see them again. We both went in pursuit of them and found them across the prairie. I caught my horse and Louis's horse followed, but Louis stayed back among the trees until I returned with the horses to the fire and tied them. He seemed to crowd one into the danger, if there was any; however, I did not have any great fears.

After his return we looked about us but could not see or hear anything, and were about to lay down again when we again heard them, and they seemed to be going round us. I took my station beside a large tree to observe any movement that might be made, while Louis laid down near the high grass, rolled snugly in his blanket—so snugly that you could not believe it contained a man. While I was standing by the tree, Louis observed to me that I had better take my hat off if I remained there much longer; it being of straw, he thought it would be a good mark to shoot at. I asked him if he thought that there was much danger. He replied that there were many Indians who did not know anything about our business and they might think us spies and intruders, and would take advantage of our situation; and he said that there were occasionally some of the Sauks and Fox Indians wandering about in this part of the country, and from them we could not expect much mercy.*

I concluded to lie down and therefore prepared myself against the encroachments of the troublesome mosquitoes. I placed my saddle for a pillow, wrapped myself in my blanket, and placed the well loaded gun at my side for a companion, and tried to sleep, but could not. The moon had risen in all its glory, and I thought I never witnessed such a clear and beautiful night. I got up and looked about me and I could not but admire the moonlit scenery, so bright and splendid. I could not help thinking of that beautiful night that I visited Niagara Falls to see the lunar rainbow. The same feelings came over me, and it seemed to me that I was standing on enchanted ground. All around me looked wild and terrible, yet lovely and romantic and beautiful, and I really felt happy to think that I was so far from home and in such a

---

*The Black Hawk War of the preceding year was still fresh in Louis Ouilmette's mind. He had reason to be apprehensive of wandering Sacs and Foxes.

place, surrounded by the western savages, and perhaps even then in danger of suffering from the power of their tomahawks and scalping knives. There was something so novel and romantic about it that it dispelled every fear, and I really believe it would have been dangerous to the Indians if a dozen had attacked us.

In the forepart of the night we heard a good many strange noises, which Louis said might be Indians, for he said that they made all kinds of noises, and mocked almost every wild animal. However, the noises might have proceeded from some strange birds that we were not acquainted with, and from foxes and raccoons, and I have no doubt but what they did.

I was up and down very often in the course of the night. I could not sleep for the noise, although it was very still near us, save the stepping of our horses and their eating. But at a distance we could distinctly hear the lonely howling of the wolves. Sometimes they would seem to be very near, and the owls accompanied them with their hoarse hooting, and so it continued until daylight. By the way, the mosquitoes joined the music of this interesting night with their eternal singing, and I was obliged to keep time for them. I never before have passed a night so interestingly, and so pleasantly....

〰〰〰 IN 1952 THE CHICAGO HISTORICAL SOCIETY PURCHASED A manuscript called "Journal to the 'Far-Off West'" from Hamill and Barker, a dealer in paper collectibles. Nowhere in the document was there any identification of the author, but its significance was obvious, as explained by Paul Angle and James Getz, who later edited the journal for publication: "Reports of journeys from New England to what was then called the 'West' are common, and our traveler's narrative of this part of his trip, while detailed and lively, could hardly be called superior to several that have been long available. But the second part, dealing with the Indians of southern Wisconsin and northern Illinois on the eve of their removal beyond the Mississippi, was and is unique." (The annotations in this selection are those of Angle and Getz.)

Getz had traveled to New England to track down possible leads mentioned in the manuscript, but his assiduous labors were for naught. As submission to the publisher neared, both editors despaired of ever determining the author, admitting that they had "exhausted our ingenuity in libraries... [and] that unless some relevant bit of information comes to light unexpectedly, our author must remain without a name."

This state of affairs did not sit well with the editors, especially Getz, who made one more trip. A detailed account of his sleuthing appears in the edition's preface, but suffice to say here that he had made enough progress to feel confident that the author was one Stephen Kendrick Jr. of Lebanon, New Hampshire. This was based in part on what was revealed in a collection of newspaper clippings from the 1870s written by Colbee C. Benton. But finding the reminiscences "interesting… [and with] plenty of time," Getz continued his perusal of the material until he encountered a key paragraph that referred to a trip Benton himself had made in 1833. Details in the articles, as well as from other sources, jibed with those of the journal, and thus Benton's authorship was determined beyond a doubt.

According to a local obituary, Colonel Colbee Chamberlain Benton (1805–1880) was born in Langdon, New Hampshire, but spent most of his life in Lebanon. Starting with $147.50, he amassed enough wealth through his store to retire "in the prime of manhood" to devote "the remainder of his life to his favorite studies, congenial pursuits, and travel." Among his long-time interests were "the study of nature," including botany, horticulture, and geology.

Benton reached Chicago on August 18, 1833. The following day, he learned that government dispatchers were to contact local Potawatomi to inform them that their annual payment would be distributed on September 10. Benton immediately decided he wanted to accompany one of the runners on the mission, as he "had a great curiosity to see the Indians at home and far away from the whites." The Indian agent warned that he "would find it rather a hard journey," but upon Benton's insistence he was teamed with Louis "Wilmot," a young man of French and Potawatomi heritage whose surname was really spelled Ouilmette. (His father, Antoine, is the person for whom the city of Wilmette is named.)

The two men departed Chicago on August 21. On the following evening, they camped in an Indian village near the present town of Mettawa (Lake County, Illinois), named in honor of the chief Me-tai-wah. Benton found the inhabitants "very lively and sociable," and he "began to like them very much." Adding greatly to the quality of the journal is this openness and honesty, most evident when he expresses the carnal twinges he seems to feel for some of the native women he meets along the way. (Throughout the journal he comments on the physiognomy of the females he encounters.)

Source: Colbee Benton, *A Visitor to Chicago in Indian Days: "Journal to the 'Far-Off West'"* (1833), ed. Paul M. Angle and James R. Getz (Chicago: Caxton Club, 1957).

Talk of the Town
Robert P. Saunders

᭑᭑᭑ The first cranberries ever seen in the Chicago market were brought here in the fall of 1839 by myself and three other young men. We gathered them in the Calumet marsh, where they grew in enormous quantities. We went into the marsh with a covered wagon drawn by oxen, and took along a winnowing machine, such as was used in those days to separate the chaff from the grain. We used it to clean the berries. We hauled into the city 8 loads of 25 bushels each. We succeeded in selling most of the lot at 50 cents a bushel. Afterward, on a trip home to England, I took some of those same cranberries with me and displayed them in the Covent Garden market, where they were looked upon with astonishment. The English people had only been accustomed to cranberries from Holland which were about the size of small currants.

〰〰〰 THIS ESSAY IS FROM AN UNNAMED AND UNDATED NEWSPAPER clipping. It is an interview with Robert P. Saunders, who came to Chicago in May 1839 and set up a business on the South Water Street Market as a wholesale merchant with a specialty in game and local produce.

A Summer Journey in the West
Eliza Steele

᭑᭑᭑ JULY 7TH.—I fell asleep, and when I was awakened at dawn this morning, by my companion, that I might not lose the scene, I started with surprise and delight. I was in the midst of a prairie! A world of grass and flowers stretched around me, rising and falling in gentle undulations, as if an enchanter had struck the ocean swell, and it was at rest forever. Acres of wild flowers of every hue glowed around me, and the sun arising from the earth where it touched the horizon, 'kissing with golden face the meadows green.'

Robert P. Saunders, "Talk of the Town," in Charles Harpel, scrapbooks on "Chicago Men and Events," vol. 16 (1880s and 1890s), 205.
Eliza Steele, *A Summer Journey in the West* (New York: John Taylor and Co., 1841), 125–27, 133–35.

What a new and wonderous world of beauty! What a magnificent sight! Those glorious ranks of flowers! Oh that you could have 'one glance at their array!' How shall I convey to you an idea of a prairie. I despair, for never yet hath pen brought the scene before my mind. Imagine yourself in the centre of an immense circle of velvet herbage, the sky for its boundary upon every side; the whole clothed with a radiant efflorescence of every brilliant hue. We rode thus through a perfect wilderness of sweets, sending forth perfume, and animated with myriads of glittering birds and butterflies.

> "A populous solitude of bees and birds,
> And fairy formed, and many colored things."

It was, in fact, a vast garden, over whose perfumed paths, covered with soil as hard as gravel, our carriage rolled through the whole of that summer day. You will scarcely credit the profusion of flowers upon these pra[i]ries. We passed whole acres of blossoms all bearing one hue, as purple, perhaps, or masses of yellow or rose; and then again a carpet of every color intermixed, or narrow bands, as if a rainbow had fallen upon the verdant slopes. When the sun flooded this Mosaic floor with light, and the summer breeze stirred among their leaves, the irredescent [sic] glow was beautiful and wonderous beyond any thing I had ever conceived. I think this must have been the place where Armida planted her garden, for she surely could not have chosen a fairer spot. Here are

> 'Gorgeous flowrets in the sun light shining,
> Blossoms flaunting in the eye of day;
> Tremulous leaves, with soft and silver lining
> Buds that open only to decay.'

The gentle undulating surface of these prairies, prevent sameness, and add variety to its lights and shades. Occasionally, when a swell is rather higher than the rest, it gives you an extended view over the country, and you may mark a dark green waving line of trees near the distant horizon, which are shading some gentle stream from the sun's absorbing rays, and thus, 'Betraying the secret of their silent course.' Oak openings also occur, green groves, arranged with the regularity of art, making shady, alleys, for the heated traveller. What a tender benevolent Father have we, to form for us so bright a world! How filled with glory and beauty must that mind have been, who conceived so much loveliness! If for his erring children he has created so fair a dwelling place, how well adorned with every goodly show, must be the celestial home reserved for his obedient people....

A line of trees proclaimed a river near, and we soon dashed through the Au Sable, the horses dancing with joy, as the clear cool waters curled about their feet. The sight of a house upon the opposite bank, seemed quite a novelty, as we had not seen one since leaving Joliet, at nine o'clock, and it was now one....

Our landlord spoke of his prairie land with the greatest enthusiasm. The ground is very hard to break, generally requiring several yoke of oxen, while beneath that the mould is several feet in thickness without stones, requiring no manuring and apparently inexhaustible. Some of the old settlements, where farms had been worked for twelve years, it was still as fertile as ever— giving the tiller very little trouble, and yielding rich crops. The oasis, or 'oak openings,' upon the prairies are very beautiful. We passed through one this morning. It presented the appearance of a lawn, or park around some gentleman's seat. The trees are generally oak, arranged in pretty clumps or clusters upon the smooth grass—or in long avenues, as if planted thus by man. From their limbs hang pretty vines, as the pea vine—lonicera flava, honey-suckle— and white convolvulus. While our carriage wound among these clumps, or through the avenues, it was almost impossible to dispel the illusion that we were not driving through the domain of some rich proprietor, and we almost expected to draw up before the door of some lordly mansion. Our afternoon drive from the Au Sable to Ottowa was through a treeless prairie, looking very much like a vast lake or ocean. So much is this appearance acknowledged by the country people that they call the stage coach, a prairie schooner. When the sun shines brightly over the landscape, its yellow light gives the prairie an azure hue, so that one can scarcely see where the earth ends, and the sky begins. The undulations are a very singular feature in the landscape. This is best seen at early morning or sunset light—the summit of every little swell is illumined, while the hollow between lies in shadow, thus making the ground a curious chequer work. We saw many prairie hens, or species of grouse this afternoon, but no wolves or deer, much to my regret. The road is so much travelled that they avoid it and retire to more sequestered places. Birds innumerable, were sporting in the sun's light among the flowers, and butterflies clad like Miltons angels, in 'purple beams, and azure wings, that up they fly so drest.'

━━━━ I CAN FIND NOTHING ABOUT ELIZA STEELE, BUT SHE WAS obviously possessed of an adventurous and tolerant spirit. (Even today travelers are often disappointed when they discover the gulf between their desire to experience exotic locations and their assumption that they can do so without some sacrifice of the creature comforts.) In her preface she says, "This little book assumes to be nothing more than a note book of all that passed before the observation of the author, during a summer tour of four thousand miles." Her purpose is to provide accurate information "to future tourists and emigrants."

Berrien and Cass Counties
G.F.

⌘ ⌘ ⌘ Mr. Editor—I take the liberty of sending you a brief sketch of my remnicencies [sic] of Michigan, in the year '31, and contrast that it presented with the year '42.

When travelling through the rich and lovely Peninsula of Michigan, in the spring of '31, upon a tour of discovery with a young friend of mine, we were struck with the wild, beautiful, and picturesque scenery which exhibited itself in almost wasteful profusion upon every side. Having recently left one of the most wealthy and populous States in the Union, in order to scan the resources of our western territories, and supposing it to be like most other new countries, we were ill-prepared for the agreeable surprise that followed our peregrinations in this now flourishing and interesting State. The numberless streams that then moved in lone and solitary grandeur from the heart of this American Paradise, to water and fertalize [sic] its soil, the bright and sparkling lakes, with their verdant banks, that dotted its surface in almost every direction, the majestic prairies, clothed in living green, waving their soft and luxuriant folds gently with the breeze, were scenes altogether new and striking. We gazed with rapture on the prospect before us, and the encomiums we bestowed upon it were as profuse as the country was beautiful. Yet still, they were merited, for never was my imagination, and in fact all my senses, so bountifully entertained as on the present occasion. It seemed as if Nature, on some joyful holiday, had decorated and fitted up the land for her own peculiar enjoyment. We had traversed the whole interior of the State, and were now about entering the county of Berrien. As we moved slowly along, we took note of that which was most interesting and striking, and as we neared the old "St. Jo," as she is familiarly styled, our eyes were occasionally greeted with

---

G.F., "Berrien and Cass Counties," *Niles Republican*, June 30, 1842.

the sight of the cabin of some enterprizing [sic] pioneer who was just making an opening in front of his domicil [sic], from the prolific soil of which he formed undoubtedly vast ideas of future wealth and happiness. The orchard-like appearance of the oak openings, contrasting strangely with the thick and timbered land, afforded another source of enjoyment and fruitful topic of conversation to young farmers (as we then were); but it was soon broken off by the appearance of smoke curling gently from the several hamlets in the valley below. Before us the beautiful St. Jo rolled in peerless grandeur between her high and towering banks. The small group of houses below was the village of Niles. We put up at a small and comfortable inn, and after supper, strolled up the river along its bank in search of some of those antiquated remains for the ancient race that once was supposed to have peopled the western part of this continent. We had not rambled far before we came across the remains of an old fort, the appearance of which indicated that it was near a century and a half old. Upon the higher bank, on the opposite side of the river, were to be seen several small mounds, apparently the burial place of our red brethren. A few rods from the fort, and higher up the bank, we found one solitary mound surmounted with a rude cross bearing no inscription. We seated ourselves upon the mound and indulged in loose conjectures as to the occupant of the narrow dwelling beneath us. The old fort, mounds, and the general appearance of the country around, argued strongly with us that death had been striving there. The sun was just sinking behind the western horizon, as we took an Indian trail and pushed for the village, at which place we arrived in the space of twenty minutes.

The next morning, our horses being in good spirits, we determined on a visit to Cass county, which was highly recommended for its beauty and fertility, and through the southern part of which we crossed on our route to Niles. We found this county, like Berrien, very sparely populated, and like her possessing every natural advantage that can make a new country valuable. In this county we touched upon no less than five beautiful prairies, all bearing the appearance of once having been the domain of some half civilized lord, and who, in the plentitude of his powers, sought to assist nature in her gambols. The heavy timbered land and openings were of the choicest kinds; the soil wore the appearance of being unusually rich and productive. Upon a small prairie, near the centre of the county, my friend determined to locate, and, after making the necessary preparations, I left him. I had intended to settle with him; but circumstances of a private nature forbade. I now journeyed towards my eastern home, where I arrived in the month of August.

One thing was remarked in my tour through this county: I had not seen any appearance of a village. Now, readers, look to the year '42, I am here again. The county presents a far different appearance. The wild and almost uninhabited county of Cass, is now one of the richest, though not the most populous, counties in the State; but for a county of so recent growth, she is well populated, having near seven thousand inhabitants. Since my last visit, the handsome and tasteful villages of Cassopolis, Edwardsburgh, Lagrange and Adamsville, have sprung up and become ornaments to the county, and are filled with worthy and industrious inhabitants.

Sketches of My Own Times
David Turpie

ᔉ ᔉ ᔉ [The death] of Mr. James Whitcomb ... in October, 1852 ... caused a vacancy in the Senate, to be filled by a successor chosen by the legislature, for a seat in which I had become a candidate. . . .

Travel in this campaign was made altogether upon horseback. A large part of the country to be canvassed lay in the Grand Prairie. There were miles of uncultivated land, wholly treeless, without even a bush to make a riding whip, but the growth of the grass was so rank and close that a wheeled vehicle would have been a useless encumbrance. I met, and accompanied to their appointments in the two counties, the Whig and Democratic candidates for Congress, who traveled together on horseback, and who also had joint discussions. This was the usual method of making the campaign, and continued to be such for many years afterward. . . .

It was during this campaign of 1852 that I became really acquainted with the prairie and its people. The country was very sparsely settled; there were few roads, and the traveler might ride for hours without meeting or seeing any one; he directed his course by the sun, or, if it were a cloudy day, by the distant groves, which looked like islands in this vast expanse of grassy plain. Sometimes he traveled in solitude a tract where he could not see timber at all, like the sailor out of sight of land; the landscape in every direction was bounded by a horizon wherein nothing appeared but the green below and the blue above. The surface was generally level, broken only by slight

David Turpie, *Sketches of My Own Times* (Indianapolis: Bobbs-Merrill Co., 1903), 110–15.

undulations, and had the monotony of an ocean view with the same pleasing variety—whenever the wind blew, the tall grass rippled, fell and rose again in marvelous similitude to the sea. When the sun was not to be seen, and the weather was so hazy that the groves were not visible, the stranger had better retrace his steps; to be lost on the prairie was by no means a pleasant experience.

The most notable plant in these great natural meadows was the blue-joint grass, so called from the color of its stalk and leaves, which was dark green with a bluish tint near the ground. It was indigenous to the prairie, not found in the woodlands. The blue-stem ordinarily grew to the height of a man's shoulder, sometimes so tall as to conceal a man on horseback. Cattle, sheep and horses were all fond of it; during the whole growing season and until late in the fall it was tender, juicy and succulent; cut and cured as hay, it was by many thought to be as good as the best varieties of the cultivated grasses. It was not at all like the swamp or marsh grass, being found only on rich and comparatively dry land. The acreage of this wild meadow growth was co-extensive with the prairie.

Although the range was pastured by numerous and large herds, there were many miles of blue-stem that seemed never to have been grazed upon save by the deer. When the deer, tempted by curiosity more than appetite, made a visit to the fields and clearings in the timber, a chase followed. As long as the pursuit was confined to the woods he might be overtaken or brought to bay: but when the stag reached the open prairie he ran no longer; he jumped, he leaped twenty or thirty feet at a bound; the hounds entangled in the long thick grass soon lost both scent and sight, and the game escaped. The prairie was a grand resort for game, both great and small, but it was hard to draw the cover.

The blue-stem was a free-born native of the soil. It would endure burning and thrived lustily after its cremation, but it could not bear captivity. It scorned inclosure, resented being too often trodden under foot, and brooked not cultivation in any form. Thus when fields and fences came into vogue it soon disappeared and has now become almost extinct. It was this grass, the blue-stem, which furnished fuel for the prairie fires. In the fall it ripened, becoming very light and dry, changing its color to a grayish white. Here then were thousands of acres of highly combustible material, awaiting only the touch of the torch. Sometimes the fires originated by accident, sometimes by design. A herdsman intending to burn off a certain space to improve the pasturage, set out fire for that purpose; but if it escaped from his control, and were carried by the wind, it spread with amazing rapidity, and became then what was called a wild fire. A wild fire on the open prairie was a magnificent spectacle, combining all the elements of terror and grandeur. It compared with a fire in the woods or in a city as Niagara compares with the waterfall of a mill-dam. In advance of it was heard a loud roar, sullen and incessant; volumes of smoke arose from its burning front, obscuring the light of the sun, clothing the whole landscape at midday for miles in the somber hues of twilight; huge masses of flame, in startling form and figure, leaped high into the air; innumerable glowing sparks, as if from a furnace, flew before it and fell on the dry grass like flakes of fiery snow.

When this blazing peril threatened a farm, the neighbors mustered a fire-brigade in hot haste. They came, riding at full speed; they had seen the signal at a distance and knew by the course of the wind what place was threatened. These firemen were each equipped with a pair of buckskin gloves and a bundle of long twigs made into what was called a brush or fire broom. A sort of skirmish line was quickly formed between the premises in danger and the coming fire. These skirmishers rapidly set out fires along their whole line, which spread toward the place to be protected; but these fires were kept carefully under management until they had consumed all the grass in a space of sufficient width, when they were whipped out with the brush. The firemen, taking their stations at intervals along the inside of the edge of the space thus burned off, waited for the coming of the wild fire. The heat from it became intense, the smoke was dense and stifling; but they remained at their post. When the wild fire reached the outer line of the "need-burn," as the burnt-off space was called, it halted. Having nothing to feed upon it died down, and the flames gradually subsided. The only danger then was from the sparks, which, borne by the wind, now risen to a gale, were sometimes carried clear over the need-burn into the dry grass beyond. The skirmishers, at this time, did lively work. They watched where the sparks fell, and wherever a blaze appeared they whipped it out. After a while the fire in front ceased burning, the sparks coming from it were all black and dead, the danger point was passed, and the neighbors dispersed to their homes.

This service of the prairie fire-brigade was one of hard work and of some danger, requiring a quick eye, rapid movement, presence of mind and much endurance. The thirst, aggravated by the heat and smoke, became almost intolerable. Having been once or twice engaged as a volunteer in such a contest, I can speak somewhat as to its character. I have, in some sudden emergency, heard even of women working on the fire-line, but usually they were sent with the children to a place of safety.

Charred remains of a great prairie fire were, to the beholder, more impressive than the ruins of any other conflagration. Let him take his position near the center of a burn of three or four thousand acres and look round him. He might well fancy that the whole earth was hidden beneath the pall. Here and there, rarely, he might see a white spot on the blackened surface; this was a small patch of the dry blue-stem which, by some inconceivable caprice of the wind, had been left untouched by the flames. Such a sight was more remarkable, as it was more unaccountable, than that of the famous Ogden house among the fire ruins of Chicago. The escape of either seemed a miracle, but that of the tiny grass-plot was the greater.

〰〰〰 DAVID TURPIE (1828–1909) SERVED AS ONE OF INDIANA'S U.S. senators during 1863 and then again from 1887–99. His earliest years were spent on a farm in southern Indiana, but his parents moved to Lafayette because they felt that only a sizable town could provide adequate education for their children. Even there, however, books were scarce.

Turpie entered Kenyon College as a second-semester sophomore, having tested out of his first year and a half of courses, and received his degree two years later. Although his goal was to become an attorney, he first needed a job and so he decided to try the teaching profession. His good penmanship won over the school trustees, but all they could provide was the school building. If he wanted a salary in addition, it was up to him to go house to house "with my article of subscription, soliciting pupils." He eventually signed up fifty students.

Turpie did find a local attorney who agreed to be his mentor, and after finishing three terms as schoolmaster, he applied for admission to the bar. Upon successfully passing the three-hour exam, his name was entered on the rolls of those who could practice law in Indiana. The year was 1849. On the advice of his mentor, he moved to Monticello, the county seat of White County, just south of Jasper County. Four years later, he was elected to the General Assembly.

During one losing campaign for the U.S. House of Representatives (he lost three times to the same person), Turpie was to debate his Republican opponent in an outdoor venue near a railroad station. As Turpie launched into his opening remarks, an engine pulled in and let loose with its whistle so as to drown him out. Exasperated, he looked for guidance from the local

Democratic chairman, who urged him to continue. When it was the Republican's turn, another engine appeared, which then proceeded to whistle just as loudly as the first.

In 1887 Turpie was elected to a full term in the U.S. Senate, barely defeating Benjamin Harrison on the sixteenth ballot. In those days U.S. senators were chosen by the state legislature. (It is easy to forget how much distrust the framers of the Constitution had toward most of the populace, for it wasn't until the ratification of the Seventeenth Amendment in 1913 that senators were elected by popular vote.) To gain an advantage, the Republican-controlled state house unseated a Democratic member, which only led the Democratic-controlled senate to unseat a Republican. This went on for several rounds until both sides conceded the futility of the maneuvers. Then a group of four house members opted to support a third independent candidate, thereby denying either Turpie or Harrison a majority. Throw in a lawsuit regarding who should be president pro tempore of the state senate (a position with much influence on how the process would transpire), and the election wasn't completely settled until months after Turpie had already been sworn in. One has only to recall the 2000 presidential election to realize how far the practice of politics has advanced in this country since the days when fires fed by native grasses still raced across the midwestern plains.

Source: David Turpie, *Sketches of My Own Times* (Indianapolis: Bobbs-Merrill Co., 1903).

Rats! Rats! Rats!

Carl Schurz

⁀⁀⁀ At twilight we entered the "grand prairie" which occupies the northwestern quarter of Indiana. The movement out of the forest on to the prairie is comparable to that out of a stream into the high sea. On both sides of the railway track the woods recede farther and farther, just as the stream opens out into the bay; and as you gaze ahead ever on the level, endless prairie meadows, your eyes find no point of rest save the sharp straight streak which the horizon makes; here and there perhaps a straggling clump of trees or a small farmhouse, which stands forth like single islands in the waveless

Carl Schurz, "Rats! Rats! Rats!" (1854), ed. Paul Angle, *Chicago History* 8, no. 2 (Winter 1966–67): 47.

grass sea. Finally, the forest banks right and left disappear, and wherever you run your eyes they see only the unbroken inexorable dead plains. I believe there can be no profounder sense of abandonment than to be alone upon a great prairie. The sea is much more alive than the prairie. There at least the waves shift grandly and the horizon changes with their movements, but even a storm leaves the prairie still. It must be a remarkable sight to witness from a distance a train rolling over the prairie. Flowers are abundant and of many colors, but when one regards the prairie as a whole its flowers are forgotten. The grand prairie has a rich soil and in some localities is already studded with farms; but however much I am compelled to love the west, at least what I have seen of it, I should not like to live upon a great prairie.

～～～ ACCORDING TO PAUL ANGLE, CARL SCHURZ (1829–1906) "WAS one of the most remarkable men of the nineteenth century." It is a pity that his writings on the midwestern landscape are too short to justify going into detail here about his amazing life. Suffice it to say he was active in the losing side of the German Revolution of 1848, an accomplished pianist, a U.S. senator from Missouri, an organizer of the Republican Party, a U.S. minister to Spain, a major general in the Union army, secretary of the interior under Rutherford Hayes (where he advocated fair treatment of native peoples), and editor of several major newspapers.

Sources: Carl Schurz, "Rats! Rats! Rats!" ed. Paul Angle, *Chicago History* 8, no. 2 (Winter 1966–67): 47; Claude Eggertsen, "Schurz," in *The World Book Encyclopedia*, vol. 16 (Chicago: Field Enterprises Education Corp., 1962), 160.

Prince of Wales in Canada and the United States
Nicholas A. Woods

ᖆᖆᖆ It is comparatively easy for any of my readers to imagine a prairie—it is next to impossible to describe one. Leave Dwight behind you, and walk out to the east till all sight and sound of the little village is lost in the distance, and then look round you. There is a huge, undulating ocean of long,

Nicholas A. Woods, *Prince of Wales in Canada and the United States* (London: Bradbury and Evans, 1861), 297–307.

rich grass and flowers, which the warm, soft wind keeps in a gentle ripple. There is not a sound but the shrill chirping of millions of crickets, not a shrub or a bush to break the dead level of the distant horizon—nothing to vary the wide-spread area of verdure but its own masses of bright wild-flowers, over which gorgeous butterflies keep always skimming on noiseless wings. This is the prairie. About a mile or so a-head is a slight, but very perceptible rise in the ground, and you push on for this to get a good look about you. There is, of course, no track, and your way lies through the prairie grass, in autumn little more than breast high, but in the spring almost over your head; you stride through clumps of resin and compass weeds, through patches of blue, yellow, and purple flowers, through thyme and long rich grass with tall, tufted, reedy plants in the midst, which attract your notice at once. It is the rattlesnake weed, always most plentiful where this deadly reptile abounds, and the root of which, with immense doses of corn whisky is said, under certain favourable contingencies, to have averted fatal results from the bites of small reptiles of this species. Where the snake-weed is plentiful, beware and look out well for the snakes too. You can't walk far through the prairie on a hot morning without hearing the dry sharp hissing rattle of one of these deadly serpents, as with his tail so quickly vibrating that you can scarcely distinguish its end, and with the lean, hungry-looking head erect, it moves sluggishly away in search of a place where it may repose and bask undisturbed. Such dangerous occupants of the grass are very common in the prairies, and may with prairie wolves and sometimes deer be seen within a stone's throw of the houses of Dwight itself. But all this while you are plodding through the grass, turning aside for one minute to look at the little prairie crabs which burrow down their holes some fifteen feet to the level of the water below the gravel, and into which they drop at once on the slightest sound of alarm, or else you watch the coveys of prairie hens as they rise with a whirr right and left, and go skimming along like grouse a little a-head of you. At last you gain the summit of the gentle rise, and can look around you for miles on miles in all directions, yet you are almost disappointed to find that you have gained nothing by your walk—that the same tremendous extent of wild meadow land, clothed with a rich luxuriance of grass and flowers, stretches away on every side till deep green fades into brown in the distance, and a line of blackish-blue on the ocean, far, far out, marks where the horizon meets the sky. Yet the land is not all level. It has a series of gentle undulations—of low, long sloping ridges, as if an inland sea, when slowly moving with a quiet, regular swell, had on the instant been changed to rich and fertile land. The prairie of which I write this is known as the Grand Prairie, from the extraordinary fertility of its land—for its length is only 150 miles by 60. But in a south-easterly direction from Dwight one may journey for more than 300 miles and never once quit their long, shallow ridges—never see anything but the external expanse of deep green grass, perfumed with the gum droppings of the resin weed. The southern prairies are broken here and there by water-courses, of clumps of cotton-wood and groves of locust trees. Occasionally, though at rare intervals, a little line of locust trees, looking like rocks in the great ocean of grass, mark where pools of water may be found. These varieties, however, are but few, and after a journey in the great wilderness a tree almost startles you as something out of place in the huge soft green meadow-sea, where the long coarse silky-looking grass bears nothing stronger than a resin weed among it, and where a breath of wind ripples its whole surface into breakers of verdure, which even in the calmest days gives such an aspect of life and animation to these silent and deserted lands. One might write for days and days on prairie land and prairie life, and yet give but a faint idea of either to those who have not seen them. It is the wild, the overflowing abundance of animal and vegetable life which fills these great reservoirs of nature, the knowledge that the thousands of square miles of soil over which you travel is the richest and most luxuriant in the world; and yet, in spite of this, the utter desolation and absence of the trace of any human being which surprises you, one time with gratitude that there is much land to spare, and the next moment with regret that its great riches should be so neglected and forlorn. Travel on for miles and miles, for days and nights pass from Illinois across the broad turbid waters of the Mississippi, into the slave State of Missouri—journey for hundreds and hundreds of leagues, as you may do then, yet not quit for a single day those monstrous grassy wastes, those perpetual land calms, in which a silence as great as that upon the sea seems always to remain unbroken.

The inexperienced person, however, must be very careful how he ventures on these luxuriant steppes alone. Let him but lose sight of his faint landmarks, and make one or two incautious turns, and he will instantly find himself engaged in a game of blindman's buff on a most extensive and unpleasant scale, and must catch what way he can back again. In vain you search for the track you have made through the long grass. A breath of wind is sufficient to conceal it from your inexperienced eyes, though a week afterwards an Indian runner could follow it up with as much ease as if it were a paved road. You push forward in what you think a straight course, but it is ten to one that you only make huge circles round the place from which you started, and it is then that the eternal solemn silence of the great plains becomes not only impressive but almost alarming, when every fresh effort to strike a track increases your weariness, and you feel yourself a helpless prisoner in these huge, bright smiling solitudes. Fortunately, none go on the prairie for the first time without being shown, in case of such mishaps, the groups of compass weed, which abound all over the plains, and the broad flat leaves of which point due north and south with an accuracy as unvarying as that of the magnetic needle itself. And thus with the aid of these useful little weeds and the sun's course, you may make tracks across the broadest prairie with the most unerring certainty.

The great danger to which travellers on the largest prairies are exposed is fire. Scorched during one or two months of summer by an almost tropical heat, the grass shrivels up into a coarse brown-looking hay, and while in this state is constantly lit accidentally by the carelessness of travellers or hunters, or by flashes from the terrific lightning storms which are always sweeping over the plains. With a brisk night wind in the height of summer a prairie fire spreads over the whole plains with awful rapidity, and, unless well mounted, woe betide the unlucky travellers, who, roused by the smoky heat from their slumbers, see the great horizon of orange-coloured flame in the distance, like a vast semicircle of fire bearing down rapidly towards them. On the small prairies instant flight is the only chance of safety. On the great and wilder prairies flight is useless, and the only expedient that offers any hope of safety is riding madly with the wind some ten miles in advance of the fire and lighting the prairie before you at two or three points. As the wind bears the flame rapidly a-head, the travellers, after a short interval, are enabled to follow along the scorched track comparatively out of reach of the flames coming up behind, which of course stop on the margin of the burnt ground for want of fuel. But even this dangerous expedient will fail if the fires take place in June when the grass is very high. So much scorched embers then remain behind, that no horse can venture in, and no rider could live in the dense stubble smoke. In such desperate straits the only chance is to slay and disembowel the horse, and literally creep into the raw cavity till the flames have passed, and as there are instances of this resource having sometimes saved the lives of Indians and hunters it is perhaps not too much to conclude that it has often been tried with less successful results. In fact this is the last resource, and must be prompted by the same desperate clinging to life which induces a sailor to hold on to a plank when shipwrecked in the middle of the Atlantic. What chance is there of escape for the man who survives suffocation from the passing flame and emerges alone and on foot in the middle of a vast burnt prairie? On the first Monday of the Prince's visit he was so fortunate as to see one of these tremendous prairie conflagrations. The day had been very hot and sultry, and the royal guest was still out shooting with the Duke of Newcastle, Captain Retallack, and Mr. Spencer. The rest of the suite had taken different directions widely apart, and were still absent on the prairie as night fell. With the darkness came an almost deeper gloom as huge masses of dense thunder clouds rose up into the angry sky. Before any of the parties could reach Dwight a dreadful storm raged, and the wind, after moaning and roaring about the plain like a hurricane at sea, would suddenly cease, and a portentous silent darkness reign over the whole scene—a silence so intense that the vivid flashes of lightning, noiseless as they were, seemed almost to break as the great livid streaks darted down and went flickering over the plains in all directions. While watching the dreadful solemnity of this storm in such a wild, I could not help noticing three dull red, copper-coloured banks of clouds at different parts of the horizon, and asked my kind host Mr. Morgan to explain what they meant or were. The explanation was given in five words, for the instant his quick eye caught the distant tinge, he exclaimed, "The prairie is on fire." And so indeed it proved to be. Whether it had caught from some smouldering gun wadding, or, as was thought far more probable, had been ignited by the incessant flashes of lightning during the storm, it was hard to say. Only one thing admitted of no doubt whatever, and that was that the grass had caught in three distinct places.

At first it seemed probable that a short, quick flood of rain which fell after the storm, and which for two or three minutes was heavy enough almost to extinguish anything, would check its progress, as for a time in fact it did. But the fire had obtained too firm a hold, and as the rain ceased the wind rose, and the smouldering red patches on the verge of the horizon grew brighter and brighter, spreading along with an angry rapidity that brought each separate conflagration closer and closer every minute. The wind was away from the village of Dwight and its rich belt of corn fields, and turning the flames westward, over the mass of prairie; and as these fires, especially at that season of the year, do the land much good, the progress of the conflagration was watched with perfect indifference. Soon the sky, from reflecting a narrow strip of red, lit up with an angry glare as the mass of fire spread beneath it—the little patches of flame began to crest the undulations, and ragged columns of dense fiery smoke streamed away in lurid masses as if it would carry the flame and heat up into the clouds themselves. An hour more and the three fires had apparently joined, or, at least, were so close together that they formed one huge belt of flame that covered the earth and lit up the sky for miles and miles. The fire was at least eight or nine miles distant from Dwight, and from there it looked comparatively a small space in the immense horizon of and around, and only by the bright orange flame in the distance, and the mass of fiery sky above, could one judge adequately of the real area occupied by the burning plains. Not so, however, as after a long ride you approached them from the windward side. For a mile and more before you reached the edge of the fire you were in its bright orange light, which made everything as visible as if it were noon day, and the sun was shining fiercely through a blood-coloured haze. You could hear the sharp barking howl of the prairie wolves as they rushed away for the darkness, and see the prairie hens fluttering and fluttering from place to place, turning in their wild terror full into the smoke, when they fall and perish instantly. At last you gain a little rise and look beyond into such a scene as nothing but a prairie fire can show. It spreads out a sea of red smouldering ashes, glowing for miles in all directions, while the deep white ridge of flames a-head mount the slopes with awful rapidity, and flap their heavy tongues up into the air with a hoarse roaring noise that fills you with astonishment and almost terror. Hour after hour you may stand, fascinated with the terrible beauties of the scene, as the mass of red sultry ruins grows and grows each minute, till your eyes are pained and heated with its angry glare and you almost dread the grand, fierce sheet of fire, which has swept all trace of vegetation from the surface of the prairie. On Monday night, when near twelve o'clock, the wind changed a little and turned the flames nearly back upon the ground they had already devastated, and this at once checked their progress. On the next day, however, they sprang up afresh and raged faster and faster than before, and the whole extent of prairie east of Dwight was hidden under such a dense cloud of yellow smoke as I never saw before. And on the last night the glare was tremendous—as if the world itself was burning.

The first day's sport of the Prince was far more successful than was anticipated. The prairie hens resemble English grouse (except that they are larger—almost the size of cock pheasants), and the sport of shooting them is followed in the same manner with pointers. There was a bet as to which of the three parties would return with the heaviest bag. The Prince with the Duke went east. General Bruce, with Colonel Grey, Major Teesdale, and Mr. Wilkins, went west; and Viscount Hinchinbrooke and the Hon. Mr. Elliott steered due south. Dr. Ackland went out with a gun in his hand and a pencil in his thoughts, and, as usual, after one or two shots fell to making beautiful sketches of the prairies. In the evening when the three parties returned, there was, considering the lateness of the season and the wildness of the birds, rather a good bag. More than fifty brace of prairie hens, exclusive of such other game as plover and quail, was the result of the united day's sport. Of this number eleven and a half brace had fallen to the Prince's gun—eclipsing the Duke of Newcastle's sport by three birds. The dollar bet therefore was won by His Royal Highness, who in this as during other days' sporting showed himself to be a crack shot, and the best of the party.

On the following morning the whole party started at six o'clock to a place called Stuart's Grove, on the edge of the prairie some thirty miles from Dwight, and one of the most celebrated covers for quail in the country. Here there was a regular battue from about eight in the morning till twelve in the day, when the heat became great, so a halt was called in a shady little nook between the brushwood, and the Prince and the Duke rested themselves and had lunch, and afterwards slept for a couple of hours till nearly four o'clock. Shooting was then renewed with redoubled vigour and the united bag of the whole party amounted to ninety-five and a half brace of game, twenty-eight of which, with some rabbits and plovers and a brace of prairie hens, were brought down by the Prince. Again therefore he had the honour of beating all the party by several birds. The skill and rapidity with which he knocked over the quail perfectly astonished the prairie sportsmen who were with him.

He was certainly most fortunate in his visit, for, for the time of the year, he had most unusual sport; he saw a prairie thunderstorm, a prairie fire of immense extent, and, above all, a prairie sunset. The latter took place in all its supernatural glory—a glory which can never be described or understood by those who have not seen it—while the party were shooting the quail the night before their departure. As the sun neared the rich green horizon, it turned the whole ocean of meadow into a sheet of gold which seemed to blend with the great firmament of reds and pinks, pale rosy orange hues, and solemn angry-looking crimson clouds above till not only the sky but all the land around was steeped in piles of colour as if the heavens were reflected from below, or as if the sinking sun shone through the very earth like mist, and turned it to a rainbow. The immensity of stillness which lay in the prairie then—a stillness as profound and vast as the green solitude itself, while not a breath stirred over the whole horizon as the great transmutation went slowly on, and the colours over the land turned from rosy to pink, from pink to orange, orange to red and crimson—darkening and darkening always as the tints ebbed out like a celestial tide leaving fragments of scarlet clouds over the heavens—the embers of the fire which had lit the prairie in a flame of glory. There was such a quiet unspeakable richness in this grand farewell of day— such a terrible redness about the sky at last that one could almost fancy some supernatural phenomenon had occurred, that the sun had gone forever, and left a deep and gory wound across the darkening sky.

◆◆◆◆ NICHOLAS WOODS WAS A CORRESPONDENT FOR THE *TIMES* OF *London*. The only fact I have found about him is that he also covered the laying of the first transatlantic cable for the newspaper. Dwight, Illinois, is located in Livingston County, just two miles south of the Grundy County line.

Breeding Habits of *Ardea heroias* as Seen During a Visit to Crane Town

B.

🐦 🐦 🐦 Every year the Great Blue Heron return from their winter sojurn [*sic*] in the South and generally seek a suitable place for their nests. Being gregarious, a nesting place often assumes vast proportions, covering acres, and is called "Crane Town" or "Rookery." The town in question, is about 20 miles from Valparaiso, in the Kankakee Marsh; which in the spring, is invariably covered with water, four or five feet deep. This portion of the swamp was visited by a fire which destroyed many trees, and left only high, black stumps, with a few charred branches.

To the uninitiated, Crane Town is inaccessible. We were favored, however, with guides; old trappers, with whom arrangements had been perfected by letter.

We left Valparaiso very early in the morning, but found, on reaching the landing, that our men had been patiently waiting for us, for more than two hours.

There were two long, narrow boats, that they called "pointers," and I might give you a "pointer" right here, that they were very long, very narrow, and very easily upset.

B., "Breeding Habits of *Ardea heroias* as Seen during a Visit to Crane Town," *Hoosier Naturalist* 1, no. 6 (January 1886): 81.

Our luggage was placed at one end and we [were] directed to sit on the bottom of the boat, at the other end. The narrowness of the boat, however, made this a very unpleasant task. The marsh being passed, we made rapid progress up the muddy Kankakee.

For wildness and novel grandeur, the river and surrounding forest excel every thing of the kind I ever saw.

After ascending the river for two or more miles, we approached a small island; the only one for miles around, on which was a low hut, built by the trappers.

With difficulty we extricated ourselves from the boat, and entered the hut. The walls were black with the smoke and grease of a dozen winters. On one side was a tier of bunks, stea[m]boat fashion, made of rough poles, and covered with blankets; later, we found them comfortable sleeping places. In one corner was a stove, with necessary culinary apparatus, and strewn around rather promiscuously, were dogs, traps, hides, guns, and fish-spears. The place was so extremely odoriferous, that we were compelled to beat a hasty retreat.

I suppose our olfactory nerves becam[e] paralyzed, for in a short time we did not notice the "odors warious," and ate our dinner with a keen relish.

After dinner we reluctantly entered the boats, and were poled through many winding passages, until, unexpectedly, we reached Crane Town.

It was a wonderful sight. As far as we could see, were great, charred, forest monarchs, patiently supporting in their blackened arms, from one to six nests. The water under the nests was covered with rubbish that had been dropped during their building.

Our first shot caused such a din and commotion among the residents as we had never heard before. It was deafening. It seemed like all the heron of the universe had assembled to give us battle, and their wide, outspread wings, as they circled round and round above us, literally hid the sun from view. Recovering from our surprise, we fired rapidly, and although the birds soon rose out of reach of our guns, yet we obtained five fine fellows in a very short time. After this fusilade they became wary, and it was with difficulty one could be approached. Becoming tired of this, we directed our attention to the nests, which were sixty and seventy feet from the water, and resembled huge inverted brush heaps, many of them being more than eighteen feet in circumference. The upper surface of the nest was flat, with a slight depression in the center, which was sparsely lined with twigs.

Four eggs generally comprised a set, they were greenish-blue, unspotted, and varied in form, from elliptical to oval. Their average length was 2.75 inches.

Many of the larger trees had five or six nests in their branches and were inaccessible without climbers; so, to reach the eggs, climbable trees had to be selected and even this was dangerous and fatigueing work. During the afternoon, seven nests were examined, three of which were empty, the remaining four had four eggs each, and from their rareness (?) abundantly paid us for our trouble. Their market value, among naturalists, is about $2 per dozen.

The Shooting Clubs of Chicago—No. XI: The Grand Calumet Heights Club

Emerson Hough

☙ ☙ ☙ The man was nearly right who said of the Chicago clubs that there is not "any one of them alike." It is a fact that each has some features of interest not possessed by any of the others. It is this fact which has made the visits to the clubs so exceptionally pleasant, and which is the main foundation of the belief that so long a series of articles may still retain some interest, even those not members of the clubs themselves.

Of all the clubs, none has more unique features than the Grand Calumet Heights Club, and none will better repay the curious seeker after novelty or pleasure.

This organization, not yet two years old, but already numbering well on toward a hundred members, is practically an offshoot of the old Lake George Club, which, like itself, lies in the famous Calumet region, by many thought to be the greatest duck country in the West, and equal or superior, in natural advantages to the famous Fox Lake district. The Tolleston, Lake George, and Grand Calumet Heights clubs, are all kindred "sandhill" or "Calumet" clubs, and lie not far apart in that favored region....

The main Calumet stream connects Calumet Lake with Lake Michigan. Its mouth is now built up with the wharves and docks of South Chicago. Calumet Lake is feebly initiated by Hyde Lake, Wolf Lake, and Lake George,

Emerson Hough, "The Shooting Clubs of Chicago—No. XI: The Grand Calumet Heights Club," *Forest and Stream*, March 21, 1889, 172–74.

which lie between it and the big water. It was in this marsh district, supplied fully with all sorts of feed, and inhabited by countless myriads of ducks in the good old days, that the old Kleinman homestead lay. Kleinman *pere* kept the wagon hot, hauling ducks to Chicago, and if the boys didn't have another wagonload killed by the time he got back there was blood on the moon. That was where the Klienmans learned to shoot.

Out of this district two more absent-minded streams start forth, but these run in a general direction, exactly opposite to that of the main Calumet. The Little Calumet starts out for the middle of the State, changes its mind over by Blue Island bluffs, where it takes in the "Feeder" from the "Sag," and then whirls around four or five times and wobbles in an entirely drunken and disreputable fashion over toward Lake Michigan, which it reaches no one knows where. It passes though Tolleston Marsh, and is the basis of the Tolleston Club, which has been described in an earlier article. The only object the Little Calumet has in going out into the country in this way is to get up in the world. Near Miller station the Little Calumet is by the Government survey shown to be 19.3 ft. higher than the Grand Calumet, which is only three miles distant from it. A canal cut across this neck—as is the proposition now before a certain improvement company—would drain Tolleston marsh dry as a bone apparently. It would not affect the Grand Calumet marshes, which are always of just the same level as Lake Michigan, since they lie right along the shore, and would be maintained by seepage in some localities, even if the crawling current of their main waterway had more ambition than it has.

As may be seen the Grand Calumet, the remaining one of these three erratic rivers, is necessarily a slow, deliberate, tortuous and torturingly crooked stream. It doesn't go inland, and it doesn't go north. It just strolls off among the sandhills and pine barrens toward the foot of the lake, sometimes running within a half mile of Lake Michigan, and then again changing its mind and taking a while over in the opposite direction. Its general appearance is that of a long, crooked valley of rice and cane, running between low wooded banks and stretching out from half a mile to three-quarters of a mile or more in width. Somewhere in this winding marsh, hidden by what a poet would call the lush and dank sedges of the marsh, creeps the deliberate Grand, 10 to 15 ft. deep in much of its channel, a lake creek rather than a river, and a darling for ducks.

History does not quite agree whether it was Mr. Lloyd and Mr. Booth or Mr. Lloyd and Mr. Cleaver who first explored the Grand Calumet to its end. The first two gentlemen were old Calumet Lake habitues, stopping at

Florence Benner's old house, and the latter was one of the old Lake George Club. They bethought them it might be well to seek fresh fields and pastures green, and the result was an exploring voyage, which started at Wolf Lake, went through the cut-offs into the Grand Calumet and down. The first camp was made a mile below the village of Clarke. From there on the explorers found themselves in what is very nearly a complete wilderness. Huge sand-hills, covered with scrubby oaks and gigantic pines, rose up in serrated ridges. The river meandered and meandered, often nearly blocked by floating islands of matted vegetation. At last they got through, and their little craft caught sight of the blue waters of Lake Michigan, nearly at the extreme foot of the lake. Fifteen miles up the beach, not a house and not a soul in sight. Four, five, eight—nobody knows how many miles back from the lake—and not a soul or a habitation for one, except the old Negro who was found dead in his cabin, and the dugout of the horse thieves who, undiscovered for months and months, were finally arrested in Indiana without requisition by South Chicago parties, and justly, though, non-technically, sent to the penitentiary. Eagles were seen in the woods, and can be seen to-day. One night a wildcat came down to the beach. Wolves are still sometimes seen. The country is to-day as wild looking as any you would see in the middle of the North Woods. Picturesque, rough, a little gloomy and forbidding, a stranger set down in the middle of it might think it the foothills of the southern Rockies. It seemed only pleasing to our explorers, who hastened to spread their sails, and so sped in their hunting boats across the foot of the lake, north and northwest, until they landed at Sheffield on the shore, and, so again passed by a narrow chan-nel into the system of lakes from which they started.

This was two years ago. This was within thirty miles of Chicago. No other city has anything of consequence which Chicago has not also, and Chicago has some things which no other city has. I submit there is no city in America which can show so strong a string of sporting organizations and none which has as good and immediate foundation for their existence and growth. There is no more notable a sporting rival in this country, unless it be that on Chesa-peake Bay, and it is probable that a familiarity with that section would lead one to give it second place.

Very well, our explorers had seen the new land, and pronounced it good. The result was the organization of the Grand Calumet Heights Club, the leasing of 1,500 acres of land lying between the lake and the Grand Calumet River, the securing of a tidy flag station twenty-eight miles down the B.&O., and the expenditure of considerable sums in buildings and improvements.

The club was put on a good footing. Memberships were placed low, at $25, although none but proper persons were ever allowed to enter the club. Annual dues were placed at $5.... The membership was limited to 100, and, although the club is only in its second year, its membership is now seventy odd....

The social element is strongly developed in the club. The club house, situated directly upon the shore of Lake Michigan, is a most delightful retreat in the summer. The spot is always breezy, entirely free from mosquitoes, and possessed of that interest which always attaches to a wild and unfrequented locality. Last summer parties of good numbers were an almost constant occurrence. Not a week passed without seeing some one down at the club house, and on one occasion ninety guests, ladies and gentlemen, clamored for a place at the dinner call. Fishing, boating, bathing and exploring are all possible at the Grand Calumet Heights Club, which has, indeed, a far greater diversity of interest and amusement than is possible at any club of the city. Trap-shooting and target shooting with the rifle are possible under particularly favorable conditions. The blackbird traps throw the birds directly into the water of Lake Michigan, and those which are not hit are recovered unbroken, for the water is only a foot or so deep on the gradual beach, and is found not to damage the birds thus dropped into it. There is probably not any other trap ground in the country where this unique feature obtains. The big lake is useful as well as beautiful. Its presence renders the use of the rifle a thing not to be constantly guarded and dreaded, as must be the case at most clubs. No target butts are necessary, and the bullets have the whole lower end of the lake for an untenanted range. Long-range naval ordnance would be harmless there, for it is a portion of the lake rarely visited by vessels, and the pleasure craft of the club members show almost the only canvas seen there.

The lake fishing of the Calumet beach is good. Two ladies have taken as many as 100 perch in a morning's fishing, and although perch fishing does not afford very exciting sport, it serves to pass away an idle day of gentlemen, as well as ladies. The Grand Calumet, distant only half a mile inland, often shows a good day's fishing for bass and pickerel. The woods near at hand have a good many rabbits in the winter time, and the close law has developed a pretty fair supply of ruffed grouse. There is always sport of some sort to be had in the daytime and at night the great veranda or the big reception room invites to a dance if the season is such that the fair sex is present; or if it is in the ducking season, the tired hunters may cock their feet upon the balcony rail, or blow blue clouds and tell portentous yarns around the mammoth stove in the gun room. It is the just boast of the Grand Calumet men that one may visit their club every week in the year and amuse himself differently each time.

The policy of this club is not a close one, and there is no suspicion of the cloven foot in any of its management, such as might suggest that the club was mostly an affair to glorify or gratify a few. There is no rule keeping out such visitors as the members choose to bring down, and the more the merrier seems to be the liberal and hospitable precept. Gaming is not allowed in the club house, there being one rigid rule, that money shall not be staked at cards.

Leaving aside all social and summer pleasures, however, let us consider what the club may claim under its primary object, that of a ducking club. Let it be borne in mind that this Calumet country has always been a magnificent duck country, and even to-day is by no means shorn of its glory, since the natural feed and the natural resorting places for the ducks still exist in much of their former excellence. Let it be remembered also that this club lies almost directly at the foot of Lake Michigan, with nothing between it and Mackinac Island. The lake flight—and it is a very considerable one—naturally coasts along until it goes near to the heel of the lake. It may be that the bluebills and redheads find some resting place of the big reef at Kensington, or some feed in the lines of city sewage which run out into the lake yet higher up; but, if this is so, it only tends to establish a shore flight up and down the lake. Indeed, such a flight does exist, and often yields good tribute to the gunners of this club. Upon the other hand, if the ducks wish to leave the lake they may naturally be expected to do so over the narrowest possible strip of land dividing the lake from the inland feed; and they may also be expected to establish a flight back and forth between these localities. This is precisely the case at Grand Calumet Heights Club grounds. The birds work backward and forward between the lake and the marshes of Little Calumet and Grand Calumet. Harried too persistently on the great Tolleston marsh they fly over and drop into the big lake, or stop on the quiet river. Pounded at too severely on much-beleaguered Calumet Lake they essay a wearied flight between the devil and the deep sea, and drop into the little-hunted bends of the Grand Calumet, where, alack the day for them! they find the little boat houses of our club, and meet the puff of the expectant gun, as deadly, if less frequent than before.

The boat houses, and the little abode of old Blacki, the odd character who, with his ten-year old son Gus, takes charge of them, lie on the bank of the river, about half a mile from the club house, and a few hundred yards from the little railway station building. A crooked trail winds through the sandhills over to the boat house, marked by an occasional blazed tree. This sandhill country is very puzzling, the different little eminences being so much alike and the character of the wooded growth monotonous. I have rarely seen a country, outside of the sandhills of the Cimarron—which this section much resembles—where one could so easily get "turned around." It is a joke of the club to misdirect a new member as to the path, in which case he may be two hours or so in going the half mile. There is a rumor to the effect that one member has ordered a half mile of wire, which he proposes to stretch along the path.

Of the boats used on the big lake, there is space to describe only two, Mr. Cleaver's Merganser, and Mr. Loyd's Calumet Turtle, both of which offer something new in duck lore. The Merganser is a vast low-lying scow, with a great water-tight cockpit nearly 8 ft. high perched like an elongated turret upon it. The scow is worked out a quarter of a mile or so into the lake, and anchored with a 200 lbs. iron anchor. The cockpit is surrounded by brush nailed fast to it. The fleet of decoys is let out down wind from the corners. The great boat rides the heaviest waves with ease. The ducks apparently mistake it for an island, for they do not pay much attention to it, and come right into the decoys. Very good bags of mergansers and also of redheads and bluebills have been made from this odd contrivance. The ducks understand marsh blinds, but they are ignorant about this one, evidently, and they decoy into it much better than to a grass blind. The Calumet Turtle is built on much the same principle, except that its cockpit is not so high, and its supporting scow or boat is neater and trimmer, being pointed at both ends, only a few inches of air chamber being left between the perfectly flat top and bottom....

Besides the varieties of sport already mentioned there is one kind of duck shooting followed at the Grand Calumet grounds which is not possible at any of the other clubs and which impresses one as a rather unique sort of fun. This is merganser shooting along the ice in the spring. There has been a large body of these birds wintering on the lower end of the lake this year, and they are always early in making their appearance. They frequent the marshes to some extent, but seem to prefer the shallow water along the bar. They are restless in their habits, and continually fly up and down the shore. The ice is packed into the lower end of the lake and crowded upon the bars by the action of the waves into a huge field of hillocks and rough ridges, which is this spring three-fourths of a mile wide. A warm day and an off-shore wind breaks the ice into long seams and threads of water, and often moves it all out except about 100 yds. of rough little ice cliffs that hang on the bar. The hardy duck hunters walk out upon this ridge or cross to it in boats—in once case a hunter poled out on a cake of ice—and put out their decoys in the strip of water lying beyond the edge of the icebar. No blind is used, the shooter lying down on the edge, or seeking shelter in some cave or crevice. The birds work back and forth along the ice and decoy very well indeed, and good sport is had at the hard-flying if hard-eating sheldrake [mergansers] and kindred ducks. Often a good bag of bluebills and redheads is gotten in this way. I cannot imagine any more picturesque form of sport. It was a foggy day when we went out on the ice, and in a moment we were out of sight of shore. The ice rose all around us in a thousand huge fantastic forms, glimmering dully through the mist and making the scene like the dream of the Arctic Sea.

〜〜〜 THE SKETCHES BY EMERSON HOUGH (1857–1923) PROVIDE some of the best descriptions I have read of the Calumet region. (They also provide fascinating insights into one stratum of the nineteenth-century hunter/angler.) This area supported one of the largest local wetlands outside the upper Kankakee River but was the scene of both heavy industry and extensive landfills. After being ignored for many years as an eyesore, the territory has within the last fifteen or so years begun to receive the positive attention it has long deserved. The marshland on the Illinois side and the dune and swale complexes in Indiana retain their status as being among the most biologically important portions of the Chicago area, and indeed their respective states.

I had not expected to find much information on Emerson Hough, but it turns out he was quite famous in his day as a writer who explored both western and conservation topics. (I should point out that back then the Midwest, including Chicago, was considered the West.) Born in Iowa, he was the state's first prominent writer, producing twenty-seven books and hundreds of short pieces, both fiction and nonfiction. While a student at the University of Iowa, he "played on the football team, edited the college newspaper, and graduated Phi Beta Kappa in 1880" (Johnson, 2). Two years later his first article was published (in *Forest and Stream*), and he was admitted to the Iowa bar. (He was to be one more in that proud tradition of attorneys who abandoned the law in favor of a career in writing. May the great spirit bless them all.)

A friend invited Hough to join his law practice in White Oaks, New Mexico. The offer coming right after a failed romance, Hough was in the mood to travel and he enthusiastically relocated to White Oaks, a town he described as being "half mining camp and half cow camp" (Johnson, 3). Despite the widespread lawlessness of the place and the business opportunities spawned by local industries and the newly arrived railroad, the law firm of Eli Chandler and Emerson Hough languished. Hough became enamored of bear hunting and devoted increasing amounts of time to his new hobby, but realizing he needed to make a living, he began writing for and sometimes editing the local newspaper. At the same time, he prepared stories for such outlets as *American Field* and *Outing*. One of his goals as a writer was to describe the West as he had known it to be, rather than the West as myth: "One saw there the actual old West, not the railroad tourist West, but the real West, with its population made up of flotsam and jetsam of the westbound tide of humanity" (Johnson, 3).

His mother's illness and father's financial woes forced Hough back to Iowa in 1884. He was employed there for ten months by a local newspaper as its business manager. From Iowa he moved to Sandusky, Ohio, to edit that city's newspaper, but after four years headed to Chicago, where he worked for *American Field* and later became the western representative for *Forest and Stream*.

The happiest years of his life were from 1889 to 1904. One event that contributed to that happiness was his getting married—an act that "brought stability to his life and gave a direction to his career" (Johnson, 3). Hough traveled extensively through the West, researching his many articles. On one such trip to Yellowstone, he was part of a group that discovered how poaching had reduced the park's population of bison to about a hundred head. Hough's account of the situation led to Congress enacting tough legislation prohibiting the illegal hunting, which represented the first federal law to protect bison. Hough reflected on his contribution: "I have always thought this was about as useful a thing as I ever was able to do in the somewhat thankless attempt to be of service to the wildlife of America."

Hough's first major breakthrough as a writer came in 1897 with the publication of *The Story of the Cowboy*, one volume in the History of the West series. Its financial success was modest, but it was critically acclaimed. Five years later he hit a home run with his historical novel *Mississippi Bubble*, which made it to number four on the year's best-seller list. Subsequent novels on western history were financial disappointments, so he devoted more energy writing about contemporary subjects. A series for youngsters, The Young Alaskans, addressed the joys of nature and the need for environmental protection. He was so effective an advocate on behalf of the national parks that he was offered at various times the superintendencies of Yellowstone and the Grand Canyon, but he declined each time, citing the low wages associated with the positions. A bigger concern was that acceptance would abridge his freedom as a writer.

Hough's love for the old West remained undiminished. He commented to a friend in 1921 that "I'm afraid I belong to an earlier age.... I reckon there is no place for me" (Johnson, 6). He made another foray into the realm of the West with *The Covered Wagon*. To his great joy, this depiction of "pioneer migration" became wildly popular. The motion picture version played at one New York City theater for fifty-nine weeks, exceeding the previous record set by *The Birth of a Nation*. *North of 36*, dealing with the cattle drives of the 1870s, was his last novel and was also very popular. The serial and movies rights alone earned him $45,000, a huge sum back then. Unfortunately, he never had the chance to savor the financial success that came to him so late in life, for he passed away in 1923, "just a week after he had attended the Chicago premier of *The Covered Wagon.*"

Emerson Hough evidently had a tremendous impact on those who knew him. Will Dilg, editor of the Izaak Walton League's *Outdoor America*, had the opportunity to work with many of the country's leading scientists, public officials, and writers. In a memorial article, Dilg referred to his thirty-year acquaintance with Hough and said of him that "he was the greatest American I have ever known." He goes on to quote George Horace Lorimer, editor of the *Saturday Evening Post*: "He was my best friend. I did everything I could to get him to live in the east even to the point of offering him an eighty thousand dollar estate so that I might be with him often. But Emerson Hough was of the west and the east had no call for him."

Sources: Will Dilg, "My Last Talk with Emerson Hough," *Outdoor America* 2, no. 5 (December 1923): 227; Carole M. Johnson, "Emerson Hough's American West," *Books at Iowa* 21 (November 1974), http://www.lib.uiowa.edu/spec-coll/Bai/johnson.htm.

Voices of the Dunes and Other Etchings
Earl Reed

THE WINDING RIVER

ᔓ ᔓ ᔓ To enjoy a river we must adjust ourselves to its moods, for a river has many moods. It moves swiftly and light-heartedly over the shallows, as

Earl Reed, "The Winding River," in *Voices of the Dunes and Other Etchings* (Chicago: Alberbring Press, 1912), 255–76.

we do, and it has its solemn, quiet moments in the shadows of the steep banks, where the current is deep and still. It begins, like our lives, somewhere far away, and twists and turns, flows in long swerves, meets many rocks, ripples over pebbly places, smiles among many riffles, frowns under stormy skies, meditates in quiet nooks, and then goes on.

As it becomes older it broadens and becomes stronger. It begins to make a larger path of its own in the world, which it follows with varying fortunes, until its waters have gone beyond it.

The Winding River begins miles away and steals down through the back country. It curves and runs through devious channels and makes wide detours, before it finally flows out through the sand hills into the great lake.

Along its tranquil course there are many things to be studied and learned, and many new thoughts and sensations to grow out of them. We must go down the river, and not against its current, to know its strange spirit, and to love it. There is always a feeling of closer companionship when we are traveling in the same direction.

It is best to go alone, in a small boat, carrying a few feet of rope attached to a heavy stone, so that the boat may be anchored in any desirable spot. You should sit facing the bow, and guide the boat with a paddle, or a pair of oars in front of you, and let the current carry you along.

The journey commences several miles up in the woods, where the banks are only a few feet apart. The boat is piloted cautiously through the deep forest, among the ancient logs that clog the current. The patriarchs have fallen in bygone years, and are slowly moldering away into the limpid waters that once reflected them in their stately Indian summer robes of red and gold.

Masses of water-soaked brush must be encountered, and sunken snags avoided. Fringes of small turtles, on decayed and broken branches, protruding from the water, and on the recumbent trunks, splash noisily into the depths below—a wood duck glides away downstream—a muskrat, that has been investigating a deep pool near the bank, beats a hasty retreat, and a few scolding chipmunks flip their tails saucily, and whisk out of sight. A gray squirrel barks defiantly from the branch of an over-hanging tree, and an excited kingfisher circles around, loudly protesting against the invasion of his hunting grounds.

All of the wild things resent intrusion into their solitudes, and disappear, when there is any movement. If we would know them and learn their ways, we must sit silently and wait for them to come around us. We may go into the woods and sit upon a log or stump, without seeing the slightest sign of life, and apparently none exists in the vicinity, but many pairs of sharp eyes have observed our coming long before we could see them.

After a period of silence the small life will again become active, and in the course of an afternoon, if we are cautious as well as observant, we will find that we have seen and heard a great deal that is of absorbing interest.

Larger openings begin to appear among the trees, the sunlit spaces become broader, and patches of distant sky come into the picture. There are fewer obstructions in the course, and the little boat floats out into comparatively open country. Tall graceful elms, with the delicate lacery of their green-clad branches etched against the clouds, a few groups of silvery poplars, some straggling sycamores, and bunches of gnarled stubby willows line the margins of the stream, and detached masses of them appear out on the boggy land.

The Winding River flows through a happy valley. From a bank among the trees a silver glint is seen upon water, near a clump of willows, not so very far away, but the sinuous stream will loiter for hours before it comes to them.

A few cattle, several horses, and a solitary crow give a life note to the landscape. A faint wreath of smoke is visible above some trees on the right, there are echoes from a hidden barnyard, and a fussy bunch of tame ducks are splashing around the end of a half-sunken flat-bottomed boat attached to a stake.

A freckled faced boy, of about ten, with faded blue overalls, frayed below the knees, and sustained by one suspender, is watching a crooked fishpole and a silent cork, near the roots of a big sycamore that shades a pool.

He wears a rudimentary shirt, and his red hair projects, like little streaks of flame, through his torn hat. His bare feet and legs are very dirty. He looks out from under the uncertain rim of the hat with a comical expression when asked what luck he is having, and holds up a willow switch, on which are suspended a couple of diminutive bullheads, and a small but richly colored sunfish. The spoil is not abundant, yet the freckled boy is happy.

After the boat has passed on nearly a quarter of a mile, his distant yell of triumph is heard. "I've got another one!" Paeons of victory from conquered walls could tell no more.

Farther on, the banks become a little higher, the stream is wider and faster. In the distance a dingy old water-mill creeps into the landscape. This means that a dam will soon be encountered. The boat will have to be pulled out and put back into the river below it. For this it will be necessary to arouse the cooperative interest of the miller in some way, for the boat is not built of feathers.

A crude mill-race has been dug parallel to the river's course, and the clumsy old-fashioned wheel is slowly and noisily churning away under the side of the mill. The structure was once painted a dull red, but time has blended it into a warm neutral gray. Some comparatively recent repairs on the sides and roof give it a mottled appearance, and add picturesque quality. A few small houses are scattered along the road leading to the mill, and the general store is visible among the trees farther back, for the little boat has now come to the sleepy village in the back country. There are no railroad trains or trolley-cars to desecrate its repose, for these are far away. Several slowly moving figures appear on the road. There is an event of some kind down near the mill, and the well-worn chairs on the platform in front of the store have been deserted. Whatever is going on must be carefully inspected and considered at once.

There is an interesting foregound between the boat and the mill, the reflections to be seen from the opposite bank seem tempting, and an absorbing half hour is spent under the tree, with the sketch book and soft pencil.

The curious group on the other side is evidently indulging in all sorts of theories and speculations as to "wot that feller over there is tryin' to do." It is a foregone conclusion that curiosity will eventually triumph, and soon the strain becomes too intense for further endurance. The old miller, with the dust of his trade copiously sifted into his clothes and whiskers, gets into the flat-bottomed boat near the dam and slowly poles it across. All of the details of the voyage are attentively scrutinized from the other side.

After a friendly "good morning," a few remarks about the stage of the water, and the weather prospects, he stands around for a while, and then looks over at the sketch. He produces a pair of brass-rimmed spectacles, which enables him to study it more carefully, and he is much pleased. He "haint never noticed the scene much from this side, but it looks pretty. After this is finished off you'd better come 'round on the other side, so's to show the platform an' the sign. A feller made a photograph of my mill once, an' 'e promised to send me one, but 'e didn't never do it." The long remembered incident, and the broken faith, seemed to disturb him, and he appeared to be concerned as to the destiny of the sketch. He wanted it "to put up in the mill." ...

After further pleasant conversation, the dusty miller helps to drag the boat around the dam. He waves a cheerful farewell, recrosses the stream, and immediately becomes the center of concentrated interest. The fat woman in the road waddles down to the mill, and a number of bare-headed children come running down the slope, who have peeked at the proceedings from secluded points of vantage.

As the boat floats on, the figures become indistinct, the houses fade into the soft distance, the mill, like those of the gods, grinds slowly on, and, with the next bend in the river, the sleepy village is gone.

The story of the eventful day percolates from the store off into the back country, and weeks later we hear it from a rheumatic old dweller in the marshy land, near the beginning of the sand hills. He unfortunately "wasn't to town" at the time.

"A feller come 'long in a boat an' stopped at the mill. He was 'round thar fer over an hour an' drawed some pitchers of it. He made one o' the old man with 'is pipe showin'. He was some city feller, an' had to git the old man to help 'im with 'is boat 'round the dam. The old man's got a pitcher 'e made of 'im stickin' up in the mill now. A feller like him oughter larn some trade, instid o' foolin' away 'is time makin' pitchers. Nobody 'ud ever buy one o' them dam' things in a thousand years. I'll bet 'e was spyin' fer the railroad, an' they'll prob'ly be 'long here makin' a *survey* before long."

A little farther down is a loose-jointed bridge with some patent medicine signs on it. Another sign tells the users not to drive over the structure "faster than a walk." Any kind of a speed limit in this slumbrous land seems preposterous, but the cautionary board is there, peppered over with little holes, made by repeated charges of small shot, and partially defaced with sundry initials cut into it with jack-knives. Some crude and unknown humorist has changed some of the letters and syllables in the patent medicine signs, and made them even more eloquent.

Another lone fisherman is on the bridge, watching a cork that bobs idly on the dimpled tide below. Another single suspender supports some deteriorated overalls. Possibly the freckled boy up the river was wearing the rest of the suspenders. He is an old man, with heavy gray eyebrows, and long white whiskers that sway gently in the soft wind. His face has an air of patient resignation. He wears a faded colored shirt and a weather-beaten straw hat. His feet, encased in cowhide boots, hang down over the edge of the rickety structure, and he sadly shakes his head when asked if he has caught any fish. His lure has been ineffectual and he is about ready to go home. There is still a faint lingering hope that the cork may be suddenly submerged, and the appearance of a new object of interest has decided him to remain a little while longer.

He explains that "the wind ain't right fer fishin'. I've seen fish caught off 'en this bridge so fast you couldn't bait the hooks, but the wind has to be south. Besides the water's all roily to-day an' the fish can't see nothin'. I bin drownin' worms 'ere most all day, an' I ain't had a bite, an' I'm goin' to quit."

Just after the boat had passed under the bridge, a dead minnow floated along on the current. A large pickerel broke water and seized it. His sweeping tail made a loud swish, and the water boiled with commotion as he turned and dove with his prize.

Instantly the dejected figure on the bridge became thrilled with a new life, and a torrent of profanity filled the air.

"Now wot d'ye think o' that! The gosh dangled idjut's bin 'round 'ere all the time, an' me settin' 'ere with worms fer 'im. They's a lot o' fish in this 'ere river that I'll teach sumpen to before I'm through with 'em. I'm a pretty old man, but you bet I'm goin' to play the game while I'm 'ere. I wonder where 'e went with that dam' minnie!"

The boat goes tranquilly on, and in the dim distance the old man is actively moving around on the bridge, flourishing his cane pole and casting the tempting bait all over the surface of the water, evidently hoping that the "gosh dangled idjut" will rise again.

The river now comes to the beginning of the vast marsh, through which its well-defined channel follows a tortuous route among big wet stretches of high grasses and bulrushes, winds with innumerable turns, makes long sweeps and loops, and comes back, almost doubling itself in its serpentine course. The current slackens and the water becomes deeper.

The cries of the marsh birds are heard, and muskrats are swimming at the apexes of the long V-shaped wakes out on the open water. On small boggy spots are piles of empty freshwater clam shells where these interesting little animals have feasted. As the crows seem to dominate the sand hills, the muskrats contribute much picturesque quality to the marsh. Their little houses add interest to the wet places, and traces of them appear all over the low land.

A wild duck hurries her downy young into the thick grasses—a few turtles tumble hastily from the bogs into the water—a large blue heron rises slowly out of an unseen retreat, and trails his long legs after him in rhythmic flight down the marsh—mysterious wings are heard among the rushes—immense flocks of blackbirds fill the air—there is a splash out among the lily pads, where a hungry fish has captured his unsuspecting prey, and the deep sonorous bass of a philosophic bullfrog resounds from concealed recesses.

Another bend in the channel reveals a flock of wild ducks feeding quietly along the edges of the weeds. The intrusion is quickly detected and they swiftly take wing. A sinister head, with beady eyes, appears on the surface behind the boat, and is instantly withdrawn. A big snapping-turtle has come up to investigate the cause of the dark shadow which has passed along the bottom.

Some open wet ground comes into view around the next curve, and some lazy cattle look up inquiringly. After their curiosity is satisfied, they turn their heads away and resume their reflections.

The Winding River has its solemn hours as well as those of gladness. Heavy masses of low gray clouds are creeping into the sky, the shadows are disappearing and a moody monotone has come over the landscape. Deep mutterings of thunder, and a few vivid flashes, herald the approach of a storm.

Some thick willows, which can be reached through openings among the lily pads, a short distance from the main channel, offer a convenient shelter, and from it the coming drama can be contemplated.

The big drops are soon heard among the leaves, the distant trees loom in ghostly stillness through veils of moving mist, the delicate color tones gently change into a lower scale, and the voices of the falling waters come. The reeds and rushes bend humbly, and there are subdued cries from the feathered life that is hurrying to shelter among them. The rain patters and murmurs out among the thick grasses and on the open river.

There are noble beauties and sublimities in the storm, which those who only love the sunshine can never know. Truly "Our Lady of the Rain" weaves a marvelous spell, and her song is of surpassing beauty, as she trails her robes in majesty over the river and through the marshy wastes. Her pictures blend with her measures, for a song may have other mediums than sound, and there are many symphonies that are silent. The prelude in the lowering clouds, and the melody of the loosened waters, bring to us a sense of unity and closer communion with the powers in the skies above us.

The sheets of flying waters have gone on up the marsh, a long rift has appeared in the clouds beyond the hills, a bright gleam has come through it, and the end of a rainbow touches a clump of poplars far away. The storm is over and the little boat is piloted out through the lily pads, to resume its journey on the tranquil stream. It finally reaches the sand hills. The river narrows and runs more rapidly as it leaves the swamp. Another sleepy little town, with two or three bridges, appears ahead. There are more still figures on the bank, watching corks on lines attached to long cane poles, which are stuck into the earth and supported by forked sticks. The labor of holding them has proved too great and natural forces have been utilized to avoid unnecessary exertion. The anglers appear much depressed and are soaking wet. A nearby bridge would have provided a refuge from the recent rain, but possibly their intellectual limitations did not permit of advantage being taken of it.

A friendly inquiry as to their success evokes sleepy responses, and looks of languid curiosity. "The fishin' ain't no good. I got one yisterd'y, but I guess the water's too high fer 'em to bite."

We have now come to the end of the Winding River. Its waters glide peacefully out and blend into the blue immensity of the great lake. Like a human life that has run its course through the vicissitudes and varied paths of the years, they have ceased to flow, and have been gathered into unknown depths beyond.

There are many winding rivers, but this one has numberless joyful and poetic associations. On its peaceful waters many sketch-books have been filled, and happy hours dreamed away. From the little boat wonderful vistas have unfolded, and marvelous skies have been contemplated.

The heavens at twilight, flushed with glorious afterglows in orange, green and purple—the clear still firmament at mid-day, lightly flecked with little wisps of smoky vapor—the lazy white masses against the infinite blue, and the billowing thunderheads on the horizon on quiet afternoons—the stormy array of dark battalions of wind-blown clouds, with their trailing sheets of rain—and many other convolutions of the great panoramas in the skies, have been humbly observed from the little boat. The Winding River has reflected them, and the picturesque sweeps and bends, the masses of trees on the banks, with the silvery stretches of slowly moving waters, have given wonderful foregrounds to these entrancing prospects.

Fancy has woven rare fabrics, and builded strange and fragile dreams among these glowing and ever-changing symphonies of light and color. The little boat has been a kingdom in a world of enchantment. The domes and vistas of a fairyland have been visible from it. The Psalm of Life has seemed to float softly over the bosom of the river, and mingle with the harmonies of infinite hues in the heavens beyond. The lances of the departing sun have trailed over the waters, and dark purple shadows have gently crept into the landscape. Manifold voices are hushed, and the story of another day is told.

Nature, seemingly jealous of other companionship, yields her spiritual treasures only to him who comes alone into her sweet solitudes. Before him who comes in reverence, the filmy veils are lifted, and the poetic soul is gently led into mystic paths beyond.

In her great anthems of sublimity and power, she fills our hearts with awe, and appals us with our insignificance, but her soft lullabies, which we hear in the secluded places, are within the capacity of our emotions. It is here that she comes to us in her tenderness and beauty, and gently touches the finer chords of our being.

One may stand upon a mountain-top and behold the splendors of awful immensities, but the imagination is soon lost in infinity, and only the atom on the rock remains. The music of the swaying rushes, the whispers among rippling waters and softly moving leaves, and the voices of the Little Things that sing around us, all come within the compass of our spiritual realm. It is with them that we must abide if we would find contentment of heart and soul.

The love of moving water is one of our primal instincts. The tired mind seeks it, and weary travelers on the deserts of life are sustained by the hope of living waters beyond. There are winding rivers on which we may float in the world of our fancy, and it is on them that we may find peace when sorrows have afflicted us and our burdens have made scars. They may flow through lordly forests, and stately mansions and magic gardens may be reflected in their limpid tides. The songs of these rivers are the songs of the heart, and in them there is no note of triumph over the fallen, or despair of the stricken. They are songs of courageous life and melodies of the living things, but only those who listen may hear them.

Sometimes, in faint half-heard tones from far away, we may imagine echoes from another world than ours, and, as we enter into the final gloom, these harmonies may become divine. In the darker recesses of our intellectual life we find shadows that never move. They seem to lie like black sinister bars across our mental paths. We know not what is beyond them, and we shrink from a nameless terror. Into these shadows our loved ones have gone. They have returned into the Elemental Mystery. Their voices have not come back to us, but their cadences may be in the singing winds and amid the patter of the summer rain.

Our Ship of Dreams can bear a wondrous cargo. We can sometimes see its mirage in the still skies beyond the winding rivers, though its sails and spars are far below the horizon's rim. We know that on it are those who beckon, and its wave-kissed prow is toward us. Frail though its timbers be, the years may bring it, but if it never comes, we have seen the picture, and new banners have been unfurled before it.

Tales of a Vanishing River
Earl Reed

THE "WETHER BOOK" [BY JOSIAH GRANGER]

⚭ ⚭ ⚭ SEPT. 1ST—The meteors in my almanack did not fall in August & predictions not reliabel. Nuthing of the kind around. It is geting along toreds fall. Pidguns are around. They broke som ded lims on the iland this week whare they roosted. Thares slews of them. This is a good yeare for pidguns. I got 33 with 2 shots. They did not kno that your uncle Josiah was around with a gunn. I notis in my almanack Oisters are now in season. Nuthing of the kind around heare.

SEPT. 4TH—Soon after sunup it looked like streky black cloudds up above but it was pidgun flocks coming south. Pidguns are all over now. Big droves roosted around last nite. I must salt down som. They are in the woods after the young akerns. Pidguns still going over. Cant tell if it is clouddy. Warm day thow.

SEPT. 10TH—Must get a houn pupp. Old Tike is geting wobblie in the nose & he looses his nose now & then. He is sick som & not lively. He is a good dog but he has erned his money. He is now going on 13 yeares & has ben over the country som sence I had him. S. Conkrite had some pupps last week & I must go up. They may be all spoken for thow. Must get som supplys & som backake ointmint. Hell I broke my pipe. Wether breeding clouds in the west tonite as I rite.

SEPT. 12TH—A sorel mare was stolen by 2 men & a buggy Tuesday nite from Ed Baxter who had just bote the mare. They caught these men over 18 miles off on the Hickery Top Road & they are now locked in jale. He was down at evening to see how I was & to get some eggs. The sherif & a possy was what nabbed the theves. I hear from Ed that Henry Clay died last June & that a chese facktory & brick kill are to be bilt neare West Crick. I fore see a church next. This country is geting too much setled up. Thares too dam meney pepil. It rained som today but cleared at noon. Ed had a lot of noos. He went off home by bugg lite about 9. He kep me up. I rite this on the 13th.

Josiah Granger, "The 'Wether Book'" (1852), in Earl Reed, *Tales of a Vanishing River* (New York: John Lane, 1920), 85–101.

SEPT. 14—A wolf has ben on this iland frequent & has ben after chickins & eny thing he can get. I set a trapp & he turned it over & got the bate evry time. Last nite I set it botom sid up & he turned it over & I got that cuss. He did not kno the trapp was botom upwards & he was astonished. You can not fool much with your uncle Josiah. Som drizzel in the air tonite & som colder. It is geting into fall all rite. I kno whare 2 bee trees are. Your uncle has them spotted. Thare will be honey heare in about a week. You Bet.

SEPT. 17TH—The merkery took a sudden jump & it is hot as July & August. I slep out on the grass last nite. A good mush mellin in the shade is a fine thing now. Conkrite & Baxter com yesterdy when I was not within & left a buckett they borowed Saturday to take down the river. I must put a date on that for its the first thing they ever brought back.

SEPT. 20TH—I got a cubb bear that was 1-2 in & 1-2 out of a bee tree after honey & got him home well chained with a colar. I got about 60 lbs honey. This was yesterday & the day befoar. The animil eats well & acts tame but scared. I name him Jim Crow.

SEPT. 21ST—S. Conkrite & Ed Baxter & Wife com today to see how I was & to see if I got eny honey yet. They are rite on skedule. Also they wanted to borro som small shot & to get som fouls. Ed's wife made beleve she was scared of the bear. Probly so Ed would save her from it. Conkrite says he got a wild catt over to the swamp that was 37 inches tip to tip . I got one 40 inches last winter that I spoke nuthing of. Mine was a feerce animil. Conkrite blows a good dele. The pupp I got from Conkrite houls all the time & has et his hed off up to date. Jim Crow got a peice of the pupp yesterdy when he got neare. The pupp tried to bite Conkrite & I think this shows he was treated bad at home. I asked Conkrite about pork for winter pikel but he semes to think my place is whare money dripps off the roof & shakes out of the trees. At killing time it will be different. Ed Baxter says he has dug a deeper well. His other he says is full of mushrats that com for watter in dry spell in July to qwench their thirst & now living thare. I tell him to sett & fish for them with a pole. It is now 8 P.M. & your uncle is reddy for his blankett.

SEPT. 25TH—I went after supplys. Old Josiah now has plenty of evrything. Thare is Backake Remedy Foot Ointmint Magick oil for Stif Joints & Pain Killer & 2 kinds of Bitters & Sistom Tonick & pills both blue & pink. I got Condition Powders for chickins if sick. I got som tabaco black as Egipt for those who com to borro. It is strong enough so you can pull nales with it. I got all they had and some candels. Jim Crow is well & he likes all swete things. I got Jim som stripped candy 3 sticks. The Pacific Ocean was discovered in 1513

by my almanack on this day. Funy they missed it befoar. When I com by Ed
Baxter's place last nite the boat that used to be mine got loose & com along
down with me. I find certain marks on it that I will show Ed. I reckonize my
own boat & it now seeks its home. A drizzel of mosture as I rite. I tended to a
lot of bisness today. Conkrite says the Sistom Tonick I ben buying is loaded
but does not say what with. He says mix a lot of pump watter with it & not
take to much or darkness will com.

SEPT. 28TH—The wether stays moist. Today in 1828 in the almanack the
sultan proceeds to the Turkish Camp with the sacred standard. Probly stole
from som whare.

SEPT. 29TH—These cold stormy drizzels may bring in a few ducks. Would
like som ducks. Moon full last nite but not sene.

OCT. 1ST—Sept. was a quere month without much wether ether way. Oct.
now opens clear with frost that nipped the vines last nite. Had the pupp out
for a run on rabbitts. His nose is good & he may learn. I never sene a good dog
that com from S. Conkrite's yet. Was down to the marsh yesterdy & meney
noo rat houses. They are bilding thick & high & this menes a hard winter &
high watter in the spring. All sines say a hard winter. Snipe are skitting around
& thare is a lot of mudd hens & loons in the marsh. 2 deer swum the marsh &
dove into the timber. They kno when Old Josiah has got a gunn & when he
left it home. Sam Green & his friend Wasson com in a boat tonite to see how
I was & to get som honey. The pupp bit Wasson. Tally 1 for the pupp. These
men also wanted to borro tobaco. Gave them som of the black. I tell them
smoking that kind makes me strong.

OCT. 6TH—Stormed & I stade in. Conkrite com in the rain to see how I
was & to borro powder & see if I had eny thing in my medicins for boils. He
says he com yesterdy & nocked but I was not within. I was then in the woods
traning the pupp. His noos is Ed Baxter claims he has 2 twins that com erly
this morning & I bet they look like young mushrats. He spoke of pork but old
Josiah is keaping pretty still until after the snow flys. He says of Ed's twins they
are both boys & red hedded. Thares too meney Baxters now. S.C. Says them
2 twins will be named James & John.

OCT. 12TH—In the full of the moon & on a frosty nite your uncle Josiah
goes after coons & I note this down. It will be the 27th if nite is clear. I notis
Columbus landed today in the almanack in 1492. He was the first of the for-
riners.

OCT. 18TH—Nuthing happened sence the 12th, but last nite a killing frost
& today a swizzel of rain & sleat with N.W. wind. This will bring down ducks

& gese. Stade in today & clened up shot gunn & rifel & all trapps. Saw to all aminition. Evrything all fixed up as I rite. Put all potattoes & vegitibels in sod celer & evrything all tite up to date. Cleared off som today & som ducks are coming & som gese are in the sky. Unusual wether for Oct. Gese honks all nite long as I slep. This was last nite. I got 25 lbs tobaco in the sod celer too. When I need tobaco this winter I kno whare som is.

OCT. 19—Blowing strong from N.W. Rain & sleat. Sky all speckeled with ducks & gese. They are coming in slews now. Gese honk all nite can not sleep. Active wether will come rite along now. No more lofing for your uncle Josiah. He gets on his sheap skin coat now. Take notis. He is in the field.

OCT. 20-21-22-23-24-25—I ben busy all this time. Josiah is around with a gunn. He makes fethers fly & he fetches in the birds. Fine gese & duck wether. The marsh is black with them evry morning at sunup. The Irish Rebelion was on the 23rd of this month in 1641. They begun coming heare then.

OCT. 30TH—Duck & Gese wether has stoped & ingun sumer is upon us. I fore saw this. They are around som whare but shooting is poor. No duck & gese wether for a while yet. I stoped at S. Conkrite's. I got to hav pork, but he said nuthing of pork & neither did your uncle Josiah. He has 9 squeeling around all fat in good condition.

OCT. 31ST—This has ben a remarkabel month & changabel at times as almanack predicted. Jim Crow is well. He has et well. I see hevy bunches of cloudds in west that I fore see will breed duck & gese wether as I rite. I notis in my almanack that meney thousans of pepil died of sickness in India at this time of the yeare in 1724. Thare is too many pepil. No sickness heare much at eny time. This is a helthy section only 3 died in 5 yeares. I see deer are around.

NOV. 2ND—Althow a stormy day Ed Baxter com in P.M. to see how I was & to get honey & som tobaco if I hed eny. He told all the noos of them 2 twins James & John & you would think nobody ever had eny befoar. It is all about them 2 red heds all the time how they et & how they are smart & how much they way. All the branes in the country are setled in James & John. He says he will bring them & show me. They must be som site & I will be struck blind in 1 eye probly. You would think the world had com to the end in them 2 & they was Danl Webstor. Thare was an awful famin in Italy in the yeare 450 when parents et their children.

NOV. 3RD—Lite snow bust in the nite & I found bear traks all around this morning. Som friend com to see Jim Crow probly. The pupp now sleeps with Jim in the dog house & he howld in the nite. Som rain sputtering as I rite.

NOV. 4TH—Roring wind from the North today. A hevy sky & sleat. I notis meney duck flocks & gese.

I will be busy now rite along. Must get a deer. A little venzon rite now would be fine. Your uncle Josiah has apitite for som.

NOV. 6TH—Got a buck rite on the iland. They will go poking their heds in the window to get shot if I dont watch out. This was yesterday. Jim Crow is loose now & spends time mostly on the roof & up the cottonwood. He was in the chickins Tuesday nite & today he was in the house & upsett things. Might as well be a horse loose in the house. Must put him back on chain. If you want to keap busy you want to keap a bear. He is a quere cuss & probly smells the honey. She still blows & tomorro I go for ducks. Wish I had all the lead I spattered around on that marsh in my time. Must have raised the watter som.

NOV. 7-8-9-10-11-12—Was on the marsh all these days & tired at nite. Wether lite winds & drizzeley. No finer duck & gese wether ever sene. Your uncle was among them & he shook them loose. I com in wet tonite & must sett around a while. I see traks showing sombody has ben heare. Probly Conkrite or Ed Baxter to see how I was & to borro somthing & tell me of them 2 twins. Must wrap up in my blankett & take som strong medicin. I got a cold & I got wether pains. Will stay in & rite in my wether book. On Nov. 9th in 1837 the quene of England dined at Guildhall. Good meal probly.

NOV. 13—When your uncle Josiah takes medicin he doses up. I took 4 kinds today & kep my feet hot with my watter jug. I got a good fire. Storms hevy outside but that does not hurt me eny. I read all it says on all my medicin botles & I can get nuthing they will not cure. I got Jim Crow & the pupp in the house for company now. They sleep mostly. When they awake they make troubel. I fore see that these animils must be put out.

NOV. 14TH—Somthing I took yesterdy or last nite has helped som. I slep well. Probly it was 1 of the bitters. Snow prevales outside & she falls hevy as I rite. I put Jim & the pupp out. Thare was too meney in the house. Jim has got honey coam & the pupp has got bones in the dog house so they are hapy. Nobody could want more than that unless they are crazy about money.

NOV. 15-16-17—I stade within mostly on these days. We are having a spell of wether. My bitters & my Sistom Tonick are most gone but I still got plenty of 2 kinds that I take internal & 3 kinds to rub on. Wolves howl around a good dele at nite. I keap my sasafras tea het up rite along but the bitters do most of the work. They are strong stuff & have som get app to them. Sky is full of ducks & gese do a lot of honking over the house. Probly to twitch me while I cant get out. Your uncle feals som beter but he is wise. He will not go out too soon. It would be beter for som body to go that would not be so much loss.

NOV. 18TH—S. Conkrite com today to see how I was & wanted to trade me a nice fat hogg for Jim Crow & I done this. Jim is geting a litle sassy & Conkrite's will be a good place for him. Will now hav pork to put in pikel & to smoak. He is to kill the pork & bring it & after that is to take Jim home. I fore see that Jim will make troubel. I am up & around all rite now. Must go after supplys of bitters & Sistom Tonick soon & I must get a chese. A smitch of chese helps out a meal. Looks wethery tonite & snow probabel.

NOV. 19TH—S. Conkrite com today with the pork & it is good pork. We fixed a crate to put Jim Crow in & he made a lot of fuss. Them 2 looked funy going off in the boat. Cold & freezing som & ducks & gese have lit out. Thare are deer around thow. I made soft soap today.

NOV. 20TH—Ed Baxter com in P.M. to see how I was & to hang som meat in my smoak house. When he sene the soft soap he wanted to borro som. Probly to wash them red hedded twins. S. Conkrite also com at evening & Sam Green & Wasson all with pork to smoak. I got lots of friends. My pork must pikel a while befoar it smoaks but I got to fire up the smoak house now for these men's pork. They all like this because its something for them. Ed told a lot about them twins. Thare has never ben such twins. Conkrite's noos is Jim Crow got away. The traks stade around the chickins a while & then went to the woods whare fethers were found. Lite sift of snow to nite. The Cape of Good Hope was doubled in the almanack today in 1497. Quere they wanted 2 capes thare.

NOV. 21ST—Jim Crow was up the cottonwood this morning when I went out. Him & the pupp are now in the dog house. Conkrite will probly com after Jim. She snows & blows hevy as I rite.

NOV. 23RD—My smoak house is well knone. Pete Quagno & 2 other inguns com today to see about puting things in it but I tell them I want to kno what they are. They say all sines show a hard winter coming. No danger of them inguns stealing my soft soap. Your uncle Josiah is now all well & feals fine. He was all over the iland today. He could pull up a tree or kick the chimbly off the house if it had to be. I notis too meney small animil tracks on the iland & I will now tend to these. The pupp is fine & he now goes with me. Lite snow last nite & I see a wild catt has ben across and I would like to get his fur.

NOV. 25TH—Yesterdy I stade within with my medicins as I did not feal so well. I got a stummick misry. Conkrite was down & took Jim Crow back to-day. I do not think Jim likes Conkrite. He tried to get a peice out of Conkrite when they was in the boat. Me & Jim always got along all rite. Snow is faling.

NOV. 26-27-28—Snows all the time now. She dont know when to quit. My almanack says G. Washington crossed the deleware Nov. 28th. It missed saying what yeare but he got whare he wanted to go. Moon was full on the 26th but not sene.

NOV. 29TH—S. Conkrite com with som meat to smoak today & it looks like bear meat. I fear Jim Crow is now in the smoak house. That man knos nuthing of how to keap pets. I was off in the woods when Conkrite com but I kno it is Jim all rite. He was a fine bear & affecksionet. I wish Conkrite had his dam pork back & I had Jim Crow.

NOV. 30TH—That meat is not Jim at all for Jim is back & up the cottonwood this morning. He did not want to com down but him & the pupp are in the dog house as I rite. Jim likes it around heare. Mackarel sky tonite & changing wether probabel. Nov. a remarkable month all through.

DEC. 1-2-3-4-5-6—I ben fealing porly now som time with the misry in my stummick. Tried som of all my internal medicins & feal som beter today. Hav rubbed my Rumatiziam with Pain Killer & took pills both blue & pink that are for liver complaint. Poor old Tike was sick too. I gave him the box of condition powders I got in the fall for the chickins but he quit that nite. This was on Saturday the 4th. The powders may not hav kep well or maybe not good for a dog. I lost my best friend. Bad wether now. I think animils should have no medicin at all of eny kind.

DEC. 7TH—Ed Baxter com today to see how I was & to get his smoaked pork. I promis to take Christmas diner with Ed & Wife. I must take presents for James & John. Likely a buckett of soft soap will be good for them 2. Looks gusty & snowy tonite.

DEC. 8TH—S. Conkrite & Green & his friend Wasson all com to see how I was today & get their smoaked stuff. Conkrite says would like me to keap Jim Crow a while longer for he is too meney up to his place. This I will do for Jim & me get along fine. Jim went up the cottonwood when he sene Conkrite. Thares too meney smoak houses on this iland & too much smoaking going on for other pepil. Snow storm slanting from the north west & drifting som as I rite. I fore saw this last nite. I think Conkrite is the one that is too meney up to his place instid of Jim Crow. I got wether pains in both back & legs now.

DEC. 9TH—Now she snows. Big drifts. Can not see dog house from window. I now got Jim Crow & the pupp in the house. My wether pains som worse. Must stay in my blankett.

DEC. 10TH—A soft thaw has come on sudden. A warm sun prevales & evrything all slushy. Good wether for wet feet. Your uncle still stays within.

DEC. 12TH—Both S. Conkrite & Ed Baxter com today & brought me a new almanack for next yeare. This is the first time they ever com that it was not somthing for them. They said I don litle favers for them & they would like to make me this litle present. This all shows that if you keap being good to pepil all your life some day they will bring you a nice litle almanack. Probly they will want somthing next trip. I gave them som Sistom Tonick & they liked that. Ed Spoke of them 2 twins & they are both well & awful smart. He asked if my smoak house was still in good working order & if my hens ben laying well lately & if I had plenty of potattoes on hand.

DEC. 13TH—Them 2 inguns that come heare last with Pete Quagno & his squa com today & their noos is that Pete & his squa are both sick & wanted tobaco. I sent Pete 2 pink pills. Them 2 inguns wanted me to send Pete & his squa a big lot of tobaco by them but they did not know that your uncle Josiah was setting around smoaking befoar eny of them was born.

DEC. 14TH—Last nite I read in my noo almanack. I notis it predicts worse wether for next yeare. Storms & Tempests will prevale with intense frosts probabel at times, but thare will be much changabel wether & meney meteors that will betoken war. Thare will be awful winds on Parts of the Earth. In the back are som Prophesies made by the Seventh Son, which I copy down. He says thare will be wars and rumours of wars & Turbulence & Teror will apear on evry hand & cloudds of darkest hue will hang over the World in the East. Fires will abound & Tumults & Bloodshed & Plots & Uprores in som Nations. Subject Pepils will turn & bite the hoof that holds them down. A certain Luckless King may loose his hed & something may hapen to the Pope. Armed Men may march to & fro & meney will be smitten to the Dust. Blood will be shed in Ireland. Tyrants will shake their Rods & the Torch of Discord will be hurled in Crimea. The Couch of Mortality will be spred & meney pepil will die during the yeare. Low Moans of the Oppressed will be heard in Italy. It is all bad noos in the almanack for next yeare. The 7th Son predicts that Flocks of Boobies will assale the TRUTHS OF PROPHESY. He predicts no troubels for eny whare around here. Your uncle Josiah is in out of the wet.

DEC. 15TH—Sam Green com & says his friend Wasson is sick & wants som medicin. I give him som of each kind but I ought to see the simptoms. Wasson does not kno what ales him but my medicin will probly fix him up. He probly has stummick complaint. Stedy freezing wether now.

DEC. 16-17-18—Evrything is froze tite & so is the pump. I ben out on trips & I think one ear is froze. I tended to a lot of bisness. I got supplys & same kind of almanack for next yeare that I ben having. I notis the predictions in it are not half so bad as the onc that was fetched for the litle present by Conkrite. He probly wanted to scare me into the woods. I notis he keaps the same kind I do & he gave me the other. I stopped at his place today & I saw Green & Wasson & J. Podnutt thare. Wasson got well. Those were all good medicins I sent. Their noos is timber theves are at it again down the river. Wasson hunts down thare & he wants us all to form a possy and chase them out of the country but your uncle chases nuthing these days he does not want. I tell them the owners must be notified. I do not know what them old mud turkels talk about all the time up to Conkrite's. I got som candy for Jim Crow & I paid Conkrite for his pork at a low price & Jim is now mine again. Jim is good good company if you kno how to get along with a bear. I got a noo medicin. Instant Relief for Internal Disorders. Will try on sombody that coms to see how I am & to borro medicin. It looks like a good remedy. This has ben an active day.

DEC. 20—Think I got som cold on my trip Saturdy. Am taking the noo remedy but do not yet kno what it will cure. I notis that 2 things that are on the wrapper I am troubeled with. Big snow storm now going on.

DEC. 21-22-23-24—Your uncle Josiah has felt pretty poorly for these 4 days. Hav taken my medicins stedy. Think I am now beter. Must go to Baxter's to-morro. Wether clear & cold.

DEC. 26TH—I took diner up at Baxter's & it was a good diner. We had chickin fixings & cooked appels & a grate dele of other things & pie of all kinds. I took the chickins up. We talked & smoaked & in P.M. Ed got his fiddel out & playd hoppy tunes on it. A string was busted but he done well with the rest. I got along fine with them 2 twins. Their parents hav a lot of plesure with them babys. I had them on my lap & it took me back to when I had 2 litle boys that did not kno beter than to like to be around with their pa. I wish I had them litle boys back now. They grew up & went away probly looking for better friends. It is lonesom heare on the iland with them & their mother all gone; once in a while I find somthing around they playd with & things their mother had & them things are what I got left. I must hav the Baxters down heare next Chrismas if I am around. I will cetch them twins some young rabbitts when they get old enough & som young mudturkels & pollywoggs to play with like I used to do. Full moon at nite on my way back to the iland & them 2 litle boys was asleep when I left.

DEC. 27-28-29-30—I ben too sick to rite in my wether book.

DEC. 31ST—This was the last day of the yeare & whatever hapened is now all over. It is awful cold & still outside & once in a while I heare frost cracking in the woods. The yeare is now coming to its end in a few minits. It is pretty late for me to be around but I am waiting for the old clock to strike 12. Maybe next yeare at this time I will be asleep. It is awful lonesom heare tonite & I wish I had my folks around or if them 2 litle boys was only heare or sombody. Maybe tomorro sombody will com. I notis by the looking glass that the old man' hed is prety white. He has ben frosted som. He now goes into his blankett for the yeare ends as he rites.

⌐⌐⌐⌐ EARL REED (1863–1931) WAS BORN IN GENEVA (KANE COUNTY), Illinois. His several books on northwest Indiana, including *Voices of the Dunes* and *Tales of a Vanishing River*, show him to be a wonderful writer, but he is best known as a visual artist specializing in etching. His works are held by such institutions as the Art Institute and the Library of Congress. He was also instrumental in founding the Chicago Society of Etchers.

In his beautifully written *Tales of a Vanishing River*, Earl Reed tells of being taken to the Granger family cabin late one night by his friend Buck Granger. The "decayed but well-constructed old house," located on Jerry Island deep within the Kankakee Marsh, was built around 1810 by Buck's grandfather Josiah, who had settled there with his wife and two sons. His children married and moved west, and with the death of his wife, he lived by himself in the cabin for many years. Two years before he died, however, one of his sons returned with his wife and little boy. Buck was the last surviving member of the family.

Perhaps because he had no heirs, and knowing Reed's interest in history, Granger shared the cabin's contents with his visitor. Among the items that Reed used in his book was the remarkable diary of Josiah Granger, who called the volume his "Wether Book." Of the several years covered by the diary, Reed selected 1852 as representative and published "extracts from the entries."

Alfred Meyer then included a heavily redacted version of what Reed had published as part of his 1934 doctoral dissertation, "The Kankakee 'Marsh' of Northern Indiana and Illinois" (University of Michigan). (Meyer was for many years a geographer at Valparaiso University, and his 1936 paper, based on his dissertation, remains the definitive study of the marsh.) Meyer was impressed both by the range of Granger's topics and "the animated presentation." He goes on to say that "the trapper's diary, however crude its phraseology, impresses the writer...in revealing the human soul as well as the landscape physiognomy of the Kankakee, and the way the latter looked to the native." I would also point out that intentional or not, Granger pens some absolutely hilarious comments.

The selection from *Voices of the Dunes* depicts a canoe trip down what is almost certainly the Little Calumet River. The river originates in the Valparaiso Moraine of LaPorte County and flows west to Riverdale, Cook County. Originally it looped back on itself through a northerly channel that moved east into Lake Michigan. Today, however, what had been one river is now three. The Little Calumet is the southern branch that flows into the artificial Cal-Sag Channel; the Grand Calumet River is the northern branch; and where they meet at Riverdale is the Calumet River, another artificial, or at least enlarged, channel that flows straight north to Lake Michigan.

Sources: F. M. Fryxell, *The Physiography of the Region of Chicago* (Chicago: University of Chicago Press, 1927); Alfred H. Meyer, "The Kankakee 'Marsh' of Northern Indiana and Illinois," *Papers of the Michigan Academy of Science, Arts and Letters*, vol. 21; Earl Reed material in Calumet Archives, Indiana University–Northwest.

Most Interesting Interview with Mrs. Druscilla Carr

J. W. Lester

ᖶᖶᖶ When I undertook to write in their own words the stories of Lake County pioneers and other interesting personages, I had not thought of contributing anything to local history, for it seemed obvious that the ground had been thoroughly searched for information of historic interest. Mrs. F.J. Sheehan, R.F. Knotts, J.O. Bowers, W.H. Mathews, and our friends in Hammond, Crown Point, and other places in the county had given the public accurate and intensely interesting reports of their findings.

But I have always enjoyed visiting strange places and meeting interesting people. It is to me a diversion which has become a hobby. I take in shorthand the exact words of those who have played a part in the upbuilding of the county. In recording them I hold myself unaccountable for grammatical or other errors. Should any occur I can point to the other person and say, "This article was dictated." Then, too, the one who tells the story can evade any misstatement by saying, "I didn't write it." My plan lessens the responsibility for all concerned....

J. W. Lester, "Most Interesting Interview with Mrs. Druscilla Carr," *Gary Evening Post and Daily Tribune*, March 10, 1922.

Copies of these stories are to be preserved in the city library by our secretary L.J. Bailey and they will be accessible to members of the Historical Society and others who might be interested.

—J.W. Lester

Mrs. Druscilla Clark Tells of Early Days at Miller Beach (December 24, 1921)

I was born at Gossetburg (Gosset's Mills) on Salt Creek, seven miles north of Valparaiso, in 1856. My father, William Benn, had a mill there. I was married in 1874 to Robert Carr. He died in 1903. We had seven children; four are living.... The two boys are both here and are working as fishermen.

I first came to Miller in March 1872, to my brother's at the fish house. It was near to where the Gary park is and close to the lagoon. John and Hank Granger owned the fishery and I cooked for them. I was married in 1874 and worked for Carolina Carr, my husband's mother. Then I picked cranberries for a dollar and a quarter a week. The hollows were full of cranberries and there were plenty of huckleberries. Then a man by the name of Nels Anderson (we called him Moss Anderson) pulled the moss out and shipped it to Detroit. I think to use in packing fruit trees. That is the reason there are no more cranberries here. When the moss was taken out the frost killed them.

Carolina Carr came to Miller in 1862. A boat builder by the name of Allen Dutcher, who came from Michigan City and built boats for the fisherman, had been here a good many years. Another man, John Beaubien, a trapper and duck hunter from South Chicago, batched with him most of the time. There was another old man, older than either of them, named Davy Crockett, who built a log shanty back in the cottonwoods right at the mouth of the Grand Calumet. His shack was probably the oldest house there. They called him "Colonel" for a nickname. His shack was on the west side of the outlet, north of the river. When I came here Dutcher lived in a log house that Mr. Carr built right at the blowout about three hundred to five hundred feet south of the lake and at the west edge of the Gary park.

We did all of our trading at Clark Station, west of here. Charley Kelley had a general store there. He was an engineer on the Michigan Central, and his wife ran the store.

There were no horses around here, but there was a team at Bailleytown; everyone used oxen. We had no roads, but what we called a towpath, run from the little Congregational Church in Miller, down to the Grand Calumet River. We had a kind of a road that came in a half-mile east of Lake Avenue, from the beach to Miller. The old stage line was still running when Grandma came. It ran from Chicago to Michigan City, along the beach, but it wasn't used much after the railroads were built.

People used to travel to Chicago from Michigan City, either by horseback or by team. There was quicksand along the Grand Calumet River. One night a fellow by the name of Doc Foster came along from Chicago and when he got to the river his horse got down and drowned. Doc didn't believe in anesthetics or the knife, but doctored with herbs. When my baby, Henry, was six months old, he had a cold, and I gave him an overdose of Bochero German cough medicine. It put him to sleep and he slept all day till near seven o'clock in the evening. Then I looked out and saw the doctor's horse tied over by the post office. The post office and the depot were together. I called him. He came over and said, "He's all right. Take the lids off the stove and get me a few teaspoons of soot." I gave him a cloth and he dampened it and put soot on it, and then put it under the baby's arms. In a few minutes the baby opened his eyes and looked around and laughed.

I was the only American woman on the lake front or in Miller in 1874. My husband's uncle, Albert Carr, and George Cook, the Lake Shore station agent, and my husband, were the only three American men.

My husband was a fisherman and fished for whitefish and sturgeon, but we hadn't any sale for our fish, only as the farmers came in and traded for them.

We could trade for flour, buckwheat, pork, butter, and so forth. Then, finally, there came a man from South Chicago by the name of Martin Hisner. He wanted to buy our sturgeon if we would take them over. We had a small boat and on one particular night my husband had gone to South Chicago with a load of sturgeon and was late coming back. I was sitting out on the doorstep waiting to see if I could see a sail anywhere on the lake. The wind was out of the north and blew pretty good, and as the sun went down the wind went down with it, and that left a dead sea. Sturgeon like a rough lake, especially when the water is warm. As I sat on the doorstep I saw them jumping, one after the other, and I thought, "My what a haul we could make if someone was only here to help me."

Finally David Crockett came along. I called his attention to the lake, and he said he would go to the mouth of the river and see if he could get any help. The farmers came to the mouth of the river because that's where the only road was. When he came back he found only John Bobian [*sic*], a nephew of old Mark Bobian [*sic*]. I was sorry when I saw there was only John—he was such a lazy man—I didn't want to have to pull a net with him. The three [*sic*] men made a haul. They laid out the seine while I picked up driftwood and made fires—one to go out by, the other to come in by. We pulled in the net and wound up the windlasses and when the net came in, we had fifty-seven large sturgeon. About the time we had them lined up, the boat containing my husband and his men, came in from South Chicago; and there we had a load ready to go again. They were surprised.

I always helped do the fishing. When my oldest son was a baby, I took him with me and rolled him up in a blanket and laid him on the beach. As we moved our windlasses closer together, I moved the baby accordingly. On one particular night I was on the beach helping my husband and I had laid the baby behind a log. Finally a man came along on horseback. I said, "Oh, my baby! Did you run over my baby?" He said, "No, I don't think I did." The baby was all right.

My husband was a great bee hunter. The hills were full of bees, wild ducks and geese. Robert Carr was supposed to be one of the best shots on the Grand Calumet river. In those days they did not have what we call breech-loading guns; they had muzzle-loaders. They did not need them. The game was so plentiful that they had time to load up and get all they wanted without being in a hurry. I remember one morning when it was foggy. Robert started to Clark Station for groceries, and I heard such a lot of shooting, I couldn't think what it was. He was gone about an hour and he had killed fifty-four blue-winged teal.

A man by the name of Miller helped build the Lake Shore Railroad. His child died and was buried in the cemetery here; then the town was named for him, I think. He went away when the road was finished. There were only four houses at Miller, and I knew all the people. There was Ansboro, the telegraph operator; Jim Bennett, section boss of the railroad; Albert Carr; and O'Connor, who worked on the section.

Albert Carr was a trapper. He worked just one day on the section, and came home and said it was his last, and that he could make more in hunting and trapping. He knew the Indian language well. He trapped and got lots of skunk, coon, mink, muskrat, wolves, and foxes and killed lots of deer. He was a brother to my husband's father. In them days there was any amount of game here. The wolves stood back in the hills around here and cried just like a woman. There were lots of white and blue cranes, and hundreds of bald eagles along the beach. When we went along the lake we could see an eagle on every hill; but we don't see any now.

Once I was attacked by one and had a terrible time with it. A fisherman had got it in a nest about a mile west of here, and he kept it in a pen part of the time and fed it fish. But he let it fly around sometimes. One day he got through fishing and went away, and when I went down to the beach, it tried to get at my baby. The eagle was terribly hungry, and I fought with it for quite a while. My sunbonnet seemed to keep it away from my head and eyes, but I had to fight hard to keep it away and it might have gotten the baby if it had not been for the bonnet. My husband was out in the lake in a boat looking after his nets, and when he heard me call, he came in and used his paddle on the bird and drove it off. Afterward it attacked a man by the name of Olander who lives in Gary and he shot it at the dam a mile east of here.

The Tippecanoe River

Theodore Dreiser

☙ ☙ ☙ With all the emphasis that I can summon I wish to applaud the inspiration and energy that in northern Indiana has prevented the further desecration, if not the complete obliteration, of the Tippecanoe River—as poetic and restful a bit of inland water scenery as I know anything about. As a boy in Warsaw, Indiana, I became familiar with a portion of this stream—the very portion, as I only recently learned, that was afterwards totally obliterated by the cutting of a very practical drainage ditch, the only purpose of which was to recover certain marsh lands, then frequented by wild rice-birds, blackbirds, wild duck, mud hens, cranes and loons. Even this land, the practical recovery of which for farming purposes necessitated the complete destruction of the very lovely upper reaches of this stream, did little more than to add a few hundred or, at most, a thousand or two acres to the already ample farming area of the state. But at what a loss.

For at that time these very marshlands were among the most picturesque areas of this region. Marsh grass, sedges or cattails, wild rice and water lilies flourished agreeably and poetically here throughout the spring, summer and

Theodore Dreiser, "The Tippecanoe River," *Outdoor America* 2, no. 8 (March 1924): 24–25, 54. Courtesy of the Izaak Walton League of America.

fall. And throughout the winter when the cold turned the marsh pools to ice the picturesque brown speared cat-tail was to be seen about as far as the eye could travel. And to these fields in the fall when they were frequented by the birds most desired as a delicacy came the fowler with his gun. I can hear the popping of the guns and see the rising covies even now.

But far more painful to me than the loss of the open reaches of marsh grass which somehow suggested the preserves of a great estate and lent color and charm to the entire region—appreciatively recalled by many since, was the loss of this same little stream, the edges of which these sloughs of marsh grass paralleled in places for so great a distance as a mile or more. The poetic facets of that little river. Its nooksy, winding coves and pools. The water lilies and clustering blue marsh flags that starred its shallow bays and inlets. How in a rowboat, alone or with a girl or some fellow schoolboy I have gathered them—great armfuls to ornament the jars and vases of those who cared for them.

The Tippecanoe completely encircled the northern and northwestern environs of Warsaw, the very pretty county seat of Kosciusko County. It provided for the youth of the town and the country adjacent one of the loveliest play grounds imaginable—at least three sandy and shimmering swimming holes into which one could toss a white rock or doorknob from the shore and see it lying on the stream bed at the center. Upon its dry and grassy banks in places one could sit or lie, pole or book in hand, and watch the placid fresh river fish reconnoitering the bait or the grassy hiding places below. Oaks, elms and beeches sentineled it in groves at points. Even rocky gray banks of stone appeared in places and it could be viewed rippling over clean bright pebbles or those long green streamers of water grass that sway so gracefully at the bottom of clear streams. For a distance of not less than three or four miles within the immediate vicinity of this pretty little town it constituted such an idyllic natural feature as any city or town would or should have been proud of. And, for all this distance it was navigable for canoes and rowboats. In the winter season its shimmering icy surface could be followed for nearly all of that distance by the skater. And nearly anywhere in the summer, either in a boat or upon the grassy bank, one might read and dream.

Yet when I returned to my native school town in 1914 this very lovely feature had entirely disappeared. In its place was a weedy and bushy ditch, with here and there a bit of stagnant water—nothing more. And why? Well, at the behest of some local clown or hobbledehoy of the practical variety who had discovered that by draining these adjacent marshes a bit of farming land could be recovered—to be devoted to pigs and potatoes, of course, a very practical drainage canal had been dug. And this, forthwith, had lowered the level of a lovely lake called Center, which to the east of the town offered its shores for residence and beach purposes, by so much as four or five feet. And this automatically removed its waterline from gardens already built upon it by at least a hundred feet or more, leaving an unsightly rim of mud. And this reduction of the lake level from which this charming little river had taken its rise automatically served to drain and dry the river itself in this region.

Gone, at once, the lovely vistas of shallow water that as a boy had so enchanted me here. Gone, the trees by the brink of a stream. Gone, the quaint wild vistas of marshgrass and flag, the circling clouds of blackbirds and ricebirds in season, the coveys of wild duck and mudhen. Gone, the youthful swimming pools, the lovely stretches along which one might row or skate or fish. And why? Some vast economic and supposedly social benefit that contributed to the upbuilding of the city and hence the economic welfare of thousands. Nonsense. Pish. I saw the course of the original stream as well as the town in 1914 and instead of its having been aided in any way by this very practical feature, it had been most definitely injured and made much less attractive by this presumably valuable practical development. The one-time lovely stream bed, a delightful feature for any city or town to possess, was now really a disgusting thing to contemplate. A once lovely iron wagon bridge that had spanned its course at one point had given place to a low stone culvert. In another place, where had stood another bridge, an automobile road, whose foundation was dry brown sand, was all that was left. And the advantage? Some hundreds—at the most a thousand or two acres reclaimed for farming. Can you imagine any sane community any where in the world making or permitting any such exchange?

And yet at that very time this very same city, or town, rather, was seeking to build itself up as an idyllic resort for summer visitors. It was advertising its lake and its sylvan nooks. I sometimes wonder what ails the cerebral processes of the average American anyhow. Is he deaf, dumb and blind to all that really represents the poetry and the sentiment that his orations and his school histories and his very popular sentimental novels boast?

Gentlemen and ladies of my native state wherever intelligence and refinement have blossomed within historic times men have tended to not only treasure but to celebrate the natural beauties by which they have found themselves surrounded. In some parts of the world they are few enough, indeed, it is true. Not so, Indiana. That state, as we all well know, is properly distinguished for a certain bucolic restfulness which all who have ever visited it, gratefully recall. It is not for nothing that the Glades, Brown County, the Tippecanoe and the Wabash are known. Why was it possible, then, for such a thing to happen? I am curious. I do not understand.

One thing is sure. The only thing that has cheered me in connection with all this is the recent successful fight waged against what is known, I believe, as the Matchett Ditch, the object of which was to further obliterate this poetic little waterway in order, I presume, to grow more turnips and pigs! I congratulate the state upon this evidence of an aroused artistic appreciation of the natural beauties that lie within its geographic lines.

➤➤➤➤ THEODORE DREISER (1871–1945) WOULD LIKELY BE ADJUDGED the premier literary writer of those whose works are reprinted here. At least twenty-three books have been published about him, either biographies or collections of his correspondence. There is, as well, an international society of scholars devoted to studying his works. Dreiser, called by Jeanette Vanausdall "the consummate social and literary outsider," is considered one of the foremost examples of the "naturalist movement" in American literature. The term "naturalist" has a meaning here quite different from that used elsewhere in this anthology. Vanausdall has defined the "naturalist movement" as "a logical product of the growing complexity of American society, which naturalists believed rendered the individual powerless and isolated. As a literary school, naturalism was a product of an immigrant, industrial, and urban society" (72).

Dreiser was born in Terre Haute, Indiana, to parents who once owned and ran a wool mill but were wiped out when a fire consumed the mill and left the father seriously handicapped. By the time Theodore came onto the scene, as the ninth of ten surviving children, the family was mired in poverty. At the age of sixteen, he left home to find work in Chicago. He secured a series of odd jobs before a former teacher offered to send him to Indiana University for a year. Although his formal education was limited, he read voraciously on many subjects including natural history.

Dreiser was to write for many publications over the years but is of course best known for his novels. The first of these was *Sister Carrie* (1900), which is considered one of America's greatest novels. Vanausdall writes that the book's publication was "arguably the most important literary event connected with the Hoosier State" (75). Dreiser was to author a total of twenty-seven books of which only eight are novels, among them *An American Tragedy* (1925), his most popular book.

Dreiser spent his final years in Hollywood, and it was only during that time that the literary world began to bestow upon him the praise he had sought. Among Dreiser's many admirers was H. L. Mencken, who wrote of the author after his passing: "The fact remains that he is a great artist and that no other American of his generation left so wide and handsome a mark upon the national letters. American writing after his time differed almost as much as biology before and after Darwin. He was a man of large originality, feeling and unshakable courage. All of us who write are better off because he lived, worked, and hoped" (quoted in Riggio). An alternative view of Dreiser is expressed by R. E. Banta, who wrote that "to many a Midwesterner, he seemed to be only a writer who could find a rotten spot in every apple" (91).

As for the subject of his piece, the Tippecanoe River drains a small part of Starke and Jasper counties as it flows to the Wabash. Except for the short stretch near Warsaw that Dreiser refers to, the river as of 2005 has escaped the channelization that has damaged so many other local streams. And while the basin through which it meanders is 87 percent agricultural, the coarse sandy soils that predominate are more absorbent and less likely to be washed into the channel than finer loams and clay. These circumstances have combined to protect the river's aquatic life. In fact, according to Chad Watts of the Indiana Nature Conservancy, the Tippecanoe has been rated as the eighth most important river in North America for protection of imperiled aquatic species. One example of its remarkable quality is that the river still harbors forty-nine of the fifty-seven species of mussels that have ever been found within its gently moving waters.

Sources: R. E. Banta, comp., *Indiana Authors and Their Books, 1816–1916* (Crawfordsville, IN: Wabash College, 1949); Thomas P. Riggio, *Biography of Theodore Dreiser* (2000), http://www.library.upenn.edu/collections/rbm/dreiser/tdbio.html; Jeanette Vanausdall, *Pride and Protest: The Novel in Indiana* (Indianapolis: Indiana Historical Society, 1999); Chad Watts, Indiana chapter of the Nature Conservancy, telephone interview with author, October 2005.

An Early Illinois Prairie

Albert W. Herre

⌘ ⌘ ⌘ From the end of spring in 1873, until the late summer of 1878, it was my privilege to see and enjoy the life of a great tract of virgin prairie. The prairie, as yet untouched by the plow, began west or a little northwest of Delavan, and stretched in a great arc southward and westward nearly to the Illinois river. It was six or seven miles wide and twenty or more miles long. This immense tract of land was so flat that it was too wet for the plow, and was used for grazing beef cattle. In addition large quantities of wild hay were cut from it. This hay not only gave the necessary winter feed but was also a much needed cash crop which was marketed in Pekin.

The few scattered farmers lived on the low flattened sandy ridges bordering the prairie. At that time the sand hills, which rose but a few feet above the prairie, were the only land cultivated. Here and there the prairie was sprinkled with more or less circular permanent water holes or pools, locally called buffalo-wallows. Around their edges grew a dense ring of sedges, cat-tails, and tall saw grass.

The first year we lived in a one-room log cabin with a loft above, where my uncle and his hired man slept. The cabin stood at the edge of a small grove, sixty or eighty acres in area, which was said to be the only grove on the whole prairie. This was not accurate, for the whole grove and most of my uncle's farm occupied a low ridge beside and not on the prairie.

The prairie grass grew to a moderate height as a rule, but here and there became very rank. I remember one low place in the prairie not far from the cabin where the grass was of phenomenal luxuriance, so that a man on horseback became invisible when only 25 to 30 yards away.

Two kinds of lady's slipper grew within a few yards of the cabin, the common little yellow lady's slipper being abundant, while the large white one was rare. Bluebells grew everywhere, while a showy flower called "flies" by the pioneers, but which I learned in later years was Dutchman's breeches, was very abundant. In fence corners at the edge of the wood grew a wild lily, *Lilium philadelphicum* var. *andinum*, its lovely flowers as high as my head. When my startled eyes saw the first wild lily it gave me a greater thrill than when

Albert W. Herre, "An Early Illinois Prairie," *American Botanist* 46 (1940): 39–44.

I found the giant white lilies many years later in the mountains of northern Luzon.

One of the most marvellous sights of my whole life, unsurpassed in my travels in nearly all parts of the world, was that of the prairie in spring. Unfading are my memories of that waving rippling sea of lavender when the "wild sweet William," a species of *Phlox* two to three feet in height, was in full flower. It stretched away in the distance farther than the eye could reach, while I sat entranced in the rear end of the wagon bed as we jogged slowly on to Delavan.

As the sea of phlox faded it was succeeded by another marvellous flower bed of nature's planting, and this in turn by others until mid-summer was reached. Then the great coarse perennials belonging to the *Compositae* dominated, and instead of a single mass of color there was a vast garden of purple cone flowers, black-eyed Susans, rosin-weeds, blazing stars, asters, goldenrods, and others.

I was introduced to the rosin-weed that first summer by playmates from the neighborhood. Over the prairie we went from rosin-weed to rosin-weed, our little fingers picking off the tiny bits of rosin on the great flower stalks, and popping them into our mouths until by and by we each had a mouth full of highly prized chewing gum.

Every spring and fall the prairie was covered with water, so that the whole country side was a great lake. Only the sand ridges emerged here and there. Wagon traffic came to a complete stop, and men could only get around on horseback, zig-zagging about along the low sand ridges.

All day long swarms of water birds filled the air, and far in the night their cries sounded overhead. At the first gleam of dawn vast flights of ducks dashed to and fro and great flocks of wild geese sped swiftly across the sky. Many flocks of ducks just cleared the oaks and maples behind the cabin, so that one could have sat in a chair in front of the door and shot enough ducks in a few hours to last a family for a week. Not so with the wild geese, which were always shyer and harder to get.

That first summer was also remarkable for the number of snakes visible. There were many bull snakes and "blue racers" or blacksnakes from seven to eight feet long. These are not a child's estimate of lengths, but were actual lengths of snakes I saw killed and measured. As a rule people killed every large snake seen, just as they shot owls, blind to their economic value. Then of course there were hordes of garter snakes and several other kinds too, such as "milk snakes" and "spreading adders," the last named greatly feared by all the country people.

As I grew older I learned to know and dread the common water snake, *Natrix sipedon*, which was called "water moccasin" and reputed to be the same as the venomous water moccasin of the southern states. Specimens up to eight feet in length were common along the Mackinaw, and on Salt Creek and its sloughs, in Mason county. It was not until I was seventeen or eighteen, and had studied zoology in high school that I learned our "water moccasin" and "spreading adder" were perfectly harmless creatures.

The destruction of the prairie flora and fauna began when a great machine started to eat its way through the prairie, leaving behind it a stream of water on which it floated. As the "Big Ditch" was made, illiterate English and Irish day laborers dug lateral ditches by hand. This drainage made it possible to plow the prairie fauna for the first time, and I well remember when my great uncle, broke up his first piece of prairie, an eighty acre field.

The prairie life survived to a surprising degree for several years more, and was still a source of joy and wonderment until I left Tazewell county. The advent of tile drainage early in the 80's completed the transformation of the prairie into ordinary farm land and brought in many more people. Of course the ducks and geese stopped coming, for there was neither water nor food to attract them. Migratory flocks of snipe and plover continued to come for a couple of decades, but their numbers had dwindled to a mere trickle when I left Illinois in 1900. The crawfish and bull frogs disappeared in a hurry, and the prairie chickens were destroyed by the combined efforts of the plow and shot gun. I returned to the region several summers during the '90's, but the prairie as such had disappeared, and of course its characteristic life with it. What a pity that some of it could not have been preserved, so that those born later might enjoy its beauty also. Now it is merely flat unending corn fields, and moderns may look on this article as only the iridescent childish romance of an old man.

◣◣◣ ALBERT WILLIAM CHRISTIAN THEODORE HERRE'S (1868–1962) "An Early Illinois Prairie" was one of the writings that inspired this anthology. In 1958 a colleague at Stanford University asked Herre to write an autobiographical sketch, which presents fascinating details of his early life. He was blessed with parents "who regarded music and books as fundamental necessities of life." Unfortunately, while his family was living in Detroit, his father accidentally drowned when Herre was only four years old. His mother moved to central Illinois to be with relatives. She was penniless, with three children, so she scraped by doing a variety of jobs, including keeping house for her brother, who had a farm in Tazewell County on the edge of the prairie that Herre writes about. (In this sketch Herre borrows several paragraphs from his 1940 article.) After two years, however, the brother caught "gold fever" and left to seek his fortune in the Black Hills of South Dakota. A great-aunt near Delavan, then the county seat, agreed to take care of Herre's younger siblings, while he stayed with his mother and attended "country school." Soon thereafter his mother and he moved to Delavan, where she performed domestic work for a physician.

To me the most remarkable element of Herre's growing up in rural Illinois is how a first-rate mind with insatiable curiosity prevailed in an intellectual backwater with a school system almost devoid of resources. Herre read everything he could get his hands on. He "devoured" the books owned by the doctor for whom his mother worked. During one period, he and his mother lived next door to the town barber and his wife. Among the volumes in their library were *The Tropical World* and *The Polar World*, popular natural histories that Herre read "with intense interest and which had a great impact upon" him. Later the family moved to Springfield, where the availability of reading material was much greater in both type and quantity. Thus, Herre became well-versed in astronomy, religion, history, English literature, poetry, and the classics. In addition, he feasted on the newspapers from St. Louis, Chicago, Detroit, and New York.

He became friends with the city librarian James Bryce, who without formal education had mastered a wide range of languages and subjects. After decades of being in academia, Herre could still say of him that "he was one of the most learned men [I had] ever known." Herre, being uncertain what to do after high school, asked Bryce for advice. Not surprisingly, the answer was to seek more education. Impressed by the writings of the great biologist and educator David Starr Jordan, Herre decided he would attend Stanford University of which Jordan was president. The death of his stepfather changed his plans, however, and over the next several years he married, worked as a reporter for the Springfield *Illinois State Register*, and taught high school. He attempted to further his education through correspondence courses from the University of Chicago, but both the geology (his first choice) and zoology departments refused to offer such classes. The botany department, however, did have such a program, so Herre took the three years of classes that they provided. When his wife became ill, they decided to move to California and so he did eventually enter Stanford in 1900.

Herre is best known as an ichthyologist, although he writes that he never received "any instruction" in the field. He had enrolled in an ichthyology course, but Professor Charles Gilbert simply gave him a collection of fish and told him to identify the specimens. He continued working with fish, becoming Jordan's assistant in ichthyology. He received his doctorate in 1908, but because of a falling-out with Gilbert, his dissertation was on the lichens of the Santa Cruz peninsula. His reputation was forged during his tenure as director of fisheries for the Philippines. According to colleague George S. Myers, Herre's "monographs on Philippine fish… placed him among the foremost faunal ichthyologists." He became curator of the animal collections at Stanford in 1928 and continued his research in Asia and elsewhere until the outbreak of World War II made travel impossible (at least two of his collaborators in the Philippines were shot by the Japanese). After the war he worked for the U.S. Fish and Wildlife Service and then the University of Washington's School of Fisheries. A bout of pneumonia drove him from Seattle to Santa Cruz, where he lived with a daughter. His final years were devoted to lichens.

Source: A. W. C. T. Herre and G. S. Meyers, "Albert William Christian Theodore Herre (1868–1962): A Brief Autobiography and a Bibliography of His Ichthyological and Fishery Science Publications, with a Foreword by George S. Meyers (1905–1985)," in Collection Building in Ichthyology and Herpetology, ed. T. W. Pietsch and W. D. Anderson. Spec. pub. no. 3 (The American Society of Ichthyologists and Herpetologists, 1997), 351–65.

The Kankakee in the Old Days

F. E. Ling

ᔓ ᔓ ᔓ My first acquaintance with the Old Kankakee was about fifty years ago when I was a youngster six or seven years old. My father, who owned a store here at that time, enjoyed hunting and fishing and took me with him. We would leave home early Sunday morning and walk about four miles to the boat landing at the edge of the marsh, aiming to get there about daybreak and out to the river by sun-up.

F. E. Ling, "The Kankakee in the Old Days," Bulletin of New York Zoological Society 38, no. 6 (November–December 1935): 197–204.

I sat in the stern of the boat, flat on the bottom, and by using a short paddle tried to keep the boat about a certain distance from the bank as we floated downstream. Using an eighteen- to twenty-two-foot cane or bamboo pole and about the same length of line with a Skinner or Chapman spoon, Father would throw out toward the bank under the overhanging branches of trees or along the logs, grass or moss, and draw the spoon around in front of the boat. When he hooked a large pickerel, wall-eye, bass or dogfish (grinnel) the fun was on. As I grew older I learned to stand up in a boat and push or fish. Later he built me a boat of my own, and then a chum and I spent most of the summer days on the river.

We never put "if" in the fish-catching game. It was a mighty rare occasion that we didn't get all the fish we wanted. We seldom looked for bait until we got to the river. You could catch frogs, minnows, crayfish or hoppers there. We caught our minnows by hooking the knob end of the push paddle into the moss and pulling it out on the bank and picking the minnows out of the moss, or by running the boat along about five or six feet from the bank where there was moss, stepping on the edge of the boat and holding it as low as we could without dipping in any water. Then we would hit the moss with the paddle and the minnows in their fright and haste to get away would jump into the boat.

This will give you some idea of why the Kankakee had fish in it. It had ideal natural spawning and rearing conditions for fish. One season the river was alive with small pickerel weighing about a pound and a half. A boy friend and I caught eighty-seven in about an hour. He caught thirteen fish in thirteen straight casts of about thirty feet. On the fourteenth cast he hooked a fish and lost it. He threw his steel casting rod down in the boat and said, "Damned if this is fishing. It's slaughter." Then he lit up his pipe and enjoyed himself.

You could catch these fish anywhere for miles along the river. Where they came from no one knew. One other year we had a similar condition, but not so many fish, but larger ones ranging from four to five pounds. They must have had the natural feed all eaten, for they would hit anything that moved. Father and I caught fourteen in less than a mile of river. On one cast, when one hit his spoon, it jerked the end of the long cane pole into the water and another fish struck the end of the pole, the splash having attracted it. These years were the exceptions, not the rule, but we always had lots of fish. Many times when we were kids have the kid brother and I dragged a string of fish on a dog chain which was too heavy for us to carry, from the boat landing to where the old cracky wagon and pony were left on high ground. Many and many a time have I looked over the edge of the boat and seen a school of bass or wall-eye in ten feet of water. The water was very clear and there was no pollution before dredging. It is not so clear since dredging, but has no pollution even now.

In the spring of the year, after the ice had gone out and we had had our first thunder storm, the old fishermen would go dogfish spearing. They would anchor their boats on the shallow side of the river bend on a sand bar where the water was from four to eight feet deep and where the current would carry the fish over the sand. The fish would come lazily floating down, resembling sticks of sunken wood. They would come drifting along from singles to droves of six or eight and the fishermen would spear boatloads of these fish and salt them down.

The dogfish run was on every year as far back as any of the old timers could remember, and the Indians before them enjoyed the same sport. When they dredged the river the run stopped and there are very few in the ditch now. The dogfish spearing started the season's fishing and it wound up with wall-eye fishing late in the fall when slush ice was running in the river. Live minnows were used for bait at this time of the year and the fishing was done just on the lower side of a drift, the drift of logs protecting you from the floating ice. We had several varieties of fish: dog, cat, bullheads, carp, buffalo, pickerel, croppies, bluegills, perch, suckers, sunfish, black bass, wall-eyes and some eels. Many a day have we filled a stringer with a catch of different varieties.

The swamp and marsh ranged from one-half to five miles wide and was about one hundred miles long. This expanse of territory with its sloughs, bayous, ponds, natural feed and hollow trees made it a wildlife haven. Very many kinds of trees, shrubs, flowers, grasses and aquatic plants grew in this area. Many of the bayous and pond holes in the timber and marsh were a white carpet of water lillies and the banks of the bayous and ponds were lined with marsh hollyhocks. In our upland gardens they are called hibiscus. In the spring the islands were one mass of flowers—spring beauties, Dutchman's breeches and violets. Later the May apple covered large areas with its umbrella foliage and the Indian turnip with its green calla lily bloom was seen everywhere. Squirrels, groundhogs, chipmunks, flying squirrels and many varieties of birds were numerous.

The greatest thrill of a squirrel hunt was that the whole area seemed to be alive with wild creatures. Many a time have I sat at the base of a big beech tree by the hour watching the wild life and not firing a shot. This area of from ten to fifteen acres seemed to be completely roofed over by the spreading tops and branches of the big beeches. I haven't words to describe what took place in this apartment building of Mother Nature's. When the squirrels were working on the beech nuts, the dropping of shells and nuts sounded like heavy rain. A hunter killed a mixed bag of 103 black, gray and fox squirrels in this area in one day.

The trees on the islands were beech, white oak, burr oak, red oak, black oak, hickory, pepperage, butternut, black walnut, white ash, black ash, sycamore, soft maple, sassafras, wild cherry and paw-paw. On the low lands there were pin oak, burr oak, black oak, red birch, elm, black ash, white ash, sycamore, cottonwood, soft maple, quaking aspen and willow. Acres of witch hazel, blackberries, raspberries, blueberries, huckleberries and wild strawberries were found on the islands and around the edge of the marsh. Large areas of pucker brush and devil pins were found in the lower areas around the bayous and ponds. In the marsh wild rice, smart weed, Spanish needle, celery, duck potato and many other grasses and weed seeds furnished natural feed for geese, ducks, birds and animals. In the timber they found acorns, wild rice, smartweed seed, pucker brush balls and other aquatic plants and roots.

During the migration season ducks, geese, brant and cranes were on the Kankakee area by the thousands and there were all varieties, from canvas backs to fish ducks. Canada geese fed on the prairie in the wheat and corn fields and rested on the marsh by the thousands. The tales some of the old timers tell are almost unbelievable. Sandhill cranes were plentiful but not nearly as numerous as geese. Many swans were seen on the marshes. Later in the season hundreds of herons, blue and white, nested here. At different places in the swamp the cranes had what we called "crane towns." Hundreds of them would meet in one locality in the cottonwood trees. Many times I have seen as many as four or five nests in one tree. The largest town, the one we called "lower crane town," was estimated to have 1,000 cranes and, believe me, there was some music when the young were in the nests! Thousands of jack snipe, sand snipe, plover and other shore birds lined the marsh shores and upland ponds.

The principal fur bearing animals were muskrat, mink, otter, skunk, raccoon, fox and beaver in early days, and occasionally a lynx and bobcat. In 1912 on the Degolia marsh, a tract of from 1,000 to 1,200 acres, from November 1 to December 20, the date when the marsh froze over, two trappers camping together caught 7,634 muskrats. During a period of a few days after the freeze-up they speared 1,300 more in their houses on the same grounds. This was a good rat marsh but not any exception to many others along the Kankakee. In the winter when the swamp was frozen over, fur was hunted on the ice. The party generally consisted of two men and three or four dogs. Two men could work to better advantage than one in climbing trees and throwing out the 'coon and chopping mink out of their dens. It was not uncommon for a pair of hunters to catch from $500 to $1,000 worth of fur during the ice hunting season. This was a free-for-all, the old trappers holding their trapping grounds by what they called trappers' rights, but the rights ceased when the swamp froze up.

The swamp had many hollow trees and many of these held swarms of bees. I have known one man to find sixty-five bee trees in one fall. Twenty to thirty-five bee trees were just an ordinary cutting. I have seen them produce honey in amounts all the way from the size of a walnut to two washtubs full. You never could tell from the looks of a tree what you were going to find. Sometimes when you had washtubs and boilers ready, you needed a hot biscuit, and other times you had to leave part of the honey or make another trip for it. In the winter many of the natives for miles each side of the swamp cut the timber for fence rails and wood. In the late afternoon and evening the roads leading from the swamp would be a procession of loads of wood. Timber, fur, fish, birds, animals and game weren't all there was to the Kankakee. It was a beautiful, clear, winding river with its tree-covered banks, its large drifts of logs, wild hollyhock banks and lily-covered bayous.

In the fall it presented a beautiful scene with the foliage flaunting all the autumn hues, ranging from the deep green of the pin oak to the naked branches of the ash, and every bend offering a different picture. The time I liked most to be in the swamp was when the leaves were all off, the wind blowing a gale, lashing the tops and limbs of the trees, a gray, drab, fast-moving clouded sky, and the air full of ducks trying to find a shelter from the coming storm. This kind of a scene meant you were going to have a hard push home in a storm, but it was worth it.

The Kankakee was one of the greatest wood duck nesting grounds in the U.S.A. Hundreds of pairs nested yearly in its hollow trees and stumps along the river, bayous and ponds. During the summer months, before the young were able to fly any distance, the broods were scattered all over the river and swamp wherever there was water and feed. You would see many broods of them in the summer while fishing. In August when the young were able to fly they began to congregate in certain roosting places. For about two hours before dark in certain areas of bayous, ponds or sloughs that had lots of old logs, floating flag roots and pucker brush, the air was full of wood ducks, from singles to flocks of fifty coming to roost. It was not uncommon to see 500 wood ducks in the air at one time, in these areas. They fed all over the swamp and sloughs, mostly on acorns from the thousands of trees in the swamp, but they roosted in certain suitable areas as soon as the young were able to fly well. The destruction of such good nesting and rearing areas has had much to do with the scarcity of wood ducks. I believe that drainage has done more to create our scarcity of ducks here than all other causes combined. Thousands of ducks and many geese nested here before the area was drained.

One late fall afternoon when no birds but the red headed woodpecker were stirring, a slight northwest breeze had just started to blow as I was leisurely pushing my boat up the river, aiming to get to the boat landing at dusk, having agreed to meet my hunting partner there at that time. I was about a mile from the landing when I heard someone call to me to hurry up and get to the landing as soon as I could. I looked around and saw my partner pushing as fast as he could and motioning for me to hurry on. At that time I also saw several flocks of ducks coming from the southeast, going northwest, and bent on going on about their business and paying no attention to a call. He called to me again to hurry on to the landing and waste no time. I knew he had some good reason for wanting me to hurry so I put on all steam ahead, and when I turned in at the ditch that led to the landing and boathouse I slowed down and waited for him to catch up.

By this time the air was full of ducks, all going northwest, but they paid no attention to the call, not one of them. When he ran his boat alongside of mine he said, "Hurry up, let's get to the landing and try to kill some of those ducks." We grabbed our guns, hunting coats and shells and ran about ten rods to the top of the ridge, which was maybe forty-five feet above the water. My partner said the ducks were changing marshes. They had been up east in the south marsh and were going to the north marsh. These marshes were so called from their position to the river. He said that tomorrow we could go back in the territory we had been in today and found no ducks, and find plenty of them, for they were moving in.

When we got to the top of the ridge his instructions were, when a flock came over us within range, to pick out a certain bird, follow it until it was directly overhead, jump in ahead of it, shoot one shot only, and we would kill some ducks. Did he know his duck shooting? I'll say he did!

We picked up a mixed bag of sixty-eight ducks in about an hour out of this flight and they were flying just as thick when we quit shooting as they were when we began. Then the weather man decided we had enough and sent a northwest wind with heavy, thick, dark clouds, which made it rather uncertain shooting. Rather than cripple and lose the ducks we unloaded the guns.

My partner was one of the old time real hunters. He studied and understood the habits of birds and animals. He wanted to know why they did this, and why they did that, and he figured out many of the whys. It was the rule rather than the exception for him to kill two ducks at one shot on the wing when coming in to decoys. He said that a pair or more of ducks very seldom came in to decoys without two of them crossing within gun range.

Before the drainage of the Kankakee we had hundreds of prairie chickens and many partridges. Now we haven't any. Their natural home has been ruined. In destroying the native marsh grass they destroyed the home for the chicken, it being their nesting place. It seems that certain conditions and feeds are essential for the nesting and rearing of wild birds and animals. Nature provided this, but the progressive, intelligent white man in his march of progress has failed to consider this and consequently unless there is a change of policy our wild life must go in spite of its efforts to continue with us.

Some spots of this drained land are still producing fair crops, but the part we wish to restore is practically worthless and could be made a wildlife haven by reclaiming it according to the State Conservation Department's plan. Not only would it be a wildlife haven, but a source of much revenue. Its location is ideal for recreation, having several million people within a hundred mile radius. I hope to see the circle completed—from marsh to wilderness and back to the original wonderful marsh again.

◄◄◄◄ ALTHOUGH HIS DEATH WARRANTED THE FRONT-PAGE HEADline in the Valparaiso daily newspaper, the *Vidette-Messenger*, Francis E. Ling's (1878–1951) obituary provides very little information about him. He spent the first part of his life in Hebron, Porter County, leaving to attend the Chicago Dental College. After graduating in 1909, he maintained a practice in both Hebron and Chicago until returning full-time to Porter County. Once back, Ling again alternated, this time between Hebron and an office in Benkie's drugstore in Kouts. (This tidbit is from *The Kouts Centennial Book*.) After a year he settled in Hebron, where he worked for the rest of his career. He married Benkie's daughter Lulu Mae in 1913. Dr. Ling was known for his interests "in hunting and fishing and the outdoors, and conservation and preservation of hunting and fishing lands and wildlife."

Sources: "Death Takes F. E. Ling, Area Dentist," *Vidette-Messenger* [Valparaiso, Indiana] (December 3, 1951); Kouts [Indiana] Centennial Committee, *The Kouts Centennial Book* (1965).

A Naturalist in the Great Lakes Region

Elliot Downing

DISTRIBUTION AND ADJUSTMENT

ᏬᏬᏬ The purpose of this...chapter is to show how the physiographic features of the Chicago region...determine the distribution of plants and animals, not directly but by shaping in large measure the interplay of those factors that do condition the plant and animal life. The chief factors for plants are available moisture and light; for animals, the oxygen supply, food, nest-forming materials, light, heat, currents, foes, etc. Furthermore, plants and animals are nicely adjusted in structural peculiarities and habits to the complex of these limiting factors, and it will be the aim of succeeding pages to point out also some of these adjustments.

It is a matter of common knowledge that, the world over, there is a zonation of life. The vegetation of the tropics is totally unlike that of the temperate regions, and the life of the latter zones is quite different from that of the arctic. The ascent of a mountain carries the traveler through a succession of life-zones, from the luxuriant growth about the base through scant vegetation and disappearing animals to a bleak and almost uninhabited peak.

Temperature is on the whole a very great factor in the determination of the abundance and character of both plant and animal forms. Locally its effect is to settle the distribution in time rather than fix the place in which animal or plant shall grow, for naturally there is no very great difference in temperature in the various parts of the Chicago area unless it be a contrast between the deeper parts of lakes and their surface waters. But there is a seasonal distribution of both plants and animals locally. Thus we have a distinct spring flora. Note, for instance, the early annuals of the oak woods—spring beauties, anemone, toothwort, trillium, hepatica, Dutchman's-breeches, dogtooth violet, bloodroot—these and others like them are all plants that are up and in blossom before the trees are in leaf to shade them. They get through with their life-cycle—bud and flower and fruit—here on the forest floor while the great trees overhead are just beginning to stir with the thrill of the spring awakening. These plants have found an unoccupied part of the

Elliot Downing, *A Naturalist in the Great Lakes Region* (Chicago: University of Chicago Press, 1922), 90–105.

season, and they make the best of it. They all possess underground stems loaded with stored food that enable them to make this very rapid growth, then slowly accumulate during months of shade a supply sufficient for the next spring. Moreover, in many cases their tender leaves that appear while frosts are still common are clothed in dense hair—a veritable fur coat to protect them. They are replaced later by other plants that have become adjusted to growing in the dense shade of the summer under the trees. Those mentioned need abundant sunlight to carry through their brief program of rapid maturation.

The temporary grassy ponds that result from the melting of the snows are the homes of a group of animals that appear marvelously indifferent to the low temperatures of early spring; they thrive in the ice-rimmed water. There is the so-called fairy shrimp, *Eubranchipus*, not a shrimp at all, though its airy grace and mysterious appearance make the rest of the name appropriate enough. It is a reddish-brown crustacean, a quarter of an inch long when first seen, but growing rapidly to an inch in length. It swims on its back, waving nineteen pairs of feathery legs to propel itself. The head bears a pair of staring compound eyes. The egg sacks of the females are conspicuous early, and strings of slender eggs can be seen in the semi-transparent body on their way to be discharged into the icy water. The adults soon die; the whole life-history occupies only a month or so of early spring. The eggs lie dormant in the bottom of the pond. When it dries up they dry, too. They may blow about in the wind to new locations. They freeze with the winter cold; in fact they will not hatch until they have dried out and frozen. And so they are ready to start the next generation in the ponds of the following spring...

Here, too, one may find, if low prairie is near, the spring peeper, *Chorophilus nigritus*, that marsh tree frog whose clear peep is almost like a bird note. It takes to the ponds to lay its eggs almost before the ice is gone. The eggs are laid in gelatinous clusters that fill the palm of the hand and in the pond are attached to the grasses along the margins or to sticks in the shallow places.

The frog's egg is admirably adapted to hatch on these cold days. It is covered with a transparent jelly layer that retains the sun's heat like the glass of a greenhouse. It has a black upper surface that absorbs heat like a black dress. The egg is laid so early that it avoids many of the insect larvae that would later prey upon it. It is inconspicuous, its black upper surface harmonizing well with the dark pond bottom when seen from above, and its light under surface with the clouds when seen from below, so that it easily escapes detection.

Just as there are an early spring flora in the woods and a spring fauna in the temporary ponds, so there are early spring plants and animals in the swamps, on the dunes and in other typical localities. In each place the spring types are followed by early summer types, by midsummer and autumnal species. One need only call attention to this seasonal distribution to have it substantiated by many commonplace facts. Thus we name many insects by the time of their appearance, as May flies, June beetles, the fall army worm, etc. Just as the appearance of the robins and the bluebirds marks early spring, so we think of midsummer as butterfly time, and the chirp of the cricket as the first voice of fall.

In the local place distribution of plants, moisture is of prime importance, or rather the ratio between rainfall and evaporation. If the water supply greatly exceeds the evaporation, then the plant grows in a pond or marsh and is designated a hydrophyte. At the other extreme are the xerophytes, plants growing where the evaporation tends greatly to exceed the water supply. The open dunes give an excellent illustration of such conditions. The great majority of plants grow where there is neither excess nor dearth of water and are known as mesophytes. This ratio of evaporation to rainfall is a very important factor in determining plant distribution on a large scale as well as locally.... The general coincidence of forest distribution and of the areas occupied by prairies and plains with the areas of a decreasing rainfall is very striking.

That physiographic features must affect the moisture content of the soil is evident on reflection. A poorly drained area develops swamps, bogs, and wet prairie. A rock surface is prone to be xerophytic; so too will be the steep side of a clay bluff. On the other hand, the side of a rock ravine may furnish hydrophytic conditions, for the sunlight penetrates so little that the temperature is low, the evaporation is slight, and even though water be not abundant, it may be so well conserved as to be adequate to moisture-loving plants....

Many marked structural peculiarities and adaptive life-habits go along with the ability of the plant to endure drought or excessive moisture. Since the loss of water goes on largely through the leaves, the leaf surface of the xerophyte is often reduced in proportion to its volume by the leaf being needle shaped rather than flat and thin. The leaf surface may be covered with hairs, and so evaporation becomes reduced, or the surface may be covered with impervious wax; such a weed as the mullein, growing in open waste territory, is a familiar example of the former, and the glossy-leaved plants of the ground stratum in the pine dunes—like the wintergreen, shinleaf, and prince's pine— are examples of the latter, as is also the very common field milkweed. The oak leaf and that of the cottonwood are both thicker and glossier than the leaf of the hard maple or beech that grows in the mesophytic conditions of the climax forest. The leaf stem or underground portions of the plant may be thick and succulent, storing up water in time of plenty to use in time of drought. The common cactus of the dunes is a good illustration. The plants growing where moisture is lacking often develop an extensive superficial root system to gather up the dew and the showers before the water has time to dissipate or sink deep into the parched soil. One can pull out fine stringlike roots in the open dunes that run just below the surface for hundreds of feet to the tree or bunch grass.

Many of the animals of the open dunes are in hiding during the day, some of them like the burrowing spider, *Lycosa wrightii*, in holes that run down to the cool and moist soil layers. The surface of the sand is covered in the early morning with the fresh tracks of many animals that have been out during the cool of the night to satisfy their needs, while the same areas are apparently uninhabited by day. The six-lined lizard, that like the cactus is a desert form left in this sandy oasis by the lake, excretes its uric acid in solid form, a conservation of water common to many reptiles.

The light relation determines not so much where the plant grows as its habits of growth in a locality fixed chiefly by the water supply. Thus, some plants are said to be shade loving. They are found forming the ground stratum in the forest with other plants, a shrub stratum, over them, plants less able to endure the shade, though they in turn are overgrown by a tree stratum whose members insist on getting up into the full glare of the sunlight. The entire association is due to the mesophytic conditions that all need, and probably the stratification is due quite as much to the relative amounts of evaporation as to the varying light intensity. How dim the light is in the ground layers of the forest becomes apparent in trying to take pictures in the beech-maple woods; the exposure meter indicates an intensity one-twentieth that of the surrounding open pastures; so undoubtedly the light factor is to be regarded as of great importance. This is made more evident when the fact is recognized that the same shade-loving plants found on the forest floor are also often found in the rock ravine. Thus, the clearweed, the touch-me-not, and such characteristic ferns as the beech fern, the fragile fern, and the spleenwort, *Asplenium angustifolium*, are common in both locations.

Some plants growing in the intense light of the marsh and prairie have very interesting habit adaptations that avoid the intense light and heat of midday. The upright position of the leaf in the grasses, the iris, and cat-tails accomplishes this, for the edge or tip of the leaf is presented to the midday glare, the broadside of the leaf catching the less intense early morning or late afternoon sun. The common lettuce and the compass plant, *Silphium lacini-atum*, both have the habit of turning their leaves so that they are in a vertical position at noon, the tips north and south, one edge up, the other down, so showing the edge rather than the broad side to the noon light.

No single line of demarcation in the distribution of animals is as clear cut as that between the water animals—the water breathers as they are commonly called—and the air breathers. Of course in each case the oxygen in the air is the important element taken, but in the water forms this is absorbed from the supply dissolved in the water, while the others take it directly in the respired air. Probably in the course of evolution most animal as well as plant life was aquatic in its origin. The hydrophyte and the hydrozooid are the primitive types, and these early forms lived not only where water was abundant but they lived under water. Such an existence necessitates on the part of complex forms some special device for taking the needed oxygen, for every living thing must have this essential gas since it is only by constant oxidation that the energy supply is maintained, by means of which all live and move and have their being. The simple plants and animals can absorb the oxygen and give off the carbon dioxide, together with other waste products of combustion, directly through the moist skin. But the higher types have needed to develop gills or similar structures and some sort of circulation to take the gas to the working cells. The more complex plants growing totally submerged have their leaves dissected so that they consist of numerous threadlike filaments or else they are long and narrow, ribbon-like, so that every part of the leaf is near the oxygen supply of the water. See, for instance, the leaf of water milfoil, of bladderwort, of hornwort, all common to our ponds or streams. The water buttercup at times grows submerged, again only partly so. The submerged portions have leaves that are finely dissected, the aerial portions the usual buttercup leaf. It is interesting to note that the animal gill is built on the same general plan as the submerged leaf, a series of filaments, sometimes branched. In the animal these are provided with vessels that take up the needed oxygen and carry it to the distant internal organs.

The transition from water-inhabiting animal to land dweller is seen in the life-history of our common toad. The eggs are laid in the water, hatch into tadpoles that are vegetarians, and breathe by means of gills. In time these develop legs, resorb their tails, come out on to the land, replacing the gills by lungs and feeding entirely on insect food. When the first sea worms crawled out of the water to make their burrows on land, a wealth of food was awaiting them in their virgin hunting ground, for the land was largely unoccupied by animal life. When some primitive fish crawled out on the mud banks, using its fins as legs—and there is one that does this still—it was rewarded by a lot of food which others of its kind could not get. But soon the earth swarmed with life, and even the air supported a full complement. Then, apparently some of the animals that earlier forsook the water for the land or air went back again to hunt in the water, and no more interesting series of adaptations is to be found than those enabling the land animals to live in the water. Insects likely evolved from land forms and never did live in the water until some of them took to it as a secondary consideration. They are primarily air breathers, taking the air in through numerous spiracles on the sides of the body to a system of internal, ramifying air tubes. When they took to the water they were forced to strange devices. Some diving beetles like *Dyliscus* and some of the *Hemiptera* like the giant water bug carry down a supply of air between the concave back of the abdominal portion and the overlying convex wing covers. The spiracles or breathing pores, which on most insects are on the sides of the abdomen, are now on its top, which is the bottom of the air chamber. Other animals enmesh enough air in the hairs of body and legs to last them awhile; they seem to carry a film of silver over the parts, the air film reflects so much light. The common water scorpion has a long air tube at the posterior end of the body so it can stand submerged on some aquatic plant and still breathe through its air tube, the open tip of which is kept at the surface.

Even the spiders have taken to the water. Some common locally walk on the water and count it no miracle, they even run with celerity, capturing flies and gnats out of the reach of their less fortunate kind. One, the diving spider, hunts under water, even spins its silken nest below the surface. This nest is cup shaped, the opening down, and the adult spider carries down to the eggs and young a supply of air enmeshed in the hair of abdomen and legs that it scrapes off so as to keep the cup full until the young are old enough to come to the surface for their own supply.

▬▬▬ ELLIOTT ROWLAND DOWNING (1868–1944) WAS BORN IN Boston and educated at Albion College in Albion, Michigan (BS) and the University of Chicago (MS and Ph.D.). He taught at several universities, including Columbia, Beloit College, and the school now called Northern Michigan University in Marquette (1901–11). Returning to the University of Chicago in 1911, Downing held joint appointments as associate professor of natural science in the College of Education and associate professor in the Department of Zoology. (The university never promoted him to full professor, despite his many publications and years of service to the school.) Being in both departments attests to Downing's great concern: he wanted to teach and popularize ecology and related disciplines. His best known book, A Naturalist in the Great Lakes Region, is an attempt to introduce the pioneering work of both Henry Chandler Cowles and Victor Shelford to a general audience. For eighty years, it was the only comprehensive account of this region's natural history. As a sampling of his other works demonstrates, they were also aimed at education: Our Living World, Our Physical World, and Teaching Science in the Schools.

When he retired from the university in 1934 as associate professor, he wrote this letter to the board of trustees: "I hope with the new freedom to write several more books that may serve to improve the teaching of science. I should like to incorporate in them what wisdom I have gained in my long contacts with science instruction; at least I hope it is wisdom. Thus I may discharge a part at least of the obligation I feel I owe the University for the inspiration it has been to me...."

Sources: Who Was Who in America, vol. 2, 1943–1950 (Chicago: A. N. Marquis Co., 1963), 162; "[Downing's] Letter to the Board of Trustees Regarding His Retirement," Board of Trustees Minutes, University of Chicago, June 14, 1934, vol. 24, p. 81.

A Hoosier Tramp

Samuel A. Harper

I

᪥ ᪥ ᪥ I had difficulty in making the good people of Michigan City understand I did not want the "best" road out of their fair town. They seemed to think that even a man in walking togs would want a "good concrete road." Finally I was directed out Michigan Street by a young man at the office of the Chamber of Commerce. Walking eastward over the city pavements I soon became impatient to get onto the spongy sod. A brown thrasher was caught singing in an old park, but beyond that nothing but a long stretch of city streets. Finally when I could endure the pavements no longer I plunged across the back yards to a ravine through which flowed a small stream which

Samuel A. Harper, A Hoosier Tramp (Chicago: Prairie Club, 1928), 60–78.

was little more than an open sewer. It was bordered by scattered trees, however, and even the soiled water of this little stream had attracted a water thrush. Field and song sparrows, towhees and warblers were singing, so that I felt that my day was fairly begun at last. As I sat down a moment and rested my pack against a stump a female redstart darted from a maple to an old pine right in front of me and a gray-cheeked thrush twanged gloriously in a neighboring dooryard. But the town cropped out again unexpectedly to vex me just beyond the little wood and two or three mourning doves among the scattered trees expressed my disappointment.

Following the railroad track I was surprised to hear a killdeer's call in a field partially occupied by an automobile equipment factory, and a bob-white in a patch of near-wilderness beyond whistled as ingenuously and merrily as if he had been far away in the clover meadows. A half mile down the track, just at the edge of town, a small sparrow flew out of the scant grass along the right of way and ran anxiously along the gravel track ballast. Suspecting a nest I examined a little clump of grass from which she had flown and soon found the nest made of dry grass and tiny rootlets. It contained two half grown birds and one egg, the egg apparently being addled. It was a Savannah sparrow. I sat down opposite the nest about ten yards from it and watched the little mother who also anxiously watched me as she flitted about between the wire fence bordering the right of way and the meadow beyond. She was very shy, however, and refused to revisit the nest while I was near. This little sparrow which lives almost constantly on the ground, running along in the grass very much like a mouse, is very difficult to see. Any one who can see a Savannah sparrow is very likely to see many other things in nature of great interest.

As I resumed my walk a bluebird warbled in an old dilapidated orchard by the roadside. The infinite pathos of an old orchard! Limbs broken to the ground, great holes in the decaying trunks and branches, the rotting trees scattered and few, wheat or other grain growing in the open spaces among the old unprotesting trees, the crumbling foundation stones of an old house, grown over with weeds, all tell the sombre tale of an ancient homestead long since abandoned and left desolate.

As I crossed a little stream a number of bank swallows darted up and down and all around me. I examined the banks of the creek for their nests but found none.

Laborers working along the railroad right of way greeted me in a friendly manner. One of their number, an old fat man, was resting quietly on a large granite boulder.

"Does this road lead to New Carlisle?" I enquired.

"Are you going to walk clear to New Carlisle?" he countered.

"Yes, I think so."

"Well, I can show you a much better road that takes you right into New Carlisle and South Bend—the concrete road."

"I do not wish to get there too soon, and the concrete is too fast and hard."

"O," he exclaimed in a surprised tone.

"I'm just sauntering across toward New Carlisle by country roads and lanes" I explained. I was afraid to tell him how far I planned to walk for fear he would become incredulous and insist on my taking the concrete road!

"I bet you don't worry about nothin'" he said, as I moved away.

"How could I,—walking through the country in May?" I replied.

I looked about for a chance to get away from the railroad track and soon came upon an untraveled road intersecting the track at right angles. Welcoming this opportunity to turn aside I followed the country road to the north.

After walking about half a mile I noticed a migrant shrike perched on a telephone wire. As I approached the bird it flew across the road over my head into a large grape vine which overflowed the fence along the highway. His flight into the dense vine suggested that his nest might be hidden in it, and as I crossed over to the vine it became plainly visible in the scant spring foliage. It was rather large, and unusually conspicuous because of its heavy lining of white feathers. Both the male and female bird immediately flew into the vine and commenced scolding. The luxuriant vine formed so dense a mass that it was not possible for me to break through it to approach the nest. I therefore got down on my knees and crawled under the vine and then rising immediately beside the nest looked into it and found it palpitating with six yellowish-green unfledged birds. The nest was built of sticks, coarse grass and lined with large white feathers, apparently from some neighboring barnyard. Both birds flew down very close to me and scolded and snapped their bills in their excitement. Their black eyes gleamed with anger and they approached within a few inches of my hand. They uttered a harsh grating sound and also made a snapping noise with the bill which sounded exactly like an angry owl. When making this noise they held their heads low and glared at me owl-fashion. They frequently flew down under the nest on some of the lower vines very close to my body. One of them had fresh bloody sparrow flesh in his bill with downy feathers clinging to it. He seemed to have difficulty in scolding me with his mouth full so he ultimately swallowed that which was apparently intended for the young birds.

The shrike or butcher bird has the well-known habit of slaughtering sparrows and other finches without discrimination and impaling their dead bodies on thorns, wire fences, or in convenient crotches of trees. From this charnal store-house he feeds himself and young as occasion or appetite requires.

After studying the birds under this forced excitement for some moments, I finally crawled back out of the vines into the road. The male bird immediately came and peered into the nest. He apparently found everything quite to his liking, in good order and undisturbed. He thereupon hopped aside and the mother bird came along quietly and slipped into the nest. She appeared nervous and upset, however, and hopped in and out of the nest at frequent intervals before finally resting quietly upon it.

As I continued down the road a prairie horned lark ran ahead of me in the dust with that gliding, deviating movement peculiar to the horned lark. The sight of this bird in the road was proof to me that it was indeed a country road because, like myself, the horned larks have refused to follow the concrete roads. They love the dirt roads and in season may more often be seen there than anywhere else. This interesting little bird builds its nest on the ground sometimes as early as February. I once found a horned lark's nest in a clump of snow-laden grass. The bird is sometimes called shorelark because in winter it is frequently seen along the shores of lakes and streams.

This country road also boasted of many charming hedgerows and all the fence corners were full of beauty which the rigid efficiency of the concrete highway has long since destroyed. Bordering the little ditch by the road was a beautiful group of young green willows in which a Northern yellow-throat showed his brilliant plumage to advantage.

A few steps beyond the road jogged through a piece of thick woods splashed with dogwood blossoms and here a wood thrush was singing gloriously even at mid-day, and an indigo bunting was sunning himself on his favorite perch at the edge of the woods.

II

By this time I felt quite ready for a real farmer dinner. I had passed several houses that did not look clean or inviting, and at each house one or more dogs had saluted me in more or less hostile fashion. One such house I approached for the purpose of making inquiry about dinner. Getting no response to my knock at the front door I looked through the windows and found the interior to be very dusty and unclean. Going around the corner of the house I discovered an untidy young lady with bobbed hair sitting on a stump and dressed in overalls. I did not disturb her. The dog at this place ran out of the barnyard and followed me some distance down the road. He seemed particularly unfriendly, apparently because I had not made friends with the family.

The next house along the way looked neat and well kept and I stole in through the front gate but found the house deserted. Morel mushrooms were growing in the rank green grass in the front yard. I looked about for some water for making coffee, but the pump was inside the barred kitchen door. A little farther down the road on the other side was another house where the dogs again barked and the pump was again inside. Three little girls were playing in the back yard and the noise of the dogs occasioned by my approach brought to the back door a fat woman in a soiled dress. I handed her my drinking cup and asked her if I might have a drink and she sent one of the little girls into the house for it. The woman did not know the name of the next town along the road. In view of the woman's appearance, I made no inquiry, for obvious reasons, about getting dinner.

The next farm building was a neat cottage with flowers growing in a well kept yard. I went around to the back of the house where a pale, slender lad of about twelve greeted me at the door.

"Is dinner over?" I asked.

"Yes, it is."

"May I have some water for making coffee?"

"You may have a cup of coffee if you come in," he replied.

I passed through the kitchen into the dining room and found an old Lithuanian just finishing his meal at the table. I sat down and the boy soon brought me a cup of coffee.

"Would you like anything else?" he enquired.

"Can you cook?" I asked.

The old Lithuanian: "Sure, fine." It was plain that the old man was fond of this lad. I noticed home-made bread and butter and boiled potatoes on the table.

"I would be very glad to have some eggs," I said.

In the meantime the old Lithuanian withdrew to the back yard and in response to my enquiries the boy told me all about him.

"Who is the old man?" I asked.

"O, he came to our house a few years ago, and wanted a home. He said he had no uncles or aunts, and no place to go. He was sick and we thought he had been paralyzed. He stayed with us, and he is very well now, and a good worker. We don't pay him anything—he is just one of the family."

"Do you eat home-made bread?" he asked anxiously. "That's the only kind we have."

"That's the kind I like best," I replied, "I don't often get it."

"Well, we are all out of store bread. We don't go to town very often, and if we buy enough bread to last until we go again it dries out."

Going out to the kitchen he returned in a few moments and said naively:

"Will you excuse me, but I can't find the salt, but I just run into the pepper."

"I'm getting along fine," I said, laughing.

"What is your name?" I asked.

"John Mankowitz!" he answered.

"How long have you been out here."

"We moved out here last fall, but I have only been out here for a week during the winter. I have been going to school in Chicago. I am out here to stay now, and to help my mother."

While eating my meal and talking to the boy his mother came in. She was clean and neatly dressed and about forty years of age. She at first appeared somewhat surprised at seeing me at the table, but the boy spoke to her briefly saying that I was a "traveller" and she smiled and immediately showed me that I was welcome without saying very much about it.

She also apologized for the home-made bread!

"I'm afraid the bread is not very good," she said, in the conventional house-wife manner.

"Tastes good to me," I replied.

"Well, my stove here is different from the gas stove I had in Chicago. When the bread was in the oven I waited and waited for it to come and then I found the damper was turned the wrong way, and that made all the trouble. It's not so good," she persisted.

As a matter of fact the home-made bread and butter were delicious, at least to my hungry palate.

I was greatly refreshed and pleased with my dinner and the unexpected cordiality of the little family. John refused to charge me anything for the meal but I insisted upon paying for it.

The delightful uncertainty about the reception one is to receive at farm houses along the way adds greatly to the interest of the journey. One never knows whether he is to receive a gracious reception with food and lodging or whether he is merely to be bitten by a dog. The dog is always there to receive one, whatever his mood may happen to be. It is occasionally friendly, usually belligerent in appearance only, and sometimes even vicious, but it is all a part of the day's adventure.

When I got out on the road again a man came along behind me in an automobile. He slowed down as if to stop and ask me if I wished to ride and when I smiled and shook my head he seemed greatly shocked as he looked at my fifteen-pound pack. Verily what is the sense in walking when one may ride in a Franklin car! Just as he passed me in the road I saw a beautiful male goldfinch fly into the hedgerow which I doubtless would have missed as a passenger in the Franklin.

III

The right of way of the electric railroad which I had left during the morning seemed now to pass through more attractive country and I therefore turned at the next crossroad and went south about half a mile to the railroad tracks. Near this crossing was a small cluster of trees "where beeches with their boles of Quaker gray" lent grace and beauty to an otherwise commonplace grove. The beeches were in almost complete summer dress, while the scattered white oaks were still covered with small pink buds.

At the railroad crossing there was a little shed which served as a station for the infrequent patrons of the electric road at this point. The shed was about ten feet long, eight feet high and six feet wide, open at the front and covered with a sloping roof. The board seat along the back was supported at each end by a rough post which extended some eighteen inches above the seat. On top of one of these posts in the corner where any waiting passenger might lean against it was a robin's nest freshly built but containing no eggs. This nest placed in this unusual position clearly indicated that the passengers at this wayside station were few indeed.

Leaving the shed I was surprised to see the name "Springville" printed on a board fastened to one of the poles which supported the electric wire. There were no houses anywhere about and nothing to indicate the town except the little shed and the name on the board.

The country road at this point wandered off to the southeast in such an inviting manner that I abandoned my purpose of following the electric railroad and took the winding dirt road instead. A foreign looking woman with a little girl was just approaching the gate opening into a meadow where a drove of cattle had congregated. Evidently they were bringing the cows home for milking, although it was only about three o'clock in the afternoon. The old lady could not speak English and did not understand my inquiry about the next town. Following the angling road to the southeast I came upon an old wooden bridge and the inevitable pheobe was sitting in a little willow tree beside the stream which flowed under the bridge. Feeling sure this was the sort of a bridge to give shelter to a phoebe's nest I climbed the fence and crept along the abutments. I found a nest on the edge of the beam supporting the floor of the bridge. It contained four white eggs. While I was examining the nest the phoebe exhibited great anxiety and flew past me under the bridge. A song sparrow perched in a bush near the end of the bridge was singing vigorously. The wind, carrying the song over the water and under the bridge so intensified the melody as it passed through this tunnel that it became exquisitely beautiful. The song and its echoes careened through the channel beneath the bridge with such delightful abandon that I stood entranced by the delightful harmony.

Finally leaving the worried phoebe and her little nest I resumed my walk along the winding country road as a pair of quails hurried across the dusty highway ahead of me.

In a neighboring field a man was driving a corn planter. I waited for him to pull up to the fence bordering the road.

"How far is it to the next town?" I enquired, still vaguely longing for Springville.

"Eleven miles to Rolling Prairie and nine miles to La Porte" he replied,— not a word about Springville!

"Well, I probably will have to sleep in a haystack tonight." It was then about four o'clock.

"You seem to enjoy the prospect," he said smiling.

"Well, since you mention it, I believe I would like the experience. That's one thing I've never done yet. I don't want to miss anything."

"You will find plenty of barns along the way where you could sleep."

"O, I've slept in a barn. That wouldn't interest me so much."

The information received from this farmer as to the distance to the next town and the fact that Springville appeared to be nothing more than a shed by the railroad track and a board nailed to a post left me a good deal in doubt as to where I would spend the night. This was a subject that did not require immediate consideration, however, and did not worry me in the least.

I continued down the winding road and presently came unexpectedly upon an old fashioned water mill with barn swallows flying about it. A very large cottonwood tree grew near the door, its gnarled limbs hanging over the decaying roof of the mill. Among the branches an oriole was singing.

"My oriole, my glance of summer fire."

The dusty miller stood in the doorway. Like the oriole in the cottonwood he was also singing, but unfortunately stopped upon noticing my approach. An old horse, with head drooping low, was dozing in the shade of the tree. A lazy half-swish of the tail now and then seemed wholly to exhaust his ebbing energy. Opposite the mill on the green bank of the mill pond a small boy sat fishing. He proudly hoisted a string of small bluegills out of the water in silent reply to my inquiry about his luck.

As I passed along beside the mill the old moss-laden wheel revolving in the stream of water which tumbled over it was plainly visible from the road. The ancient timbers and sagging roof of the old building were weatherbeaten and stained with flour dust and in many a cranny there hung a bit of tender moss.

In a few short years this picturesque survivor of a by-gone age will pass away forever, and its like will be known only in story or song:

> "Only the sound remains
> Of the old mill;
> Gone is the wheel;
> On the prone roof and walls the nettle reigns."*

Tucked away on this quiet by-path where the main currents of travel come not to disturb or destroy it lingers on for a brief season graciously to remind us of far-off happy days, blessed in their simplicity, which will return no more. The yard of this old mill seems to have been formed by a genial spreading of the road as it gently passes by, in like manner as the mill pond has been formed by a widening of the creek above the dam. Across the yard the road after thus paying homage to the mill again returned to its normal width in order to pass over an old wooden bridge which spanned the mill race. It then continued southward in its ambling, sleepy fashion.

This old wooden bridge over the mill race appeared to be the favorite resort of a flock of barn swallows who darted about and under the bridge and occasionally perched upon it. They seemed very tame as I reclined on the grassy bank of the mill race watching them. Squatting down on the protruding ends of the old bridge timbers they resembled whip-poor-wills dozing on a fence rail. There was also the usual phoebe flying about uttering his plaintive notes, but as the bridge was built low over the water it was difficult

---

*Edward Thomas.

to examine it carefully to determine whether or not it sheltered a nest. A cardinal whistled in the willows along the race below the bridge as I arose to follow the course of the little ambling road.

> "The little road, like me
> Would seek and turn and know;
> And forth I must to learn the things
> The little road would show."*

In a ploughed field by the roadside an old man and woman were engaged in planting seed by hand. The method which they followed was conclusive proof that they were not American-bred planters because the old man was dropping the seed and the old woman was using the hoe to prepare the ground and also covering the seed after it was dropped by the old man. In other words, the woman was doing all the heavy work in true Continental peasant fashion. A number of kingbirds scattered about over the field kept them company.

Altho the old, lazy road moved slowly it soon wandered through a charming wood which transformed it into a beautiful woodland path, and darting about in the trees were many sparkling migrant warblers, including the magnolia, the chestnut-sided, the black and white, and a number of flycatchers joined them in their quest for midges floating in the sunshine. As I paused in the road to make a few notes, two dogs came racing up from a farmhouse nearby and barked incessantly for five or six minutes as I stood quietly writing. They seemed incensed because I took no notice of them and ran around me back and forth in the road. When I resumed my walk they became greatly excited and charged about me barking more loudly than ever, but their noise and activity indicated excitment rather than a disposition to be vicious. The road was an unfrequented one and the appearance of a hiker with a pack was doubtless highly stimulating to the canine imagination. The farmer, who was repairing the fence along the way, seemed to regard conditions as quite normal and made no attempt to silence the dogs, and the moment I passed them in the road they became quiet and immediately forgot all about their excitement.

A good many bluebirds were observed at various points during the day and I watched them all very carefully, but none at any time disclosed his nest.

---

*Josephine Preston Peabody Marks.

The bluebird is often careless in keeping this important secret. If one of the birds is watched for a few moments during the nesting season it usually reveals its nest by going to it.

As I passed over another little wooden bridge the usual phoebe flew out from under it. Every old bridge on this unimproved country highway so fortunately neglected by modern motor traffic had a phoebe for a tenant. If modern concrete road construction continues the phoebes will be obliged to change their habits and hang their nests exclusively upon roots and other supports along the banks of streams and these interesting little birds will then be missed along the country highways.

As I passed thru a nursery of young trees set out in orderly rows by the roadside a scarlet tanager flitted down the aisles of the trees as if to light up the shadowy passages. In such a place

> "I always go on tiptoe down the aisles
> Feeling it a sacrilege to disturb that silence."*

Approaching the concrete road running on to eastward, (so hard for the hiker to avoid and sometimes equally hard to find if one is motoring), I met a man driving a horse and buggy.

"Is there a good dirt road going east" I inquired.

"Yes, there is one a mile farther down, but it is about as much traveled as the concrete road."

"What is the first town east?"

"Springville," he answered, much to my astonishment.

"Just where is this town of Springville, pray tell?"

"About a mile and a half east on the concrete road."

"How large a place is it?"

"Oh, a coupla houses and a garage," he answered.

I consulted my pocket map and back among the last few pages there was given a list of towns in Indiana. Looking anxiously through the list I came to "Springville," and found that it was given no population whatever. Because of this surprising circumstance I was able to understand why it might well be found in several places or in a number of places all at once. It seems that population is essential to give any town a definite situs, and as the map gave Springville no population I was not surprised that I had heard of it at several

---

*Edna Becker.

points and was only anxious that in going down the road eastward I should not wholly overlook it.

A map held in the hand, as one wanders about the country, is a sort of dream thing, and none of its nebulous points becomes a reality until one walks up and puts his foot on it. The journey thus becomes a long voyage of discovery and a constant checking of the real with the unreal, like taking the latitude or longitude of a boat at sea. And the distressing thing about the maps of civilized man is that they never show the points of greatest interest or importance. They give lines for railroad tracks, and dots for cities of steel and stone, but of all the wonders which I had seen along the way my map gave no sign. One must seek and find them with no guide save his love for the beautiful.

Before going on to the village and learning the truth about it I loitered along a little stream winding through a wet meadow still yellow with marsh marigolds. Reclining on the bank of the stream with my pack for a back rest and the late afternoon sun shining in my face I rested and enjoyed the songs of the many birds in the meadow, in the hedgerows and along the little stream. The air was musical with the calls of red-winged blackbirds, bobolinks, meadow-larks, a veery, a number of song sparrows, and a catbird kept up a constant scolding in an apple tree pink with bloom. As I reclined on the bank of the little stream the view led across a natural amphitheatre, the green banks on my left forming one side, the hedgerows along the stream on the right forming the other side and across the low meadow in front a ridge of wooded sand hills, flanked at either end by groves of trees, their spring foliage glistening in the quiet afternoon sun.

I loitered here for an hour enjoying the rest and quiet of the scene for it was late in the afternoon. I had walked far enough for the day and intended taking my chances on accommodations for the night at the elusive village of Springville. As I waited and enjoyed the beauties of the declining day a brown thrasher and a bob-white added their notes to the bird chorus and a goldfinch pursuing his undulating flight over my head called:

"Perchickeree! perchickeree! perchickeree!"

This is the usual method employed by this tiny bird to attract attention to his diminutive self.

As I walked on toward the village a number of killdeers running about in an old wet cornfield raised their voices in that ringing call so characteristic of spring. The cornfield was bordered by a wonderful row of wild crab apple and wild cherry trees, the former in vermillion pink buds, the latter in lacy white blossoms. These alternating trees of pink and white formed a fence row which was more difficult for me to pass than any barbed wire obstruction I had ever encountered.

Resuming the road I came at last into what appeared to be the village of Springville. I called at the very first of a small group of houses to make sure that it was this long sought village and that I should not pass it unawares. I was reassured by an untidy woman in a soiled apron while the inevitable dog barked loudly at my heels. I inquired where I might find food and lodging for the night for I felt sure from the appearance of this woman that I did not wish such accommodations at her home. Fortunately she directed me "to the first house beyond the bridge." The "bridge" was a railroad viaduct which spanned the highway at this point.

As I proceeded up the road in accordance with directions, the dogs continued to bark and awakened more dogs in every dooryard along the way,—airedales, collies, hounds, big and little dogs of all varieties, colors and qualities of voice,—tenors, basses, altos. The house to which I was directed was well nigh hidden by a hedge which was draped its entire length with a glorious trumpet vine. The house and farm buildings were old and weather-beaten, but another wonderful flowering trumpet vine covered the sloping roof and the sides of the porch and the old orchard trees in pink and white bloom overflowed into the yard, and I felt that I could rest amid these scenes of rustic beauty. In response to my knock a slender and neatly dressed woman came to the side door.

"May I have supper and lodging for the night?" I inquired.

"We have finished our supper," she replied hesitatingly—

"But if you will come in I think I can get something for you."

Without further encouragement I promptly stepped into the little kitchen.

"Your neighbor down the street told me you sometimes kept boarders," I said.

"Yes, we do," she replied. "I think we can take care of you."

Here I was in Springville at last, with homelike accommodations in prospect.

While supper was being prepared in the kitchen I rested in an old-fashioned sitting room where the natural hunger which I had brought with me to the house was further stimulated by the pleasant odors from the kitchen. In order to divert my mind from my gnawing hunger I made some notes of the day's walk and, upon summing up, dogs seemed to take a very important place in the day's adventures.

I was not disappointed in the meal prepared for me. It consisted of home-made bread and butter, cottage cheese, fresh country eggs, fried potatoes, rhubarb, and buttermilk. I am fully persuaded that I have never tasted such buttermilk. It creamed over the top of the glass and was flaked with bits of golden butter. Mrs. Dunham, who presided over this country home, ex-plained to me that she churned twice a week and made from forty to forty-five pounds of butter weekly. It is a rare experience in these modern days to get buttermilk and cottage cheese from any place nearer to their natural sources than a creamery....

Before preparing to retire for the night I went out into the yard and enjoyed the stars and listened to a whip-poor-will calling from the quiet woods across the road. Returning to the house I surreptitiously inquired of Mr. Dunham if he knew where his wife kept that wonderful buttermilk. He obligingly took the hint and leading me into the kitchen gave me all I could drink.

Fresh country buttermilk is the nectar of the pastures! It is as fundamental as the hills where the cattle graze. It has an elemental tang like juice from wild grapes. Sweet milk from which it comes is new and green but butter-milk has the ripe maturity of rare cheese or old wine. Its ingredients are made and mixed by Nature herself, and it is happily free from any of man's in-genious adulterations. It is sufficient unto itself. Man cannot successfully imi-tate its qualities, nor counterfeit the alchemy of Nature which produced it. The chemistry of man cannot compete with the processes which Nature em-ploys in making a beverage that pleases the taste, strengthens the body and warms the heart.

~~~~ SAMUEL ALAINE HARPER (1875–1962) WAS A MAN OF MANY accomplishments. He was an attorney by vocation, and indeed was one of the foremost authorities on the law of workers' compensation. Harper was to be the principal architect of Illinois's workers' compensation statute, which, when it went into effect in May 1912, made Illinois the thirteenth state to en-act such a law. His *The Law of Workmen's Compensation: The Workmen's Com-pensation Act with Discussion and Annotations, Tables and Forms* appeared in two editions, 1914 and 1920.

Besides the two legal texts, Harper wrote at least four other books. Three of these dealt with natural history topics (*A Hoosier Tramp, My Woods,* and *Twelve Months with the Birds and Poets*), but his 1956 *Man's High Adventure* was described by the *Chicago Tribune's* Harry Hansen in his July 8, 1956, column as being "a view of the world's needs and possibilities in the light of modern science." He spent most of the last two decades of his life on Merritt Island, Florida.

In May 1924 Harper embarked on a ten-day hike from Chicago to Orland, Indiana, a trek that covered 125 miles. The first short leg of his trip was via the electric train that enabled him to avoid the "stone, steel, and concrete" of the city, so Tremont was the point where he abandoned any further reliance on mechanized transport. Tremont, in Porter County, is in the heart of the Indiana Dunes, through which he slowly meandered over the first three days of his trip.

The selection of his book that I have included covers day three, where he walked from Michigan City on the west to Springville on the east. This country was far from wilderness when he was there, and it is far different today than it was in the 1920s. It was in a short-lived transitional period just after automobiles appeared on the landscape but before they became the dominant "life-form." Farm families offered food and lodging to visitors, and apologized to patrons that the only bread available was homemade rather than store bought. The vegetation and birds that Harper encountered are largely gone from this fast-developing area (except where specific efforts succeeded in preserving floristically diverse tracts). One striking example is the loggerhead shrike (Harper's migrant shrike). It has virtually disappeared as a nesting species in the upper Midwest, and the occasional bird still recorded in northern Indiana is almost certainly merely passing through.

Sources: Samuel A. Harper, *A Hoosier Tramp* (Chicago: The Prairie Club, 1928); Samuel A. Harper, "Germany Has Unique Plan to Protect Worker," *Chicago Tribune*, November 27, 1910, E4; Samuel A. Harper, "Compensation Law for Workmen; England Furnishes Best Example," *Chicago Tribune*, November 20, 1910, E8 .

The Sand Dunes of Indiana
E. Stillman Bailey

ᘓ ᘓ ᘓ A dune is a pile or ridge of incoherent sands, orderly or fantastically fashioned by prevailing winds. Sand dunes are common the world over; they are inhospitable spaces on the earth.

E. Stillman Bailey, *The Sand Dunes of Indiana* (Chicago: A. C. McClurg, 1917), 15–32.

The lure of the dunes is kin to the lure of the desert and the desire for great waste spaces. They have, however, a uniqueness of their own which admits of no comparison. They breathe mystery and romance, and appeal to the imagination. They have a temper and charm which you come to know and feel. At times they are in the area of fierce storms, with desert winds and blinding sands, and at other times when calmness obtains, one can, paradoxically speaking, listen to the voices of their silence. A hundred theories may explain their appeal, but, after all, the spell they cast becomes your own to unriddle, and you answer the riddle in your own way. The dunes are alluring and fascinating, and with the varying seasons their changing forms and colors weave for you a wealth of fact and a wreath of fancy. Their secrets are many, and their wild beauty is a pleasure to the eye and a joy to the memory.

Perhaps it is the out-of-doors, the unconstraint, the farewell to bricks and traffic, that fascinate you at the first visit, for fascinated you will be. Maybe it's the beach, where the sky, the water, and the sand make a long blue line, with miles of freedom to enjoy. Perhaps it is the place where work and care are tossed upon the crest of the outgoing wave, and time is immersed in forgetfulness. Maybe it is the exchange of the plain or prairie for the hills, for before you is the skyline of the heights, a hundred or more feet above you, of changing sands, changing forms, changing views, and exchange of colors with a decoy that tempts you to tread an unbroken path in the mobile sand and climb to the top of the unstable crests. Whatever it may be, visits to the sand dunes are unconventional and they beget and foster a wanderlust.

The dunes are mutable, the mobile forms of yesterday pass into oblivion while tomorrow's are in the making. The dunes are matchless, their class is all their own; they are restless, changing, billowy sands.

Freedom pervades the whole atmosphere of the dunes. The voices there seem to say, "You do not trespass, you are welcome, you are no bother, you must take us as you find us today, for this is Utopia." If you care to sit and dream, the foundations for your air castles are here. If you enjoy the sudden change from your home to this wilderness, you will return again and again, and in time be recognized as a regular dune commuter and an all-the-year-round visitor.

To the query as to the best time to see the dunes, I reply, the best time is at your own convenience. To some, the dunes are uncomely—yellow, bothersome, changing sands, and nothing else. The gift of imagination was not bestowed on these people. They are to be pitied, for they miss much of life's finer pleasures.

To others, the dunes are fascinating at any and all times. Perhaps, as the "old commuter" said to me recently, the dunes are at their best in the spring; but the same enthusiast must have a short memory, for last fall, while his arms were laden with American holly and bunches of bitter-sweet for his home holiday decoration, he told me that the fall of the year is the best possible time for these excursions. So you see any time is right if *you* choose it.

If you are feeling fit, and are warmly clad, you will welcome a trip to the dunes even during the winter's snows and gales, and find a rich return for your venturous jaunt.

If you are the Indian you think yourself, you will on a summer day take a fifteen-mile hike on the beach, hatless, and unconventionally plunge at your own will into the lake for a refreshing swim, and later you may seek a resting spot to watch the sunset; or, after a night in a shack or under a bent tree for shelter, with a blanket for a cover and the warm sinking sand for a bed, you may see the silhouette of the sentinel pines against the morning sky, conscious of the beginning of another day of moods among the dunes.

If you are hopelessly practical, a one-day visit will suffice for you, but if you are thrilled by something new or gifted with imagination, or if you are curious and temperamental, the call of the sand dunes will be resistless.

Some day you may happen at the dunes when the silence will be the charm. The wind has ceased, the little lisp of the water as it caresses the sand will be the only sound. The surface of the lake will be molten. It is a calm that may cast the spell of littleness of self, and suggest the largeness of nature's forces. The calm at the dunes is to be felt; it cannot be spoken. The sky, the water, and the land meet and proclaim the peace of heaven and earth and sea.

The next day it may be the fury of the wind that will cast its spell upon you. It is well to be prepared for all emergencies, for the sands, when driven by stormy winds, will cut and sting you, and though you may enjoy it all, you may be driven to shelter. You may think stormy days will be dreary in the dunes, but not so; both storm and calm are but echoes of your own self; you acquiesce when all is still, and you thrill when all is moving.

It is quite unnecessary for you, in order to enjoy the dunes, to be a geologist or a botanist or a zoölogist, or to know about the formations or historic relationship of glaciers, ice-fields, sedimentary rock, granites, beach lines, or ocean beds. It is all the more interesting if you happen to possess this fund of information, but if you can see and feel the real dune pictures, with their high lights and contrasting colors, the shadows and shades of the purples, blues, grays, yellows, and greens, whether of the lake, the sands, the clouds, the mists, or the receding paths into tempting woodlands, or views where land and water meet, and if the vistas spread before you charm your eye and rest your brain, then the dunes are for you.

Every branching tree or wild vine, every print of foot of tern or turtle, every lee of rotting wrecks or upturned roots, every little ripple on the beach or on the sanded crests, every beating wave when the storm is wildest, or whisper of wind in the forest depth, or note of bird or tread of hoof, every sandy trail to flower or fern, every echo from surf or camp will cast its mystic spell and make a memory web. If you see these views or hear these things, or feel it all, you may be certain that the dunes are for you.

At your feet even the rounded smooth stone may suggest a story. Maybe it is a wanderer from some far-distant home, where other sands have fashioned its form and brightened its face. If you pause and gaze at the sentinel pines that stand, even when dead, pointing their tops above the advancing sand; if you see the polish on the fallen trunks and roots of trees, made by nature's sand blast of prevailing winds and diamond sand dust, then, too, the dunes are for you and are yours to enjoy. If your camera takes what your eye sees, the pleasures will double as you review the pictures from time to time in your leisure hours.

The call of the sand dunes is the call of your measure of fancy and of caprice. There is always the riddle of varying attractions, and each journey to the dunes has a reward equal to that of a new discovery. If your first visit be in the fall of the year, the views of the autumn will outrival those of the spring. Then the dunes are a blaze of colors which are heightened by the white of the sands and mellowed by the contrasts of evergreens and ripening vines. The rushes and grasses may have withered, but the autumn trees will command your attention, for so red are the maples, so brown the oaks, so yellow the aspen, and so green and feathery the tamarack, that one draws near to see it all. If the frost should have completed the ripening of the woodbine and ivy on the stumps of trees or over the rotting fences, the near view will be the more charming one. The fall of the year, too, offers a chance to walk and scuff through the fallen leaves or to pick the wild grapes, whose color is now deep purple and whose bloom is now complete.

The Indian Summer days are rare days in the dunes, with their purple haze, the curling smoke of camp fires, the calls of the wild fowl returning to the lake and to the marshes, the thrill in the atmosphere, and then the stillness over it all!

The dunes are just "different"; their language is beginning to be a different tongue, their call is irresistible. You may have come impelled by curiosity, but the wild beauty invites you to remain. And why should the despoiler come? It is time now to change the waste into a park, and the desert into a playground for millions forever to enjoy.

〜〜〜〜 ELI STILLMAN BAILEY (1851–1926) IS ANOTHER AUTHOR about whom there ought to be more information available. He received his bachelor's degree from Milton College (Wisconsin) in 1873, and an MD at Hahnemann Medical College in Chicago in 1878. After a decade of being a general practitioner, Bailey concentrated on gynecology, a field he taught at Hahnemann Medical College, then a center for homeopathic medicine.

While Bailey was studying, teaching, and practicing medicine, he obviously also spent time indulging an interest in natural history. (Bailey's situation is just the opposite of fellow physician Joseph Lane Hancock, whose vocational activities have been overshadowed by his avocational pursuits.) But, unfortunately, apart from his book on the Indiana Dunes from which the selection is taken, I have been unable to find anything about that interest.

Sources: William Harvey King, *History of Homeopathy and Its Institutions in America*, www.homeoint .org/history/king/4–03.htm; *Who Was Who in America*, vol. 1, 1897–1942 (Chicago: A. N. Marquis Co., 1943).

Botany

Our Native Grasses

George Vasey

⁓⁓⁓ It would seem to be the appropriate work of farmers to make themselves acquainted with the natural or wild grasses of the section of country where they reside, and to make investigations as to their adaptation to the purposes of profitable cultivation. By doing so, it is probable that many species would be found worthy of the attention of the agriculturist, and perhaps, some be found as useful as the universally employed timothy or herds grass.

With the hope of directing the attention of farmers and others to this subject, I present a few remarks on the indigenous grasses of Northern Illinois, as they have fallen under my observation. I hope to be excused for using the scientific names of these grasses for two reasons; first, because in many cases there are no common names known, and second, because common names are not in all cases distinctive, the same name being frequently applied to several different plants, and different plants being often called by the same name. I may remark that I adopt the names as employed in Dr. Gray's Manual of Botany.

I do not propose even to mention *all* the different kinds of grasses which the botanist may detect in this region, for they would exceed one hundred, but I will notice only those which form a principal part of our native gramineous vegetation.

It is well known that wild hay varies much in its quality and productiveness. This difference depends partly upon the time of cutting and manner of curing, and partly upon the kind or species of grasses of which the hay is composed. A very general distinction of wild hay is into slough and upland. The coarsest and least valuable slough hay is chiefly composed of several kinds of sedge or cut grass, belonging to the genus Carex. In *very wet* bogs and sloughs we have Carex lanuginosa, C. stricta, C. filiformis, C. bromoides, C. paniculata, and C. intermedia. The fruit of these species is matured

George Vasey, "Our Native Grasses," *The Prairie Farmer* 19 (1859): 50.

early in the season, and as the grass is usually cut late, we seldom find fruiting specimens among the hay. Growing with these we frequently find several species of Juncus, a Dulichium, and an Eriophorum; and also some of the true grasses, as the Phragmites communis or reed grass; which, if cut early, would probably be a very good and productive grass.

On bottom lands and low grounds, *not boggy*, we have the Calamagrostis Canadensis, sometimes called blue joint, a tall and valuable grass, also Spartina cynosuroides or cord grass, a long, tough grass with singular heavy fruit spikes, which remain until late in the season; Glyceria nervata, a slender grass with handsome spreading panicles, the glumes finely nerved or striped with purplish lines; Elymus Virginicus and C. Canadensis, both called Rye grass; and Poa serotina, a long, slender, soft-stemmed grass, with a heavily seeded open top, very valuable and deserving of extensive cultivation. It is sometimes called wild Red Top, from having a panicle similar to the Agrostis vulgaris, which is called Red Top in the Eastern States, and Herd's grass in Pennsylvania, but it is a more valuable and productive grass and will flourish on upland as well as on lowlands....

Our high and dry prairies are now so generally brought under cultivation, that it is rare to see any hay raised on such locations. Ten and twelve years ago, it was frequently brought to market, and in some places may still be procured. It is composed chiefly of the Sporobolus heterolepis, Stipa spartea, sometimes called wild oats, with long needle-like and twisted barbs attached to the seed; several species of Panicum, Carex straminea, C. festucacea, C. scoparia and C. Meadii. Intermixed with these are usually many herbs which have no more relation to grasses than has clover, which is frequently called a grass.

On prairies with a thin soil, on gravelly knobs, and skirting the timber, we find Koeleria cristata, Andropogon scoparius, Bouteloua curtipendula, one of the neatest and prettiest of grasses but not productive; Triticum caninum, and several species of Panicum; but seldom any of them in sufficient abundance to repay the expense of cutting for hay.

At another time I may describe more at length several species of the grasses I have mentioned, which appear to me worthy of some attention for general cultivation.

～～～ GEORGE VASEY (1822–1893) WAS THE FIRST SCIENTIFIC BOTANist in northeastern Illinois. Born in England, he arrived in the United States with his parents a year later. They settled in Oneida County, New York, and although he attended school only to the age of twelve, he was an excellent student who had a grasp of both algebra and Latin. Sometime over the next year, an event occurred that sealed his future: he read Almira Lincoln's *Elements of Botany*, "a little book well known to the older botanists of the country" (Canby and Rose, 171). He had to have the book, but because he couldn't afford it, he hand-copied the entire volume.

Vasey worked as a clerk, and one day saw a customer approach the store but then stop to pick a flower from the pavement. The man asked young Vasey if he knew what the plant was. To the man's great surprise, Vasey identified it correctly as *Ranunculus acris*. The customer was Dr. P. D. Knieskern, a physician and botanist of great repute. Knieskern took Vasey under his wing and introduced him through correspondence to such future luminaries as John Torrey and Asa Gray. The three young botanists were to become lifelong friends.

Later Vasey went back to school, graduating from the Oneida Institute and then on to three courses of lectures at the Berkshire Medical Institute, topped off by a few weeks at the College of Physicians and Surgeons in New York City. (Medical training was evidently accelerated back then; perhaps there was less to know. The truth is medical education in this country wasn't standardized until the 1920s.) In 1848 he and his wife and child moved to Ringwood, Illinois, where he practiced medicine for eighteen years. (And even operated a store to help support his six children.) Later, with the opening of a railroad line, he was able to offer medical services to patients as far away as Elgin.

Vasey arrived in this region just as the prairie wilderness was ending but while the native landscape still retained much of its original character. The herbarium specimens that he collected are testaments to the full richness that once endowed this territory. According to Floyd Swink and Gerould Wilhelm's *Plants of the Chicago Region*, the prairie dandelion (*Agoseris cuspidata*) is "known in our area only from McHenry County, where it was collected by Vasey in a prairie at Ringwood on May 20, 1858." And Charles Sheviak says that the only confirmed regional record of the white adder's mouth orchid (*Malaxis brachypoda*) "is the Kane County specimen collected by George Vasey in the hanging fen at Elgin." (The site is part of Trout Park.)

Many local naturalists have savored these old gems scattered through the literature. But Ed Collins did more. Mr. Collins is a restoration ecologist who works for the McHenry County Conservation District and lives in Ringwood. He embarked on a lengthy search to find out all that he could about this botanical pioneer, gathering Vasey correspondence in collections across the country.

Mr. Collins has generously shared with me the materials he so patiently acquired. Unfortunately, from my standpoint, Vasey rarely showed emotions in these writings. However, in one letter to his longtime friend M. S. Bebb dated July 25, 1861, he did express himself with unusual feeling, probably because he was still awash in the joy of a new daughter: "I have the pleasure to announce the recent arrival of a very interesting young stranger—a little daughter—making the third of that kind, and, including the boys, completing my half-dozen. We welcome her to our fireside—that means we are not sorry she has come, and we will do as well as we can to care for her, and prepare her for the race of life."

His statement on the importance of science and nature education has a surprisingly modern ring to it: "Ought we not in some way to issue a 'journal of Science' which should reach the inquiring youth of our State, and teach them how to understand and love nature? It seems to me that the State could not do itself better service in the work of public education than to provide the thousands of public schools in the State with some such medium of knowledge."

Vasey's stay in northern Illinois was terminated by the sickness of his wife. To facilitate her recovery, they moved to Washington County, Illinois, in 1866, but she died soon thereafter. Within a year, however, he was married again, to the very accommodating Widow Cameron, who apparently gave her blessing to Vasey's joining John Wesley Powell on a six-month exploration of the Colorado River. He returned to Illinois for a short while before taking the family to Washington, D.C., where he was to become the botanist for the Department of Agriculture and curator of the National Herbarium. In those roles he dramatically increased the holdings of the herbarium, became a leading authority on North American grasses, and established the Grass Experimental Station at Garden City, Kansas.

Sources: Wm. M. Canby and J. N. Rose, "George Vasey: A Biographical Sketch," *Botanical Gazette* 18 (May 1893): 170–183; Charles Sheviak, *An Introduction to the Ecology of the Illinois Orchidaceae.* Paper XIV (Springfield: Illinois State Museum, 1974); Floyd Swink and Gerould Wilhelm, *Plants of the Chicago Region,* 4th ed. (Indianapolis: Indiana Academy of Sciences, 1994).

Prairie Woodlands

Ellsworth Jerome Hill

When pioneers began to settle in our primeval forests the natural impulse to plot in right lines led to the clearing of rectangular spaces, so that the surviving pieces of woodland are mostly bounded by straight lines. Time has, however, modified and beautified the abrupt and naked forest-borders that skirted the newly cleared fields. The taller trees along the margins have been overturned by the wind, lower ones have grown up with rounded tops and limbs which spread out to reach more light; an undergrowth of shrubs and herbs has sprung up by the enclosing fences, so that an unbroken bank of foliage stretches from the ground to the tree-tops.

In the woodlands of a prairie region these sloping borders are characteristic, and the bounding lines naturally curve with the windings of streams and valleys and the outlines of timber-clad hills. These masses of timber are often surrounded by treeless prairie or cultivated fields, the woods being left to supply the adjoining country with forest-products. The large trees have been cut off, but the ground is left to grow up with timber again, and under the stimulus of self-interest a kind of rude forestry is practiced. If the wooded areas are too large to be embraced with profit in adjacent farms, they may be divided into portions of a few acres, and be owned by several farmers living within easy distance. These small holdings in the groves are bought and sold with the main estate, and this also tends to their preservation and keeps them in larger tracts, so that one may sometimes follow belts of unbroken woodland for miles along a stream. Fed by springs which issue from the bases of the bordering slopes, even the small streams become perennial in the shade of the woods, though when followed away from the forest a dry bed may mark their course through the prairie in summer. The farmer thus becomes a conservator of the woodlands which help to preserve moisture for the soil, as well as to form one of the most pleasing elements of the landscape; for this region is monotonous as a whole—a plain with no striking features in the way of hills and mountains.

These wooded areas are generally used for pasture, frequently to their injury; but cattle and horses being chiefly kept, the undergrowth is not as closely

Ellsworth Jerome Hill, "Prairie Woodlands," *Garden and Forest* 7, no. 347 (October 17, 1894): 412–13.

cropped as when sheep have their range. Portions of the adjoining prairie are often included in the pasture, especially on the hills or along swampy lands. Points of timber jut out from the main body into the prairie, running down the hillsides or along ravines and watercourses, or out into swamps into which drier land projects. An occasional tree or small group of trees stands apart from the rest, still further varying the outline of the border, while larger groups, like islands in a sea of prairie, lie apart from the main body. The use of the tree-covered ground and the adjacent prairie for grazing often saves these isolated trees and groups; for where land is devoted to the plow they are apt to be cut down, as may be observed where the rich soil of the prairie comes close up to the woods. If in wettish grounds the trees may be spared, because they do not encumber a meadow from which hay is taken, and which may be pastured only a part of the year. The limbs of these detached trees or groups come down low about the trunks, the tops are round and spreading, and a sturdy and symmetrical habit has been developed by the free play of light and air on every side.

A pleasing feature of these woods is the way in which the border merges into the green of the prairie. Belts of trees lining low and swampy prairies, or the sloughs with which they are interspersed, become attractive, or even beautiful, notwithstanding many of their unpleasant surroundings. A continuous mass of foliage joins the green below with that at the tops of the tallest trees. Willows and Alders, the Swamp Rose, Button-bush and Osiers of various kinds are followed by Viburnums, Sumachs and Sassafras. These and other tall shrubs or low trees furnish a gradual or undulatory slope from the rank Grass, Reeds and Rushes of the swamp to taller Elms and Swamp Maples, and the Oaks and Hickories of the drier background. The wide variety in the tints of green foliage makes these borders attractive in summer, and when autumn kindles its fiery colors in the leaves they are fairly radiant with beauty. Various shrubs connect the dry upland woods with pasture and meadow. The leaves and glandular twigs of the Hazel are not agreeable to browsing animals, and it becomes common along the margins of woods and fields, or forms outlying masses between the field grasses and the foliage of Oaks and Hickories. In lower and richer ground the Hazel is replaced by the Crab Apple, the Wild Plum and different species of Thorn.

The prevalence of Oaks with smooth and glossy leaves gives to many of these woodlands a peculiar distinctness under the glow of the summer sun. A shimmering light plays upon them, as well defined, although not as bright, as that which glances from a water-surface, and these bright areas catch the eye from a distance, and in the broad sweep of the green landscape these forest-masses rise out of the general level as sources of clear mellow light.

➤➤➤➤ REV. ELLSWORTH JEROME HILL (1833–1917) WAS ONE OF THE greatest of the region's early botanists. He was born in Le Roy, New York, where he lived until the age of twenty. His first interest in natural history was directed toward geology. Just before he was to attend college, he suffered an ailment of his knee that rendered him unable to walk. As he convalesced, his attention turned to botany. Armed with crutches, his initial attempt to walk outside was to collect flowers from an orchard. By the next year, he was healthy enough to move to Mississippi, where he taught for three years. To further his education (one would suspect that the impending war was also a factor in his decision to return), he moved back to New York to study at Union Theological Seminary. Upon graduation in 1863, he was offered the pastorate of the Presbyterian Church of Homewood, Cook County. Hip problems forced him to give up his position at the church, so he became a teacher once again. He taught at Kankakee High School for four years and then at Englewood High School in Chicago for eighteen.

Hill loved spending time in the field, even if being ambulatory was often painful and required the use of crutches or canes. Agnes Chase relates how "he carried his vasculum over his shoulder and a camp stool with his crutch or cane in one hand. To secure a plant he would drop the camp stool which opened of itself, then he would lower himself to the stool and dig the plant."

Hill botanized throughout the region, making thorough collections of the areas he worked. One of his richest sites proved to be Langham Island, twenty-five acres of bedrock in the middle of the Kankakee River. Hill first visited the island in 1872, and it was on June 29 of that year (perhaps that initial trip) when he came upon his most extraordinary find: the Kankakee mallow (*lliamna remota*), a plant not known to be native anywhere else in the world. For the next twelve years, Hill continued to scrutinize the island, and the list of rarities he compiled was amazing: leafy prairie clover (today federally endangered), buffalo clover (*Trifolium reflexum*) (today state endangered and one of only very few local records), and the lobed violet (*Viola palmata*) (no longer found in Illinois). While an able-bodied adult in a year of average or low flow can easily reach the island by wading, I can't imagine how Rev. Hill managed. Even getting in and out of a small boat must have been difficult for him. (The answer may lay in the fact that, with less runoff due to development, the river was much shallower back then.)

As time progressed, Hill devoted himself increasingly to in-depth studies of particularly difficult genera. These included the *Potamogeton* (pondweeds), *Carex* (sedges), *Quercus* (oaks), *Salix* (willows), and *Crataegus* (hawthorns). He corresponded with many of the leading botanists of the time, eager to share materials or information to aid others in their work. Agnes Chase writes that his herbarium comprised roughly 16,000 sheets and he published 162 articles. Hill's oak (*Quercus ellipsoidalis*) and Hill's thistle (*Cirsium hillii*) are named for him.

He left teaching in 1888 because of health problems and declining vision. For the next dozen years, over which time his health improved, he spent as much as four days a week botanizing. Chase describes those outings: "His enthusiasm and love of beauty were as fresh as a boy's, while his mature judgement and ripe knowledge made it a rare privilege to be in his company. He never lost his early love of the Greek and Latin classics and often he had a copy of Virgil in his pocket to read aloud during the resting periods."

Sources: Agnes Chase, "Rev. E. J. Hill," *Rhodora* 19, no. 220 (April 1917): 60–69; Joel Greenberg, *A Natural History of the Chicago Region* (Chicago: University of Chicago Press, 2002).

The Plant Societies of Chicago and Vicinity

Henry Chandler Cowles

ॐ ॐ ॐ A. *The ravine.*—No topographic forms lend themselves so well to a physiographic sketch of the vegetation as do those that are connected with the life history of a river. Beginning with the ravines, which are deep and narrow, because of the dominance of vertical cutting, we pass to the broader valleys, where lateral cutting becomes more pronounced. From this stage on we have to deal with two phases of river action, the destructive, which is concerned with the life history of the bluff, and the constructive, which has to do with the development of the flood plain.

Henry Chandler Cowles, "The Plant Societies of Chicago and Vicinity," in *Bulletin of the Geographic Society of Chicago*, no. 2 (Chicago: Geographic Society of Chicago, 1901), 13–22.

Wherever there is an elevated stretch of land adjoining a body of water, such as a lake bluff, one is apt to find excellent illustrations of the beginning of a ravine.... An embryonic ravine ... may be seen frequently along the clay bluffs between Evanston and Waukegan. A ravine of this type is essentially a desert, so far as plant life is concerned. The exposure to wind and to alternations of temperature and moisture is excessive. The lack of vegetation, however, is due chiefly to the instability of the soil; this instability is particularly great in the case of clay bluffs such as these, where the seepage of water causes extensive landslide action. No plants can get a foothold in such a place, unless it be a few species that may be able to make their appearance between periods of landslide action; among these plants annuals particularly predominate. The perennials that may be found in such places are almost entirely plants which have slid down the bank.... Ravines of a similar type may also be seen at many places inland, and wherever found the poverty of vegetation on the slopes is the most striking character.

As a ravine extends itself inland the conditions outlined above may be always seen about its head, but toward the mouth of the ravine the slopes are less precipitous. Torrents cut down the bed of the ravine until a depth is reached approaching the water level at its mouth. From this time on the slopes become reduced and the ravine widens more than it deepens, by reason of lateral cutting, landslide action, and side gullies. After a time a sufficient stability is reached to permit a considerable growth of vegetation. If the erosion is slight enough to allow a vegetation carpet to develop, a high degree of luxuriance may be attained. In fact, ravine conditions are usually extremely favorable for plants, after the initial stages have passed. In a comparatively few years the vegetation leaps, as it were, by bounds through the herbaceous and shrubby stages into a mesophytic forest, and that too, a maple forest, the highest type found in our region. Nothing shows so well as this the brief period necessary for a vegetation cycle in a favored situation as compared with an erosion cycle.

Of such interest are the facts just noted that it is worth while to mention some of the characteristic ravine plants. Perhaps the most characteristic trees of the Glencoe ravines are the basswood (*Tilia Americana*) and the sugar maple (*Acer saccharinum*), though the ash, elm, and other trees are frequent. The most characteristic undershrub is the witch hazel (*Hamamelis Virginiana*). The herbaceous plants are notoriously vernal forms, such as Hepatica, Thalictrum, Trillium, Mitella, Dicentra, Sanguinaria (Bloodroot); mosses abound and liverworts are frequent.... We can explain this flora only by regarding it as having reached a temporary climax. Ravine conditions are more favorable for plants than those that precede or follow. The instability and exposure of the gully have gone; in their place there is protection from wind and exposure. The shade and topography favor the collection and conservation of moisture, and as a result there is a rapid development into a high-grade forest, as outlined above.

Rock ravines are much less common in the Chicago area than are those of clay, since the underlying limestone rarely comes near the surface. Excellent illustrations of stream gorges are to be seen at Lockport, and also in various tributaries of the Illinois river near Starved Rock. A striking difference between these rock gorges or cañons and the clay ravines is in the slope of the sides. The physical nature of the rock excludes landslide action, hence the sides are often nearly vertical for a long time. Lateral cutting is also relatively slow as compared with clay. Thus the conditions for vegetation at the outset are much more favorable than in a clay ravine. Rock-bound gorges are very shady and often dripping with moisture, hence liverworts and many mosses find here a habitat even more congenial than in the clay. Among the higher forms are found the most extreme shade plants that we have, such as Impatiens, Pilea, and shade-loving ferns, plants whose leaves are broad and remarkably thin....

The stages of development pass much more slowly in cañons than in clay ravines, largely because the primitive conditions of shade and moisture remain for a long period of time. Nor do the steep slopes permit the development of a wealth of trees and shrubs, since a secure foothold is not easily found. However, as the cañon broadens out and the slopes become less steep, shrubs and trees come in, though a typical mesophytic forest is rarely seen. The Starved Rock ravines are cut in St. Peters sandstone, those at Lockport in the Niagara limestone, yet the vegetation in the two places is essentially alike; at any rate, the resemblances are greater than the differences. Much has been written on the physical and chemical influences of rocks upon the vegetation. The facts seen here seem to show that the physiographic stage of a region is more important than either. The flora of a youthful topography in limestone, so far as the author has observed, more closely resembles the flora of a similar stage in sandstone than a young limestone topography resembles an old limestone topography. A limestone ravine resembles a sandstone ravine far more than a limestone ravine resembles an exposed limestone bluff, or a sandstone ravine resembles an exposed sandstone bluff. We may make the above statements in another form. Rock as such, or even the soil which comes from it, is of less importance in determining vegetation than are the aerial conditions, especially exposure. And it is the stage in the topography which determines the exposure.

All of the preceding statements as to topographic stages, whether young or old, refer not to times but to constructional forms. Two ravines, equally youthful from the topographic standpoint, may differ widely as to actual age in years or centuries, since erosion is more rapid in one rock than in another. In our region, however, elements of actual time are not very important, except as between rock and clay, since the limestone is less soluble and the sandstone is more easily eroded than is often the case.

B. *The river bluff*—As a valley deepens and widens, the conditions outlined above undergo radical changes. From this point it will be necessary to discuss two phases in the growing river, the bluff phase and the bottom phase. We have left the clay ravine bluffs in a state of temporary climax, clothed with luxuriant mesophytic forest trees and with a rich undergrowth of vernal herbs. More and more the erosive processes are conspicuous laterally, and widening processes prevail over the more primitive deepening. As a result, the exposure to wind, sunlight, and changes of temperature increases; the moisture content of the slopes becomes less and less. The rich mesophytic herbs, including the liverworts and mosses, dry up and die. The humus oxidizes more rapidly, and a xerophytic undergrowth comes in. In place of Hepatica and its associates, we find Antennaria, *Poa compressa* (Wire grass), *Equisetum hyemale* (Scouring rush), and other xerophytic herbs; Polytrichum also replaces the mesophytic mosses. The first signs of the new xerophytic flora are seen at the top of the ravine slope; indeed, the original xerophytic plants may never have been displaced here by the ravine mesophytes. As the ravine widens, the xerophytic plants creep down the slope, often almost to the water's edge....

After a few years have passed, xerophytic shrubs appear on the bluff in place of the witch hazel and its associates. And it is not long until xerophytic or semi-xerophytic thickets prevail, in place of the former mesophytic undershrubs. Among the more characteristic of these shrubs are the hop tree (*Ptelea trifoliata*), bitterweet (*Celastrus scandens*), sumachs (*Rhus typhina* and *R. glabra*), chokecherry (*Prunus Virginiana*), nine-bark (*Physocarpus opulifolius*), wild crab (*Pyrus coronaria*). Two small trees are common on stream bluffs, the service berry (*Amelanchier Canadensis*) and the hop hornbeam (*Ostrya Virginica*); this last species is perhaps the chief character tree of river bluffs, and is rarely absent. Perhaps the best examples of xerophytic stream bluffs near Chicago are along Thorn creek. One of the most interesting things about these bluff societies is the frequent presence of basswoods and sugar maples. Doubtless these trees look back to the mesophytic associations that have otherwise disappeared. As would be expected, the last of the mesophytes to die are trees, because they are longer-lived than herbs and shrubs, and also because their roots reach down to the moisture. But they cannot be succeeded by their own kind, inasmuch as the critical seedling stages cannot be passed successfully.

The life history of the rock ravines, or cañons, is somewhat different. When the ravine vegetation is at its height, the moisture and shade are greater here than in the clay, hence the high development of liverworts and their associates. As the ravine widens, these extreme shade forms are doubtless driven out almost immediately by xerophytes, since intermediate or mesophytic conditions are seldom seen where the soil is rock. Furthermore, the xerophytic conditions become much more extreme on rock bluffs than on clay bluffs. This is well illustrated at Starved Rock, where the dominant tree vegetation is coniferous, consisting especially of the white pine (*Pinus Strobus*) and the arbor vitaé (*Thuya occidentalis*). The herbs and undershrubs here are also pronouncedly xerophilous, resembling the vegetation of the sand dunes, *e.g.*, *Selaginella rupestris* [sand club moss], *Campanula rotundifolia* [bluebell], *Talinum teretifolium* [fameflower], *Opuntia Rafinesquii* [Eastern prickly pear], etc. The entire bluff flora down to the river's edge is xerophytic, except in shaded situations.

When a stream in its meanderings ceases to erode at the base of a bluff, increased opportunity is given for plant life. Through surface wash the slopes become more and more gentle. Mesophytic vegetation comes in at the foot of the bluff and creeps up as the slopes decrease. Finally the xerophytes are driven from their last stronghold, the top of the slope, and the mesophytes have come to stay, at least until the river returns and enters upon another stage of cliff erosion. The growth of a ravine into a valley with xerophytic bluffs is rapid, when expressed in terms of geology, but far less rapid when expressed in terms of vegetation. A ravine in the vigor of youth may develop so slowly that forest trees may grow to a considerable size without any perceptible change in the erectness of their trunks.... But the activity of the erosive forces, slow as it may be, is nevertheless revealed by occasional leaning, or even falling, trees. From the above it is easy to understand that cycles of vegetation often pass much more rapidly than cycles of erosion, but never more slowly. During one erosion cycle the mesophytic forest develops at least twice—once on the ravine slopes, and then finally on the gentler slopes that betoken approach toward base level.

〰〰〰 TO PURSUE GRADUATE WORK IN GEOLOGY, HENRY CHANDLER
Cowles (1869–1939) arrived at the University of Chicago in 1895, just five
years after the school was founded. He had a diverse background, studying
not only geology and botany as an undergraduate at Oberlin College, but
Latin and Greek besides. His major professors at Chicago were Rollin Salis-
bury and Thomas Chamberlain, who specialized in the geologic forces that
sculpted landscapes. Under their tutelage, Cowles became familiar with the
moraines, streams, plains, lakes, and dunes that made up the local scenery. At
the same time, John Merle Coulter left the presidency of Lake Forest College
to create a botany department at the new school to the south. Cowles enrolled in his courses, where he was introduced to a book that would change
his life. That book, written by the Danish botanist Eugenius Warming, was
Plantesamfund (Plant community). Warming's treatise represented a radical
departure from traditional botany in that it sought to determine how the
physical environment affected the plants that were present, as well as how
plants were adapted to that particular environment. He then classified the
species that occurred in specific settings as "communities." Charles Chamberlain, a fellow student, recalled what the classes were like: "None of us
could read Danish except a Danish student, who would translate a couple
of chapters, and the next day Coulter would give a wonderful lecture on
Ecology.... Cowles, with his superior knowledge of taxonomy and his geology, understood more than the rest of us, and became so interested that he
studied Danish and, long before any translation appeared, could read the
book in the original" (Cook, 11).

Cowles took Warming's lead and applied it to the Chicago region and
particularly the Indiana Dunes. In three articles published in the *Botanical Gazette*—one in 1899 and two in 1901—he established his international stature
as a leading figure in the field of plant ecology. His great innovation regarded
the succession of plant communities: "There must be then an order of succession of plant societies, just as there is an order of succession of topographic
forms in the changing landscape. As the years pass by, one plant society must
necessarily be supplanted by another, though the one passes into the other
by imperceptible gradations" ("The Physiographic Ecology of Chicago and
Vicinity," *Botanical Gazette*, February 1901, 79).

With only minor changes, "Physiographic Ecology" was reprinted later
that year as "The Plant Societies of Chicago and Vicinity" in the *Bulletin of the
Geographic Society of Chicago*. The selection is from the *Bulletin* and addresses
what Cowles calls "The River Series."

After the publication of his landmark papers in 1901, Cowles, who spent his entire career at the University of Chicago until retiring in 1934, devoted himself mainly to teaching, scientific organizations, and conservation. He was one of the founders of the Ecological Society of America. At various times he served as president of such organizations as the Association of American Geographers, Botanical Society of America, Chicago Academy of Sciences, and Friends of Our Native Landscape. In addition, Cowles was both a key collaborator with Dwight Perkins in establishing the Forest Preserve District of Cook County and a leading advocate for the preservation of the Indiana Dunes.

Source: Sarah G. Cook, "Cowles Bog, Indiana and Henry Chandler Cowles (1869–1939)," (Chesterton: Indiana Dunes National Lakeshore, 1980).

Round About Chicago
Louella Chapin

⌘ ⌘ ⌘ This summer you have been our guest; we have taken you with us and shown you the way you should go. But you must know your lesson well, for next season you are to go forth alone, to preach the saving gospel of out-of-doors to your family and your friends and your friends' friends to the third generation. For Mother and I have missionary zeal.

Now, the only thing the o. m. claims to know even a little, is how to cultivate a child-garden, and her main implement is repetition, strategic repetition. A theme with variations is her way; for children as well as men "must be taught as though you taught them not."

This particular variation she calls—The Season of Flowers.

If there are those in your family who lack strength of spirit or of body to go to the wild flowers, do you bring the flowers to them and keep the house sweet and beautiful from early April to late October. A few hours on Saturday afternoon or, better yet, early Sunday morning will bring joy to dozens who see your trophies, and much greater joy to you who gather them; for of flowers more than of any other thing in the world is it true that it is better to give than to receive.

Louella Chapin, *Round About Chicago* (Chicago: Unity Publishing, 1907), 177–200.

But there are flowers to be picked, and flowers not to be picked, and before you can go with us you must be carefully instructed in this matter, else our present wrath will fall upon your head, and your own future conscience will smite you for your thoughtlessness, even as the o. m.'s sometimes smites her now. Moreover, there are flowers of which one may pick armfuls, and others of which a handful is better, and still others of which a single blossom is the limit of propriety; and there are flowers to be picked in full bloom, and flowers to be picked in the bud, and flowers to be taken roots and all. And there is art in knowing the best length of stem, and the right amount of foliage; and there is the important first lesson of picking one kind at a time. Oh, the hopeless flower mixtures that one sees!

And always know that the best of your trip is what you can bring back only in your memory; the picture of the flowers where God meant them to be, in their settings of meadow or prairie or woodland. We have little patience with our friends whose first instinct is for possession, and who instantly fall to picking and digging, as soon as their feet strike soft earth.

If picking flowers is a great art, arranging them is a greater. Ask the Japanese. But there is safety in the rule—keep to one kind at a time. When you come with us you are not to violate it.

Thus instructed, we will take you first for the hepaticas that early April finds peeping out on the northward slopes of the ravines on the north shore. Lakeside or Winnetka or Glencoe will be the place. You may bring a shoe box or a tin cracker box, for you are to take some home, a blue and a white and a pink-tinted one perhaps, and these with their roots and much adhering leaf mold. You must never pick the blossoms. They will only droop and wither miserably and to no purpose.

When you have your plants home, put them in low pots with their own woods earth, and cover the pots with crape paper of the same shade as the blossoms, and you will have prepared your sweetest Easter gifts. Choose the recipients with care, and admonish them to plant the hepaticas, when they have done blooming, in a sheltered corner of the lawn near the front fence; and next spring, and many succeeding springs, every passer-by who has known the country life, but especially the old people, will cry out at sight of them, and lean over the fence to gaze at their soft-tinted, waving petals, first harbingers of spring, and talk to the children about them, and go away smiling and reminiscent.

In the ravines with the hepaticas, but on the warmer south slopes you will find dancing anemones, sweet, fragile and unpickable. Let them alone.

By the time the hepaticas are gone to sleep the violets will be up in the woods along the Desplaines and at Beverly Hills, and a little later in the ravines to the north, and later still the Flossmoor woods and Stewart Ridge will be blue with them.

Take a few of the roots and do as you did with the hepaticas, and also pick many bunches of blossoms, mixed plentifully with their own green leaves. You need not fear that you will exterminate them. Next spring there will be just as many, and plenty for all who come. And take all the children you know with you, lest one of them should ever send a pang to your heart by saying, as a sweet city child of fourteen once said to me very wistfully, "But, Miss Emily, I never saw violets growing."

The most beautiful of the flowers in the ravines is the great white trillium, tall and spotless in a circle of its own green leaves; but of these you must never take but a handful, and a very small handful at that, for they are sadly diminishing year by year, as the city reaches farther and farther out. Best of all, take just one apiece to those at home, and put each in a tall clear glass vase; and for yourself, enjoy them where they stand fluttering beside the waxy May-apples, in the misty green of the maiden-hair and meadow rue on the cool shadowy slopes.

As I read my new seed catalogue lately, I was delighted to find that the white trillium is being domesticated, and that the plants can be bought. I wonder if, treated as a garden pet, it can ever be quite its own wild shy self.

While the trillium is being threatened with extinction on the north shore, the blushing arbutus and the rare sweet linnaea on the dunes are safe, because it takes long and patient search to find them, and he who will take that trouble, loves them too well to harm them.

In mid-May, the wild crab-apple blossoms begin to open, and it is best to gather the branches that are in bud, and let them spread their petals in water, for the flowers are too frail to transport. You can find them in many places: at Sag back of the churchyard, along Western avenue between Tracy and Morgan Park, at Glencoe and Ravinia, at River Forest, and best of all at Palos Park, where on every hand great pink thickets brighten the fresh young green of the landscape. Indeed, if you have not seen Palos Park in crabapple week you have missed the loveliest sight that Nature spreads for you the whole round year. And, moreover, you have no excuse, for a few hours are enough, and you could find the time somewhere in that beautiful week.

June will bring the climax of the early and delicate blossoms in the shape of the roses that bloom all about on the railroad embankments everywhere.

They too must be picked in the bud, with plenty of leaves, and put, just a few together, in finger bowls, so that the leaves and stems may show through the clear glass. Their delicate color and matchless perfume will make the whole house glad.

After the roses comes the mid-summer carnival of less exquisite but stronger, more decorative and more abundant flowers; but there are many that make the fields and roadsides showy, and yet are useless for bouquets, such as the red phlox, the puccoons and the lupines, for when picked they shed their corollas in deep dejection. The shooting-stars and gaudy painted-cups that are so effective in masses in the fields, are not strong enough in texture or color to be useful. Even the blue iris, the haughty fleur-de-lis, becomes a mass of corruption when cut and in water. I used to class the blue vervain of the vacant lots with the unpickables, but I know better now. If you put the stems immediately into water, it will hold its erect freshness and make a charming table decoration.

Our favorite flowers for decorating house or church or child-garden are the large white golden-eyed daisies. You can take them by the arm load and wagon load and no one misses them, and next year you can take just as many more. They grow at Tracy, Homewood, Willow Springs and out the "Archey Road"; but this last place I will not locate, for they are on the land of an irate foreigner who is likely to shoot you for trespassing. The daisies are killing out his timothy, and the people who come for the daisies are spoiling what is left. So, though they are a little later at Tracy, you must wait. They always come before the child-garden is scattered for the summer, and the garden is always taken out to them and they are always brought in to the garden.

As you get off the train at Tracy, look back along the track. A florist has his farm here, and at the beginning of daisy time his acres of rose-scented peonies are in full bloom, a gorgeous sight to see. Then go westward past the florist's house garden, where you may look over the fence at rare iris and pansies and what-not, and to the top of the "hill," the long ridge that extends from Beverly Hills to Blue Island, and which yields the same harvest at many points.

A sunny Sunday morning in the daisy-patch is not to be forgotten. You must take shears or a strong knife, for the stems are tough, and you want so many; and you will need strings to tie the daisies in tight bunches, so that you may carry the more. And for a while you stoop to cut them, and then you just sit down where they are thickest and pick around you. It is beautiful to see the big little girl sitting so, almost buried in the tall daisies, the white ribbon on her top-knot making her look like a larger blossom as she bends over her lapful to tie them up. The west wind sways the flowers and the ribbon. The great feather-bed clouds sail past, miles above you in the clear blue sky. Only the distant sounds from the early golfers break the Sabbath stillness.

As you go back you will find a few wild strawberries under the edge of the sidewalk, just enough to remind you what nectar is.

You have had such a season of refreshment that you find yourself pitying the church-goers whom you fall in with on your way home.

When you have eaten the dinner that has been preparing itself while you have been gone, if you have learned the gentle art of fireless cooking, you will fit your pieces of wire screening into your bowls and pitchers to make each daisy stand up free and independent, just as it grew. You can put some in everything you have that will hold water, but always very loosely disposed, and have plenty left to give to your neighbors, and for a week or two your house will be a bower of beauty. And if you like, and we always do like, by going several times you can have a continuous "daisy-fest" of a month at least.

With the daisies comes vacation, and then we can range farther and oftener. Now and then toward the end of daisy time we happen upon a lily near the daisy haunts, but to find them in abundance we must go across the ridge and into the low fields on the western side. And how abundant they are! Dozens and scores of them! They are the gorgeous, red, spotted lilies, like the tiger lilies of the garden, but brighter; those that I always see when I hear "Consider the lilies of the field, how they grow."

Pick them with long stems and be careful not to stain your garments with the deep brown pollen that the swaying stamens will dust over you as impartially as over the bills of the iridescent mites of humming birds that sometimes hover over them, almost while you pick, to take their last sip of sweetness from the lily's deep narrow throat.

The pavement dwellers that meet you returning will never suspect or believe that you are carrying wild flowers.

The lilies grow also in the moist places along the railroad between Glencoe and Ravinia, in the open fields at Clyde and Berwyn, and probably at many points between the western edge of the city and the Desplaines river.

The field lilies yield the palm but to the most magnificent of all America's native flowers, the pink lady-slipper, and her attractive but shyer cousins in yellow. The pink slippers are full two inches long and marvelously beautiful in form and tints. Years ago they grew thickest where East Chicago now is, and once I stood on a low sand ridge and counted more than two hundred bordering a swampy spot. They tossed, proud and free in the full sunshine, at the top of their tall, straight stalks, yet I picked them down to the ground, all I could carry, and they made my neighbors and my child-garden glad. Now I would not harm one on any account. I am more civilized than I was, but that does not bring back the vanished orchids.

I will only whisper to you, very privately, when and where they bloom, for now one is rarely seen, save in the clutch of some ruthless picnicker who has found and beheaded one of them, and in whose careless hand it hangs limp and pathetic.

After the lilies are faded comes a long glory of yellow! Plebeian sunflowers from the fields that any trolley line reaches, and more plebeian and plentiful black-eyed-Susans, coarse and common, but yielding place to none in decorative beauty. And then, in August, come the royal golden-rod, and the royal asters of all shades from white through lavender to deepest purple. You can go in any direction for them, but they seem purest in their coloring and most free from dust in the uplands between the ravines to the north. And for once you may mix your flowers at random, for the golden-rod and the asters never quarrel with one another, no matter what tints you find. Arm loads are in order! Pick the stems very long. Choose particularly the more plume-like golden-rods and the lighter shades of asters, and put the bunches just as you gathered them into tall, wide-mouthed jars and pitchers.

As you garner your gold and purple, the sumach thickets turning crimson in the first frosts will warn you that you must be thinking of your winter stores. Bitter-sweet you must have, and red rose hips, and the sand hills of Dune Park will yield both, and at the same time perhaps, on the edge of some slough, belated fringed gentians and one or two lonely cardinal flowers.

So late, you will get only a handful of the pure deep-blue gentians. The cardinals you will take with their roots, for in your garden they will thrive and bloom and multiply, and their intense glowing color will be many seasons' delight.

You will make one more trip to the north shore for late golden-rod and the red berries of the Solomon's seal, and one trip to Pullman for some of the cat-tails that are filling Calumet Lake.

Our brown grasses for the brown vase that would be nothing without them, we have collected early and quite incidentally at Morgan Park.

Thither we go sometimes for an hour or two, just as we used to go to South Shore, for Mother discovered long ago a charming walk that takes us all in a minute out of the suburban atmosphere into the real country air. At the end of the Morgan Park trolley line a path leads northward along the cemetery into a veritable country road, with a farm and an apple orchard on one side and tangled natural woods on the other.

Sitting by the farm fence in the shade of the apple trees we eat our luncheon and are indescribably refreshed. A busy red-headed wood-pecker is spiraling around the trunk of a gnarled oak tree in the woods across the road. Two squawking blue jays are quarrelling in another tree close by; and the farmer's dog is sniffing uneasily through the fence.

We walk on to the north and east through woods and fields and out on the main trolley line, and we make us a brown bouquet as we go, with rose-brown docks and yellow-brown grasses.

For the red-bronze oak branches that we shall hang in the dining room, and arrange in our copper bowls for Thanksgiving time, we pay a last visit to Palos Park on a mellow hazy afternoon, bringing back arm loads of branches and heart loads of thanksgiving that it has been given to us to know and love God's country, and that we have the health and strength to gain more health and strength as each summer season adds to our soul's stature.

Gentle reader, the summer is ended, and the autumn far advanced. And as the season has ripened, so in varying degree have we.

The big little girl has stretched beyond the boundaries of her summer frocks, and clamors to have her hair tied up and hid with monster ribbon bows. The big boy has symptoms of the trying time when he will roll his trousers toward his knees, be melancholy and "sporty" by turns, and refer to himself and the like as men. But Mother has nursed him through chicken-pox and measles, and will bring him safely even out of this.

Dear Mother, who used to play with her crowing babies on the floor, is still as old as they. Swiftly and unconsciously she adjusts herself to them. Her joys and her duties change and bring her growth.

Only the o. m. seems stationary, her children always young, her round of life unchanging. She knows that she is slowly crystallizing, and only prays that the crystal may be clear and many-faceted, with all its angles true.

Father has come back and says that there is nothing on earth so happy as an old maid when she "gives up."

I, the o. m., long since gave up, and I am happy. Very happy through the long sweet summer and the lingering golden autumn. But now the silver moon rides high, and the red sun seeks its solstice; and with the longer shadows comes a faint, questioning dread of the lonely winter—even the winter of life.

〰〰〰 LOUELLA CHAPIN (1865–?) WAS UNDOUBTEDLY A FASCINAT-ing person, and of all the authors about whom I can find very little, she in-trigues me the most. Born in Ottumwa, Iowa, she was a high school teacher who lived with her sister on Kimbark Avenue in Chicago's Hyde Park neigh-borhood. She was still alive in 1952, when her sister's obituary was published in the *Chicago Tribune*. That no similar notice of her death appeared in the paper suggests that she spent her last years out of this area. The only other facts I know about her relate to her activities as president of the Chicago High School Teachers' Club, where she was a strong advocate on behalf of children's welfare. Newspaper stories document her involvement in providing free lunches during the Christrnas break for destitute students and organiz-ing summer recreational programs. She was also active in the PEN group of the Chicago Woman's Club. (PEN is, according to their website, an "asso-ciation of writers working to advance literature, defend free expression, and foster international literary fellowship.")

Sources: "Asks Sympathy for the Bad Boys," *Chicago Tribune*, April 13, 1913. A6; "High School Teachers Will Dine Board Tonight," *Chicago Tribune*, November 21, 1913, 15; "Open Recreation Center in Three No. Side Schools," *Chicago Tribune*, July 14, 1929, H3; "Free Lunches to Open at Holiday Time," *Chicago Tribune*, December 14, 1931, 19; "9 Women Plan Use of Fund for Hungry Pupils," *Chicago Tribune*, October 21, 1932, 19; "Writers' Meet to Open Here on Wednesday," *Chicago Tribune*, July 22, 1945, F4.

Relic Dunes: A Life History

Frank C. Gates

〜〜〜 A relic dune is a mound of sand left for a time during the wash-ing away of a beach. It is but a temporary stage even when well protected by an efficient plant covering. The relic dunes with which this article deals are to be found along the western shore of Lake Michigan from Kenosha, Wisconsin, down to the Illinois-Wisconsin state line. In that locality the shore is being gradually washed away by the action of Lake Michigan. The dunes were studied during 1908 and 1909, during which time the complete stages of destruction of some of the dunes were observed.

Frank C. Gates, "Relic Dunes: A Life History," *Transactions of the Illinois Academy of Sciences* 3 (1910): 110–16.

To explain the formation and the stages of destruction of these dunes, it is necessary to go into the historical development of this part of the lake shore. This beach is the exposed lake bottom and sand bars of a glacial Nipissing lake which preceded the present Lake Michigan and was from 3 to 17 meters above its level.* Since the low-water period, which ended about 1896, the rising waters of the lake have been cutting into the beach in places and carrying the sand along the beach toward the southwards. From Winthrop Harbor north to Kenosha is a place where the shore is being washed away fairly rapidly. When such action commences, there is usually a bluff formed at the line of contact of the storm waves and the beach. The bluff, however, is maintained by plants, for the sand of which the soil is composed will not of itself remain in such a position....

The plant association that is all important in maintaining the bluff is the lake shore rush (*Juncus balticus* var. *littoralis*) association. From 97 to 99% of the individual plants belong to that species of rush. The rhizomes form a very dense tangle, which not only helps to protect the beach from the attack of the waves, but also serves to protect the sand from desiccation, for dry sand forms a gradual slope instead of an abrupt bluff. Back of the lake shore rush association is the silverweed (*Potentilla anserina*) association, a narrow zone, separating the lake shore rush from the grassy sand plain which stretches back for many meters. As was the case with the [lake shore rush] association, the facies comprises more than 80% of the silverweed association. The secondary species which make any showing whatsoever are horse mint (*Monarda punctata*), sand dropseed (*Sporobolus cryptandrus*), and sandbur (*Cenchrus longispinus*). Both of these grasses usually occur in the tension zone between this association and the grassy sand plain. This later association is composed of over 90% of Canada blue grass (*Poa compressa*), which, however, does not grow sufficiently thick to prevent the sand from giving the color tone to the area. Secondary species appear scattered throughout, but they are never of much importance, as they occur only as individuals here and there in the grass. Some of the most frequently occurring secondary species are:

Flowering spurge (*Euphorbia corollata*)
Blue vervain (*Verbena hastata*)

---

*It was long believed that the beach area addressed in this article was formed in the manner described by Gates. Today, however, it is thought that the beach is of more recent vintage, created when sand from the mouth of the Root River in Racine County was carried south by wave action. (Michael J. Chrzastowski and Wayne Frankie, *Guide to the Geology of Illinois Beach State Park and the Zion Beach-Ridge Plain, Lake County, Illinois* [Illinois State Geological Survey, 2000].)

Horseweed (*Erigeron canadensis*)

Common mullein (*Verbascum thapsus*)

Prairie Indian plantain (*Cacalia plantaginea*)

Horse mint (*Monarda punctata*)

Common mountain mint (*Pycanthemum viginianum*)

Panic grass (*Panicum* sp?)

False pennyroyal (*Isanthus brachiatus*)

Dwarf fleabane (*Erigeron divaricatus*)

Kentucky blue grass (*Poa pratensis*)

Field sorrel (*Rumex acetosella*)

Yarrow (*Achillea millefolium*)

Pearly everlasting (*Anaphalis margaritacea*)*

Common wood sorrel (*Oxalis stricta*)

Pale spiked lobelia (*Lobelia spicata*)

Small skullcap (*Scutellaria parvula*)

Kalm's St. John's Wort (*Hypericum kalmianum*)

Prairie cinquefoil (*Potentilla arguta*)

The most important secondary species of this association from the viewpoint of this article are trailing juniper (*Juniperus horizontalis*) and common juniper (*J. communis*), because they are the only secondary species that can form relic dunes. There are a few mats of each of these junipers at intervals in the sod. These together with the lake shore rush are the only species which, in this region, are instrumental in the formation of relic dunes.

With this brief consideration of the physiographic appearance of the region and the plant associations which occupy it, the steps in the formation of the relic dunes are now in order. In places where in every storm, not merely the more violent ones, the waves attack the bluff, sooner or later passageways or rifts will be cut through the lake shore rush association. This allows undermining of the sand plain, whose surface is sparsely covered with vegetation, the roots of which have but a very limited sand binding capacity. Consequently the plain is washed away as far as the waves have power.

---

*According to Swink and Wilhlem this is a very rare species in this region—and they list no records for Lake County, IL or Kenosha County—and one that is frequently mistaken for the more common old-field balsam (*Gnaphalium obtusifolium*).

The sand from immediately to the landward of the *Juncus* is carried into the lake in the backwash of the waves, thus leaving a mound of sand, thoroughly permeated with *Juncus* rhizomes, which become dry under the desiccating action of the wind and sun.

At first the relic dunes are elliptical in shape, with their major axis parallel with the shore line. By washing away the ends, however, succeeding storms reduce them to an approximately circular outline. In this form they may endure for a couple or more years, depending upon the violence of the storms.

The relic dune itself is a mound of sand about 2 meters high and about 2.5 meters broad. At the top is a very dense growth of lake shore rush, whose rhizomes thoroughly permeate the dunes. The outside of the exposed rhizomes, which usually form a rather dense matwork, protect the dune from both the desiccating and mechanical effects of the wind. From the attack of the waves the rhizomes are of relatively smaller value. The sides of the dunes are cut out in grooves, especially near the bottom where the wind and wave action is more pronounced. A few secondary species occur on the cap, such as sand dropseed, sand reed (*Calamovilfa longifolia*), red osier dogwood (*Cornus stolonifera*), and Russian thistle (*Salsola kali* var. *tenufolia*), but the number of individuals is so small that they are of relatively no importance.

The common juniper relic dune, of which at present writing there is only one in the area, is a poorly developed dune near the limit of wave action. The roots of this plant have not the sand binding power that those of the rushes have. Consequently any exposed sand would be blown away. As the sand is removed by blowing or otherwise, the outermost branches of the juniper sink down and cover the sides of the mound with vegetation which protects it in quite a fair measure from much further action by the wind. Toward the west or landward side the prevailing westerly winds keep piling up sand faster than the juniper can cover it up. Within the next few years, if the present rate of erosion continues, there will be seven or eight more relic dunes of both species of juniper, at which time a better understanding of this class of relic dunes will be possible.

The destruction of the relic dunes takes place through the same agencies that were instrumental in their formation. The wave action during violent storms is one of the most potent agencies of destruction, both because of its mechanical force and the ready movement of sand grains when submerged. The wind in general acts as a desiccating agent, but obtains direct action where the sides of the dunes are unprotected. As the surface dries, the outside grains no longer stick to the moister ones within, but either fall to the base of the dune because of gravity, or are blown away by the wind. This

method of destruction is very slow because the dunes are abundantly supplied with water by capillary attraction from the water table, by spray blown
in from the lake, by the quite frequent rains of this region, and by dew which
is often deposited upon them during the nights when the sand cools down
much faster than the surrounding air.

A more violent, though infrequent agency, is the disruptive power of freezing water. In instances of the action of this agency in 1909, the dunes were
thoroughly soaked by a heavy rain, which was followed immediately by a drop
in temperature from 0.5 to 12C, the result of which was the cracking of the
dunes. The broken pieces were like rocks on account of the ice, but as soon as
the ice at the surface evaporated, the wind scattered the loosened sand grains
over the surface of the beach. Several of the smaller relic dunes were thus disin-
tegrated during November 1909. The larger dunes merely suffered the removal
of 10-20 cm. of sand from around the edge of the crown. The vegetation of the
rim slipped down and now serves to protect the dune during the winter.

In the part of the area which has been under consideration, the lake is advancing upon the shore rather rapidly, so rapidly that the lake shore rush, even
with its relatively rapid means of vegetative propagation, has been unable to
retreat. It merely holds the ground upon which it had formed a zone parallel to the shore line. Consequently, as soon as the lake shore rush relic dunes
are destroyed, the rush association will become non-existent in this particular
area. The silverweed association, on the other hand, spread by seeds as well as
vegetatively. It has widened out from a strip 0.5–1.0 meters wide back of the
rushes, to large mats of 5 to 7 or more meters in diameter, which occupy the
sand area between the relic dunes and the grassy sand plain. The maximum
development occurs just below the grass. Where it overlaps into the grass the
vegetation is noticeably heavier than in either of the two associations. This is
aided by a few secondary species, especially sand dropseed and horsemint.
Canada blue grass forms a permanent vegetative covering, but the silverweed
dries up during the winter. During the growing season much blowing is prevented by the covering of silverweed, but with the removal of this in the fall the
blowing of sand commences, and during the winter the amount that is blown
away is noticeable. Much of it tends to accumulate southward of this area.

From the state line south to Winthrop Harbor the shore dips away from
so direct an attack from the waves. Here the rushes are able to retreat from
the shore line, and although relic dunes are still being formed, there is still a
reactively wide zone of *Juncus* behind them which protects the grassy plain
from being washed away.

～～～ FRANK CALEB GATES (1887–1955) GREW UP IN A HOUSE AT 4540 North Lincoln Avenue, in what was then the town of Ravenswood, now long since incorporated into Chicago. While most of the naturalists discussed in these pages acquired their interest early in life and with the help of environmental factors, Gates seemed to have had the nature bug since birth. His mother recalled that "from the time he was born, he had an incredible curiosity about the earth, its plants and their development." But, in fact, he was more eclectic than that, for birds and reptiles were also well within his purview. According to an aunt, he apparently once brought home a poorly capped shoe box full of snakes that he had just collected: "At a company dinner in the house, the snakes got loose and upset the whole party, because Frank had to go under the table and chairs till he had located all the snakes and the dinner party could go on safely. His mother could have killed him."

Not far from the Gates's home were the cemeteries of Graceland and Rosehill, where he often roamed in pursuit of specimens. And a little farther away was Lake Michigan, also a rich territory for the budding biologist. He attended Lake View High School, where his favorite teacher was the distinguished botanist Herman Pepoon.

This is a good place to state that Gates wrote copiously about his own life. There was the "document" written at the age of twenty-two that discusses the changes he experienced "during the transition from the formative to the adolescent period." He dispenses with the physical changes in one sentence—they were "effected in a comparatively short time" (a modern writer would no doubt had gone into such matters in excruciating detail)—but we learn how his previous aversion to school switched to one of great liking. Once birds became an important subject, the need to get up early during migration reduced the time he slept from seven hours and forty-two minutes to five hours. The "bird walks together with other things developed more fully the desire to be alone.... To avoid undesired company on bird trips another boy and I learned all the birds by their scientific names only and neither could nor would try to give their common names."

When Gates was twenty years old, he began a diary that he was to keep intermittently throughout much of his life. From those writings and many other sources, his son David Murray Gates (who became director of the Missouri Botanic Gardens) wrote a magnificent 502-page biography of his father and mother. It was privately printed, abundantly illustrated with family pictures, and distributed to a handful of libraries that meant something to his father. This labor of love deserves a wider readership.

Gates attended the University of Illinois and soon gravitated to Henry Gleason, one of the premier botanists of his time. It was under Gleason that he wrote his fascinating senior thesis that was published as "The Vegetation of the Beach Area in Northeastern Illinois and Southeastern Wisconsin" (*Bulletin of the Illinois State Laboratory of Natural History* 9, art. V, March 1912). He graduated in 1910 and just two years later received his Ph.D. from the University of Michigan, to where Gleason had relocated. Gates taught briefly at the University of the Philippines, Carthage College, and the University of Michigan. But two institutions were to hold him for the duration of his career: Kansas State University (1920–55), where he was professor of botany and managed the herbarium, and the University of Michigan Biological Station at Pellston (Douglas Lake), where he taught for thirty-nine summers between 1915 and 1955.

Gates's particular interests were in plant succession, revegetation, and grasses. He authored numerous publications, including such books as *Field Manual of Plant Ecology*, *Grasses of Kansas*, and *Wild Flowers of Kansas*. About the last named, A. S. Hitchcock of the Smithsonian Institution wrote: "It is one of the best state floras that I have seen." His dedication may be demonstrated by the events of March 21, 1955. He held his one o'clock paleobotany class, which that day consisted of a test. He came home, talked with his wife, and shortly thereafter keeled over dead.

One of the last sections of David Gates's biography consists of remembrances from various botanists who knew Frank. Edwin A. Phillips took summer classes at Pellston and would become a professor at Pomona College in California. He tells of the day in 1939 when Gates led a field trip to a nearby bog and took advantage of a most effective pedagogic opportunity:

We walked in towards the bog, being rather careful not to get our shoes too wet by stepping over puddles and the like. Of course, as we neared the bog, we found our feet getting wetter and wetter, finally up to our ankles. When we got to the bog, Dr. Gates walked out on the bog mat and as we followed, forming a circle around him, he began to lecture.

Gradually we noticed that we were all sinking down farther and farther, as he nonchalantly and imperviously continued to talk. Everyone was looking around very nervously at everyone else, wondering if we should call his attention to this little detail, also wondering if he was perhaps an absent-minded professor who really wasn't aware of anything being wrong.

At last we were up to our hips in water when he finally with a twinkle in his eye, explained that was as far as we would go down and proceeded to explain the whole nature of the mat.

Sources: Iralee Barnard, "The 137-Year History of the Kansas State University Herbarium," *Transactions of the Kansas Academy of Science* 106, nos. 1/2 (March 2003): 81–91; David Murray Gates, *Frank Caleb Gates: Botanist and Ecologist and His Companion Margaret Thompson Gates* (Ann Arbor, MI: privately printed, 2000); Kansas State University History Index, "Frank C. Gates," *Collegian,* March 22, 1955.

Reading the American Landscape
May T. Watts

ᔕᔕᔕ "You're walking on a history book," said the professor. "It happens, you may notice, to have a flexible cover."

"And to be extremely absorbing," added the undergraduate, using both hands to pull his other foot out of the ooze.

That was my introduction to the Mineral Springs bog—to any bog.

A member of the class broke through the flexible cover that day, up to her waist, and had to lie down on the sphagnum moss and cranberries before we could pry her out. She never did retrieve her shoes, but that was because she had, unfortunately, taken the professor seriously, on the day before the field trip, when he answered our question of "What shall we wear?" by saying "Better wear pumps."

On a later ecology field trip to that same bog, one of the male students killed a rattlesnake there.

On a still later trip, my companion, a young nonbotanical advertising man, cut a walking stick and carried it all day; and I thought nothing of it until, the following week, he called me from the hospital where he lay, his eyes swollen shut with poison sumac.

On that first trip to the bog, we all sat for a while on the end of an old black oak sand dune among columbines and bird's-foot violets while some students put on boots against the water or gloves against the poison sumac; and some put on protection against the mosquitoes (it was citronella in those

May T. Watts, *Reading the American Landscape* (New York: Macmillan, 1975), 74–79, 83–86.

days); some told tales of teams of horses that had disappeared in bogs; and inevitably someone mentioned "The Slough of Despond."

Then we crossed the narrow old corduroy road, which is said to have once carried stagecoaches from Michigan City to Chicago.

We passed the fringe of sour gum trees and took a few steps down, down among the cinnamon ferns—textured acres of them.

Those three or four steps down from the road transported us magically from that bog in northern Indiana to the woods of northern Wisconsin or northern Michigan. The trees around us were yellow birch, white pine, red maple, arborvitae. The shrubs were red-berried elder and winterberry (*Ilex verticillata*). On knolls about the tree roots we looked down upon a rich carpet of goldthread, wintergreen, bunchberry, starflower, Indian cucumber root, partridgeberry.

The two sides of the road seemed to have nothing in common, until we found, on a low branch above the cinnamon ferns, the hummingbird's nest with its two oblong eggs. That structure made a link. It was built of wool from the stipe of the cinnamon fern, and encrusted with lichens from the base of the yellow birch, and some of the food-energy for the building of it had doubtless come from the long nectar sacs of the wild columbine back there on the end of the dune where we had rested.

We swished ahead tenderly through the waist-high ferns to the point where their ranks thinned.

Then we found ourselves having to jump from the hummock at one tamarack's base to the hummock of the next tamarack. In between was black ooze. And in that ooze we found treasure—a clump of showy lady's slipper, *Cypripedium reginae*.

From this spot, the jumps between tamarack roots became longer and longer, and just as we would grab for an outstretched helpful-looking branch, someone would yell, "Look out! Poison sumac."

Between and beyond the poison sumac stretched open sunnier areas. Here we spent a joyful time jumping up and down to see the poison sumac several yards away jumping up and down with us, until the girl with the pumps broke through the surface. After she had been extricated, we still advanced gingerly, step by step, on the unstable mat. We were young and competitive, but we gave up among the cattails.

From the yellow birches to the cattails we had been reading the story of a bog backward.

Now, as we stood, where our eyes could go farther though our feet could not, we began to consider the story of a bog chronologically. This story, from open water to forest, is written over and over again. It is the story of undrained depressions.

Many such undrained depressions are kettle holes in the moraine, places where masses of icè were left behind as the glacier receded. But the Mineral Springs bog lay in a long lagoon of an ancient beach.

Probably this long inland lake started filling with vegetation like any other lake, with submerged plants attached to the bottom near the edge of the lake.

Probably the second stage, too, was like the second stage of other filling lakes—a few waterlilies attached to the bottom and floating their leaves on the surface. These first two stages, however, had long ago disappeared from the Mineral Springs bog. Their existence here could only be assumed because of their presence in other bogs.

It was at the edge of the third stage that we turned back, for lack of footing. Probably this stage had got its start long ago when, somewhere along the margin of the water, some chance surface—a floating lily rhizome, a log, a thin mat of sedges, a bit of debris perhaps—offered an adequate though shallow seed bed for pioneer plants. This group of floating pioneers soon offered a seedbed for others. Each time some root or rhizome or runner or arching branch, rooting at the tip, added a slight bit of substance to this thin mat, there was more support for more roots and rhizomes and arching branches. The mat was thickening. Bulrushes rode its wobbly surface, and cattails, and sphagnum moss, and swamp-loosestrife (*Decodon*), with its habit of arching and rooting from the nodes of the stem, and reed (*Phragmites*), and water speedwell (*Veronica anagallis-aquatica*), and water purslane (*Ludwigia palustris*), and forget-me-not (*Myosotis*). We, however, could not ride that fragile surface.

Even if we had been light-footed enough to advance farther, we should not have arrived at open water. That had long been arched over by the bulrush-decodon mat.

As soon as we had turned our backs on this insecurity, our oozing, muck-heavy footgear began to find a firmer foothold, but we felt as if we were treading a soggy spring mattress.

We had entered the fourth stage, often called the sedge, or the fen, stage. No longer were we surrounded entirely by sun-tolerant plants that can grow with their roots ever in water. On this thicker mat there was opportunity for sun-tolerant plants that do not grow with their roots always in water. We lifted our ponderous feet past a variety of plants, including such well-known water-edgers as chairmaker's rush (*Scirpus americanus*), as well as such inhabitants of dry places as aster and goldenrod. In this varied group appeared grass of Parnassus (*Parnassia caroliniana*), white fringed orchid (*Habenaria leucophaea*), grass pink (*Calopogon pulchellus*), blazing star (*Liatris spicata*), Joe-Pye weed (*Eupatorium purpureum*), blue-joint grass (*Calamagrostis canadensis*), and water-hemlock (*Cicuta maculata*). The fern of this fen stage was marsh shield fern (*Dryopteris thelypteris*).

In the fifth stage, where we had our fun jumping up and down to watch nearby shrubbery bobbing in unison, the bog began to develop a distinctive personality.

At that point we noticed that four major changes had appeared on the mat:

1. The plant society changed from an assemblage of familiar plants of a wide and general distribution to an assemblage of plants of a limited distribution, some of them found only in bogs: pitcher plant, poison sumac, sphagnum moss.
2. Shrubs became a part of the mat. Some of them were outstanding examples of a limited distribution: cranberry, dwarf birch, poison sumac.
3. Something definite seemed to be happening to leaf form and texture. Many low shrubs showed small, thick, leathery, hoary leaves: cranberry, leatherleaf, bog rosemary.
4. Carnivorous plants had arrived, a strange pair of them: pitcher plant (*Sarracenia purpurea*), and sundew (*Drosera rotundifolia*).

Each of these four changes bespoke a changing condition in the bog:

The presence of plants of limited distribution spoke of conditions not generally a part of a plant's living conditions.

The coming in of shrubs spoke of a thickening mat.

The presence of small, thick, leathery, hoary leaves would ordinarily suggest a lack of water, but here they were growing on an oozing mat resting on water. In this instance they indicated that the water, though present, was not readily available, that this situation, while *physically wet* was *physiologically dry*.

The carnivorous plants were long an enigma. It has taken the work of a succession of famous botanists to arrive at our present understanding of the ways of these plants, and of their presence in a bog. If we were to look at the pitcher plant and sundew, as early botanists looked at them, through a gradually clearing fog of mystery, to witness the unfolding of what Emily

Dickinson called "truth's superb surprise," we would see a succession of observations and experiments....

Are there any other plants that have had as long a line of distinguished plant scientists concerned in building up an explanation of their mechanism? With these investigators revealing that the plants do trap insects, do digest them, do absorb nitrogen and other nutrients from their bodies, and do profit by that absorption, the operation of carnivorous plants is no longer an enigma.

Understanding *how* they work makes it not too difficult to find a probable reason *why* they work in a bog—a place where there exists a deficiency in the nutrients that they obtain from insects.

That deficiency is caused by lack of drainage. The lack of drainage causes poor aeration, resulting in a deficient growth of bacteria, and the consequent accumulation of plant debris, and the gradual development of an acid condition. As the bacteria population is reduced, the nitrogen becomes less available; and, as acidity increases, the ability of the plant to absorb through its root hairs, by osmosis, is decreased.

Existing under these conditions there could only be a limited assemblage, plants that cannot compete with others in more favored situations, but can have a place here if they can tolerate it. Each plant must have some special adaptation to be a part of this group; the carnivorous plants just happen to have the most dramatic adaptation.

In that tough community we lingered awhile, realizing that we should not be meeting these specialized individuals for a long time, probably. We fed mosquitoes to the pitcher plants, and irritated the tentacles of the sundew. Then we dug up a plant of each to see if their roots were as insignificant as might be expected of plants that had established a new intake. They were, and the plants were easy to set back into their small soggy places.

As we continued on our way, we found ourselves not so much walking as jumping, from root hummock to root hummock. These were the roots of trees, the first tree to come in on the bog mat, the tamarack of American larch.

Each island of tamarack roots gave us a foothold that was less soggy than the area we had left behind. We were in the sixth stage of the bog. A new element was shade. On the hummocks shade-tolerant plants such as jewelweed (*Impatiens biflora*) appeared. Among the mosses, sphagnum was replaced by the moss, *Mnium*. Among the ferns, marsh shield fern was replaced by royal fern (*Osmunda regalis*).

We could see that these hummocks were formed by a surprisingly shallow root system, because the wind had recently toppled two trees over, tipping on edge a flat platter of roots, the shape of the brown pool they had uncorked in the mat.

In the low places between the root-islands, a few poison sumacs still persisted, as did several other plants of earlier stages.

We were soon through with hummock-hopping, and moving on to earth that felt firm beneath our feet. This was the seventh stage of the bog. The mat had thickened considerably, and supported a rich forest, with abundant yellow birch, white pine, red maple, and arborvitae. Shade was more dense, and shade-tolerant plants abundant: serviceberry (*Amelanchier canadensis*), white violet (*Viola blanda*), showy lady's slipper (*Cypripedium reginae*), stemless lady's slipper (*Cypripedium acaule*), wild sarsaparilla (*Aralia nudicaulis*), cinnamon fern (*Osmunda cinnamomea*) most conspicuous of all, spinulose wood fern (*Aspidium spinulosum*) and an assemblage of typical northerners. That assemblage included partridgeberry (*Mitchella repens*), starflower (*Trientalis borealis*), Indian cucumber-root (*Medeola virginiana*), bunchberry (*Cornus canadensis*).

From among the ferns and northerners we climbed up on to the road, and then to the welcome dryness of the dune. We sat among black oaks, blueberries, and sassafras, removing black muck from our footgear and looking back down at the bog.

What was that assemblage of northerners doing, across the road from black oaks and sassafras?

To find a similar assemblage we might have traveled either to another bog of our area, or to the north woods.

These bog spots of northern vegetation are considered glacial relicts left behind in the midst of conditions that the natural vegetation of the area cannot tolerate, but that the vegetation of the north is able to tolerate. Evidently plants that are able to survive when soil water is locked up by freezing can survive just as well when it is locked up by acidity.

Not all bogs lie, like the one described here, in deep depressions. In southern Michigan are several bogs that lie in shallow depressions, demonstrating that it is lack of drainage rather than depth of depression that is the essential feature for bog formation. Because those sand-locked depressions are so shallow, they do not show the successive stages of a bog in a deep depression, but have one and the same stage overall. Near Stevensville, Michigan, one may visit a bog that is in the waterlily (second) stage; another that is in the sedge (fourth) stage; another that is in the leatherleaf-cranberry (fifth) stage; and another that is already completely covered with tamarack, representing the sixth stage.

On Improving the Property

May T. Watts

They laid the trilliums low,
And where drifted wood anemones and wild sweet phlox
Were wont to follow April's hepaticas—
They planted grass.
There was a corner
That held a tangled copse of hawthorne
And young wild crabs,
Bridal in May above yellow violets,
Purple-twigged in November.
They needed that place for Lombardy poplars
—And grass.

Last June the elderberry was fragrant here.
And in October the viburnum poured its wine
Beneath the moon-yellow wisps of the witch-hazel blossoms.
They piled them in the alley
And made a burnt offering
— To grass.

There was a slope
That a wild-grapevine had captured long ago
At its brink a colony of mandrakes
Held green umbrellas close,
Like a crowd along the path of a parade.
This job almost baffled them,
Showers washed off the seed
And made gullies in the naked clay.
They gritted their teeth,
And planted grass.

May T. Watts, "On Improving the Property" (1920s or 1930s), May Watts Archives, Morton Arboretum, Lisle, Illinois.

At the base of the slope there was a hollow,
So lush with hundreds of years of fallen leaves,
That maiden-hair swirled above the trout-lilies,
And even a few blood-roots
Lifted frosty blossoms there.

Clay from the ravaged slope
Washed down and filled the hollow
With a yellow hump.
They noticed the hump,
And planted grass.

There was a linden
That the bees loved
A smug catalpa has taken its place,
But the wood ashes were used
To fertilize the grass.

People pass by
And say, "Just look at that grass,—
Not a weed in it!
It's like velvet."
One could say as much
For any other grave.

〰️〰️ MAY PETRA THEILGAARD WATTS (1893–1975), A LIFELONG native of the Chicago region, was yet one more student of H. C. Cowles to carry the torch of conservation. (She received her undergraduate degree from the University of Chicago in 1918.) There are scientists who elucidate the ways of nature, and there are those who take that knowledge and put it in a form that can be shared with the wider world. Of course, these don't have to be mutually exclusive enterprises, but for various reasons they often are.

May Watts is certainly one of this region's guiding lights in opening up nature to all. She is best known for *Reading the American Landscape.* In his memorial to her, Alfred Etter wrote that this lovely book "took the jargon out of ecology and made it approachable by anyone. For each scientific phrase removed, she substituted one with intriguing associations and quaint humor."

While Watts's book is a classic, her efforts were not restricted to writing. As a naturalist at the Morton Arboretum, where she served for over thirty years, she had further opportunities to reach thousands of people through her lectures, frequent television appearances, and field trips. And when an honorable battle needed to be joined, she was there. Even if she had never written *Reading the American Landscape*, she would be remembered for her successful fight to convert the abandoned Chicago, Aurora, and Elgin railway into the Illinois Prairie Path, a ribbon of open land winding through the sprawling suburbs of DuPage County. Watts also enlisted in the struggle to save the Indiana Dunes and tried to protect the native beauty of Ravinia, Lake County (Illinois), where she and her family lived and Jens Jensen was a neighbor. (She later moved to Naperville, DuPage County). Of that effort, she wrote: "We must decide whether trilliums, witch hazel, wild plums, and hepaticas are to be forced to join the ghosts of Ravinia. Do we want a humdrum, rubber-stamp town haunted by lovely memories?"

Source: Alfred Etter, "Reading May Watts," *Morton Arboretum Quarterly* 11, no. 3 (Autumn 1975): 38–41.

Impressions of the Warren Woods
Wendell Paddock

᙭᙭᙭ The writer's memory does not go back to the time when all of the land about Three Oaks was covered with virgin forest, but he does remember when a good deal of it was in that condition. My earliest remembrance of the home farm is that there was not more than fifteen acres of cultivated land, most of it very stumpy, a few acres of chopping, and the rest unbroken forest. The woods shut us in in all directions and they extended for miles with only occasionally a small clearing where some settler was slowly chopping out a farm. Timber was the one thing that was plentiful and every boy had to learn to take care of himself in the woods.

In the winter we frequently had to break our path to school through snow, knee deep, and many times in time of storm we were admonished to be constantly on the alert for falling trees. In the spring of the year water was a great

Wendell Paddock, "Impressions of the Warren Woods," *The Acorn* (Three Oaks, Michigan) 26 (October 25, 1917).

source of annoyance as the accumulation of the melting snows and spring rains ran off slowly; consequently roads were almost impassable and farm work was delayed. Later in the season mosquitoes and fever and ague had to be reckoned with.

Cattle ran at large in those days, so it was up to the boys to bring them home each evening. As this frequently meant a tramp of five or six miles, it was a job that was constantly in mind. So we would hurry home from school and after fortifying ourselves with liberal slices of bread and butter would start on our uncertain quest. Cattle are like people in that distance appears to lend enchantment. Any way our cows evidently found the grasses the sweetest in some distant clearing, while neighbor's cattle browsed contentedly about our door. Once I remember that after a fruitless search over the usual beat which was to the north and west a neighbor told us that they had taken an unusual direction and we finally found them enjoying the wayside pasture in the streets of Three Oaks. On these tramps we followed the cow paths through the woods from clearing to clearing finally locating our little herd by the tinkling bell.

As the country boy was constantly in the woods, he came to know each kind of tree and its uses. He could estimate fairly accurately the number of cords in a beech or a maple, or the number of bolts in a basswood. He knew many of the flowering plants, as well, as not a few of them were used in compounding simple remedies. The doctor visits our house but seldom, but when he does come he leaves some sweetened pellets called No. 1 and No. 2, and then I think of the quarts of Lobelia tea that I had to swallow for colds or the quantities of Boneset tea for ague. Once in a great while Dr. Wilcox was called and my remembrance of this kindly old man with his lean horse and buckboard and crooked yellow pipe is principally of the bitterness and copiousness of his medicines.

We knew where the blackberries grew largest, where wintergreen and sassafras grew and where all of the walnut and butternut trees in the vicinity were located. Small game was very plentiful. In the fall of the year we frequently saw large flocks of turkey and the now extinct passenger pigeon visited the clearings each fall. The barking of black squirrels was a common sound as was also the drumming of the partridge, the call of the whip-poor-will and the hoot of the owl.

The timber was in the way and it had to be gotten rid of in the quickest manner possible. While much of it was burned in jam piles and in log heaps, it was at the same time about the only source of revenue. Three Oaks was a busy place in the winter time when heavy loads of saw logs and of bolts came in from all directions to supply the saw mills and broom handle factories. Wood was used for fuel for the railroad locomotives for a good many years; consequently the nearly worthless beech and maple supplied this demand. Cord wood was delivered to the railroad by the hundreds of cords and was ranked up along the right of way at Three Oaks and at the crossings. The first buildings of all kinds were, of course, made of logs and I remember well when our well house, ash leaches, and bee hives consisted of sections of hollow sycamore trees.

Finally there came a time when it was evident that the timber supply would soon be exhausted. The question of how to meet the changing conditions was often discussed and the change came so rapidly that our family at least was scarcely ready. We came to be very careful of the final wood lot and we saw with regret the rapid passing of the most familiar object, the stately forest. I am glad now that I left the farm before the advent of the coal stove.

The above has been written for the purpose of indicating in a feeble manner how intimately our lives were associated with the woods. Our parents were the pioneers to be sure but their children arrived upon the scene in time to have had something of an experience of pioneer life.

When I drive around the old home neighborhood on infrequent visits, I miss first, of course, the once familiar faces, but to me the next great loss is the woods. The thrifty fields of grain are, of course a delight, but I always try to picture in my mind the way the country looked when I was a boy.

It was then with no small amount of anticipation and some trepidation too that I visited the Warren woods. I hoped that they would come up to my ideal, but was fearful that I might expect too much, as all too often childhood memories play one false. In this connection I recall my experience in coasting. The little hills by our home afforded us a great amount of pleasure and we coasted down them by the hour on our home made sleds. Returning home one winter after reaching young man's estate, I found my sister together with some neighbor's children keenly enjoying this sport. I was invited to join them and did so with great enthusiasm, but after one slide I gave up in disgust, for I was rudely awakened to find that the immense hill that I remembered had dwindled into an insignificant knoll.

But the woods produced an entirely different impression, for no where I am sure, have I ever seen so many immense beech and maple trees. Everything about the woods is just right and no detail is lacking. The underbrush is all there, spice brush, buck beech, iron wood and alder and no doubt in the spring of the year, there is a wealth of flowers. A woods to be just right should be provided with ravines and running water. The Galien river furnishes these details and on the bottom land I saw, just as I expected, butternut and walnut trees. An occasional tree has been partially undermined by the river till it has toppled over and leans at just the right angle over the stream. Many a time have I poked a wood chuck out of just such a tree after he had been "treed" by our dog. Here and there a tree has fallen in years gone by and now lies in a state of decay covered with mosses, lichens, and ferns. Open spaces occur occasionally along the river bottom, why I do not know, but they are always there.

How one who has lived and worked amid such surroundings revels at the sight. I know of no other locality in the State or surrounding state that can boast of such a sight as this. Surely there are no other forests of this extent anywhere in this region that have not been mutilated by the hand of man.

This forest is of untold value from many standpoints, some of which may be considered briefly:

It appeals to me most strongly from the sentimental standpoint. And if our nature is lacking in sentiment, imagination and romance, we miss much of the pleasure of life and the joy of living. I am sure that it is sentiment in connection with forests in general that gives the owner his keenest pleasure in the woods. He became a resident of Three Oaks when the settlement of the country was in its crudest stage. Unbroken forests extended for many miles in all directions and the problem of making a bare living under the hardest of conditions confronted every person. As he became an employee of the Chamberlain store at an early age he came to know every man, woman, and child intimately as he traded calico and tea for eggs, dried apples and ax handles. He participated in the heart breaking labor, the crude fare and the homely pleasures. Is it any wonder that persons who have known real woods regard this last remnant of a general condition with a feeling approaching veneration? It has been said frequently that the rising generation is lacking in this virtue. Let the young become familiar with the story of the pioneer and it will be a queer individual indeed who will not regard this woods as a priceless heritage.

Closely associated with the sentimental is the historical value. As stated above, this is probably the last important example of a perfect forest in the region. It shows the condition of the country when the white man first entered it and when the Indian and wild beasts held full sway. It is known that the Pottawatamie Indians made maple sugar from the trees in this very woods and unmistakable evidence of his crude tapping is still to be found in the trunks of some of the giant maples.

Michigan was famous at one time for its timber and for its lumber products but the necessity for tillable land on the part of settlers and careless and greedy lumbering on the part of big interests soon dissipated this resource. When we realize the very great scarcity of tracts of virgin timber the Warren woods become of immense historic value from many standpoints.

Who knows the span of life of common hardwood trees, or when they approach maturity or their rate of growth at different stages up to maturity? At what age do they begin to decline? How long may they live after maturity? What are the chemical and physical changes that timber undergoes at maturity and after? Who has studied the plant societies of which the undergrowth is composed in a mature forest, or during a period of a hundred years? Then similar questions can be asked in regard to chemical and physical changes, taking place in the soil and in the forest cover, the bird, animal and insect life, the microscopic forms of plants and animals, the movement of water in the soil in a mature forest, and a hundred other questions of like import. We do not know the answers largely because no really mature forest is at hand, at least in this region, to study.

Then there are the legion of technical questions that are of interest and value to the trained forester that this woods alone can answer one hundred years from now. It is a safe prediction to make that this woods will within a few years become a mecca for foresters, botanists, zoologists and entomologists from all over the United States. Its scientific value cannot be estimated.

The educational value of this forest includes all that has been said and many things in addition. It is not alone for the scientific man, but people from all walks of life and from all ages may gain information here. Students of forestry and of botany from the great Universities will use this woods as a laboratory, but it is equally important that the other very large per cent of our young who do not go to college should be able to recognize the common trees, shrubs and flowers, as well as birds and small animals.

The average high school student will most likely study lobsters, star fish and marine plants, which is all valuable, but he probably will hear very little about woodchucks or frogs or trees and weeds. Let a child once become interested in the common plants and animals of a locality and he will probably add to his knowledge until he has a fund of useful and ever interesting information. More than this he will absorb a reverence for Creator and things created that he will not otherwise receive. Young women who have been taught to see beauty in the wayside weed are not apt to wear dead birds as ornaments and young men will see in a magnificent beech something more than a mere place to carve his name.

The economic value of the woods is now very great. How a lumberman of the old type would revel in its possession. It would not take him long to destroy the last sapling and the procedure would add very materially to his bank account. But what has he given in return? Well, he has blotted out one of the most wonderful evidences of the Creator's handiwork and left in its place a desolated landscape. Beauty and grandeur have been destroyed and replaced with ruin and desolation. Many dollars are represented here in the value of the possible lumber which the trees would make, but they may not be realized. Instead the money value of the trees should be estimated by the student as one of his problems in forestry.

Finally if one does not appreciate the beauty in a forest or in a tree let him live for ten years in a state where there are no trees. What do you suppose the city of Denver would pay for a bit of this forest could it be transplanted to their city park? The landscape architect devotes his life to developing beautiful landscapes. He uses two principal subjects in developing his pictures and they are trees and grass. Trees and grass used with taste about a dwelling converts it at once into a home. So too, a natural landscape without trees is monotonous, drear, and uninteresting. Corporations and individuals spend fabulous sums in developing parks and grounds and all are familiar with the prominent part that trees play in this development. The keynote of all landscape art in America at least in grounds of any considerable extent, is the natural style; that is the copying of nature. If a tree or a group of trees are planted they must be so placed as to appear to have grown naturally. The Warren woods possess all of the features for which the landscape architect strives and rarely attains, so what more need be said. If one has appreciated the beauty of a woodland let him visit this place and then try to imagine how the country would look if the forest was removed.

Since visiting the woods and thinking over the whole situation, the importance of the forest grows upon me. I think the time will soon be at hand when some institution will want the privilege of establishing a sort of experiment station in connection with it, at least to have a headquarters near there so that scientific men may keep track of many of the things that are going on in the woods. This could be done without in any way interfering with a single living thing.

The citizens of Three Oaks, as well as of the state, are to be congratulated in the possession of this woods. We are assured that this tract is to be preserved indefinitely and the man who makes this possible is certain to receive much praise in years to come. But in the meantime the span of life is short, so why not show our appreciation now to Mr. and Mrs. Warren whose keen foresight and public spirit insure the preservation of this woods.

This is my humble tribute.

Wendell Paddock, Columbus, Ohio, September 27, 1917.

◄◄◄◄ WENDELL PADDOCK (1866–1953), BORN IN THREE OAKS, MICH-igan, received his higher education at Michigan State University, earning both BS and MS degrees. After early stints at the New York State Agricultural Experiment Station (Geneva) and Colorado State University, he spent twenty-eight years on the faculty of the Ohio State University, twenty of which were as head of the nationally acclaimed Department of Horticulture. Paddock was one of the founding members of the American Society for Horticultural Science and senior author of the textbook *Fruit Growing in Arid Regions*. He apparently developed strong friendships with his students, for during each of the six years prior to his death they honored him at an annual fall dinner.

Source: "Paddock, Wendell," Biographical Files, Ohio State University Archives.

An Annotated Flora of the Chicago Area
Herman S. Pepoon

THE WAUKEGAN MOORLANDS

֍ ֍ ֍ Lying north of Waukegan, and extending north beyond the limits of our area, is an extensive region of nearly uniform topography consisting of low, sandy ridges with a general north and south trend, separated by wider or narrower belts of marsh or slough that are but slightly elevated above the waters of Lake Michigan. The ridges have an elevation above the adjacent marsh varying from a few feet, in those parts most distant from the present

Herman S. Pepoon, *An Annotated Flora of the Chicago Area* (Chicago: Chicago Academy of Sciences, 1927), 3–13, 123–30. Courtesy of the Chicago Academy of Sciences—Notebaert Nature Museum.

lake shore, to twenty-five or thirty feet in the more recent sand formations immediately contiguous to the lake. The width of the region from the abrupt Glenwood Beach to the lake varies from one-half to one mile.

Near the limiting Glenwood Beach are extensive marshes, usually with much water, even in the driest summers. As the lake is approached the marsh strips are narrower, the ridges become more pronounced, and the last one-fourth to three-eighths mile is dry, sandy, and in places dune-like; about two hundred feet of recent lake beach terminates the area eastward. The region is imperfectly drained by Dead River, the lower course of which, for one mile, is a deep slough having a width of forty feet or more and lying in a depression parallel to and one-fourth mile from the present shore. The mouth, often choked by sand, discharges a small but rather constant flow of water, except after long continued east winds, which for a time effectually bar the entrance to the lake.

In the spring and early summer the moor is practically impassable from west to east in the outer or western half, owing to the depth of water in the marshes and the miry nature of the bottom soil. At such times the collector must needs travel the ridges, finding an occasional cross-shoal of sand that will enable him to work his uncertain way crosswise. In late summer it is possible with wading boots to traverse most of the area except the river and one or two small sloughs connected with it....

Most of the ridges are more or less covered with various species of trees in a rather scattered formation, the black and bur oak being dominant on higher ground, and willow, cottonwood, and ash prevailing along the marsh regions. A few tamarack still persist and in the region south of the outlet of Dead River, occupying a large area of the dry elevation just west of the shore dune, is a veritable miniature forest of conifers, of many species, the result of wholesale seed sowing some sixty years ago by Mr. Douglass. To the unobservant or the poorly informed observer, this growth has every appearance of being natural, and in fact is so for all practical purposes, and if no disturbance occurs in the next generation, the evidence will all point to Nature instead of Art as the causation of the peculiar coniferous flora.

The moor was doubtless a much more extended region in times long past, as indicated by many of the sand ridges being abruptly worn away at their northern and lake ends. This is due to the fact that the present shore does not lie parallel to the ridges, but cuts each at an angle. This wearing away of the ridges may furnish a large amount of sand for dune construction.

Botanically, the region is easily divisible on a topographic basis into *eight* distinct floral zones, as follows:—(1) the deep water of Dead River, and a few larger water bodies; (2) the permanent marshes; (3) the junction of ridge and marsh; (4) the low moist ridges; (5) the dry ridges; (6) the lake dunes and sand knolls; (7) the Glenwood basal strip; (8) the Glenwood Beach proper. In a general way the whole region is more or less a duplication of the conditions existing in the Dune Region of Lake County, Indiana, particularly that portion adjacent to Clarke, Pine, and Clarke Junction. The Glenwood element is, however, wanting, but a similar topography may be found on the steep limiting bluff of the Grand Calumet near Millers.

The soil of the drier portions is largely sand; of the moister portion much humus is intermingled with muck and peat in the marshes resting on a glacial clay formation. The Glenwood Beach is of clay, often saturated, with gravel and sand pockets, and a loam surface soil of some thickness. Many springs have their origin along the beach slopes or in the erosion ravines, that cut into its otherwise very regular face. These are invariably factors, probably thermal as much as aqueous, in plant distribution.

Like all regions of varying topographic conditions that subject plants to extremes of water and soil environment, the moor is a "final resort" for many species that have been able to accommodate themselves to the more or less unfavorable soil and water factors, and so here make their last stand before extinction, natural or artificial.

Easy of access, by the old railway grade from Beach Station or by the road nearer Waukegan, here is afforded a very favorable opportunity for the study of associations and formations of plants. Of the latter there are well marked, in different parts of the region:—(a) the black oak forest; (b) the swamp meadow; (c) the reed swamp; and (d) the meadow. Of the associations there may be studied without difficulty the following:—(a) the water lily-pondweed (Nymphaea-Potamogeton); (b) the knot-weed (Polygonum); (c) the rush-cat-tail (Scirpus-Typha); (d) the phlox-ragwort-painted cup (Phlox-Senecio-Castilleja); (e) the prickly pear cactus (Opuntia); the bear-berry (Arctostaphylos) and others. The marsh associations, as a rule, are much more marked than are those of drier lands, the determining factors in the latter being so much more numerous and complex.

Attention may now be given to some of the more interesting plant features that a careful study of the moor will disclose. As might well be expected, marsh forms are very much in evidence, and appear in great numbers. A few may be named that appeal more certainly to the eye, as well as some of possibly greater interest if less conspicuous. Marsh grasses, blue-joint (Cala-magrostis), reed grass (Phragmites), cord grass (Spartina), early bunch grass (Sphenopholis),—particularly the first two, are everywhere. The cat-tails (Typha), and an apparent hybrid, form great zones or masses; a large number of sedges (Carices) assist in the swamp meadow and meadow formations. Several duckmeats (Lemna) and Wolffia columbiana are abundant. Arrow-leaf (Sagittaria variabilis), justifying its specific name, associates with the bulrush in wide zones. Sweet flag (Acorus) is everywhere in suitable loca-tions. The rushes (Junci) are very numerous along the marsh borders, being represented by some ten species. The genus Salix or willow has many swamp or marsh loving species, the most noteworthy being silky willow (S. seris-sima), myrtle willow (S. pedicellaris), shining willow (S. lucida), and hoary willow (S. candida). It is evident to anyone who has studied willows, espe-cially where several species are associated, that numerous hybrid forms exist, that cannot be placed by the manuals. Hybrids of the hoary (S. can-dida), beaked (S. rostrata), and silky (S. sericea) willows in almost inextri-cable confusion, have been noted.

Water knotweed (Polygonum hydropiperoides) grows in densely con-gregated masses that dominate large areas; with it are the water hemlocks (Cicuta, particularly C. bulbifera, and Sium). This knotweed (Polygonum) furnishes in many places a chief food for wild ducks and other water fowl, their crops often being filled exclusively with the seeds.

Yellow water crowfoot (Ranunculus delphinifolius) is common in lo-cal communities, but is not general. Its relative, the white water crowfoot (R. aquatilis) is found in the deeper waters of the Dead River. The peculiar cursed crowfoot (R. sceleratus) appears now and then as a solitary example of strange aspect. The first crowfoot named above is showy when en masse, and like many other water plants, may propagate vegetatively by stem buds, breaking away from the plant and rooting.

The most striking plants are found in the meadow formation where the ridge land is but slightly elevated above the adjacent marsh lands on either side. These meadow strips are from two to ten rods in width and in June and July are almost a solid and glorious mass of variegated colors, due to the profusion of showy plants in bloom. In order of season we may name meadow parsnip (Zizia aurea), cynthia (Krigia amplexicaulis), betony (Pe-dicularis canadensis) and various violets, the common blue, arrow leaved, bog, and larkspur (Viola papilionacea, sagittata, nephrophylla, pedatifida, cucullata). A peculiar form nearly white, of uncertain determination but possibly an arrow leaved x hooded (sagittata x cucullata) hybrid is abundant and very beautiful. Star grass (Hypoxis) and blue-eyed grasses (Sisyrinchium), painted cup (Castilleja), shooting star (Dodecatheon), red phlox (Phlox pilosa), violet broom rape (Orobanche uniflora) of deep lavender hue, prairie phlox (Phlox glaberrima), wild onion (Allium cernuum), and others delight the eye and thrive often in soil of considerable moisture, but do so because of the good drainage of the underlying sand. Later in the season fair crops of hay are produced, but all untouched portions are gay with brown-eyed Susans (Rudbeckia), goldenrods (Solidago), asters (Aster), gentians (Gentiana crinita, procera, andrewsii), and grass of Parnassus (Parnassia).

The extensions (girdles) of the same sort of soil but with a greater water content found along the immediate margins of nearly all the marsh strips and separating them from the dry sand ridges, are the chosen homes of a number of rare orchids:—lady's slippers, large yellow, small yellow, and white (Cypripedium parviflorum, parviflorum pubescens, and candidum); rein orchis of three species (Habenaria hyperborea, dilatata media, clavellata); pink pogonia (Pogonia ophioglossoides); beautiful-beard orchis (Calopogon pulchellus); ladies' tresses (Spiranthes cernua); and green twayblade (Liparis loeselii). This entire orchid list is found in similar topographic surroundings in the Dunes of Indiana and illustrates well the influence of topography, soil, and water upon plant distribution.

A few notable plants may be named, characteristic of the sand ridges of moderate elevation. Asparagus is very abundant, thoroughly naturalized, and exceedingly vigorous. The writer has gathered many a good "mess." Coreopsis lanceolata and especially the variety villosa, are rather common. The ovate-leaved Jersey tea (Ceanothus) is present and the yellow paint brush (Castilleja sessiliflora) is very abundant on a few mid-located ridges. Lupine (Lupinus) forms great masses and shows three color forms, the ordinary, a pure white, and one with the purple replaced by a violet-pink. Choke cherry (Prunus virginiana) is occasional, but reaches its best on the dune ridges. Pinweeds (Lechea) are scattered here and there; bird-foot violet (Viola pedata) is everywhere; butterfly weeds, the orange and yellow forms (Asclepias tuberosa, forma aurantiaca, and forma aurea) are here, the latter very striking and very rare; three horsemints (Monarda fistulosa, mollis and punctata); numerous composites,—blazing star (Liatris), goldenrods (Solidago), asters (Aster), and sunflowers (Helianthus).

On the higher and drier ridges of purer sand nearer the lake grow prickly pear (Opuntia), bearberry (Arctostaphylos), red cedar (Juniperus virginiana), dwarf juniper (J. communis depressa), and in magnificent profusion the trailing juniper (J. horizontalis). This last named forms great carpets of twenty to sixty feet in diameter, and is one of the finest evergreens I have ever seen. The pine forest mentioned previously occupies a portion of this topographic formation, just south of the outlet of Dead River. The pines are of many species, white, Austrian, Scotch, Table Mt., pitch (Pinus strobus, laricio, sylvestris, pungens, rigida) and are seemingly perfectly naturalized. Mr. Douglass (the story goes) traversed the area on horseback, carrying a bag of mixed pine seeds and threw them into the wind, thus scattering the seed far and wide. This was some sixty years ago, and today some parts are covered with most vigorous forest growth. The Austrian pine has done the best, though numerous specimens of all the species are thriving.

On the lake margin of the sand ridges small irregular dunes are found, generally capped with false Solomon's seal (Smilacina stellata), marram grass (Ammophila) or sand binder grass (Calamovilfa), herbaceous forms; or when woody plants appear, sweet sumac (Rhus canadensis), dune willow (Salix syrticola), sand cherry (Prunus pumila), wild grape (Vitis vulpina), Virginia creeper (Psedera), choke cherry (Prunus virginiana), hop tree (Ptelea trifoliata), box elder (Acer negundo), cottonwood (Populus deltoides), and balm of Gilead (Populus candicans). This latter species is common and extends throughout the shore dunes the whole extent of the Chicago Area, and appears to be a genuine native.

The wave and wind swept recent beach in its higher and less wave exposed portion has a scanty growth of love grass (Eragrostis pilosa), Cyperus, sea rocket (Cakile edentula), bug-seed (Corispermum), tumbleweed (Cycloloma), and Russian thistle (Salsola kali tenuifolia).

The Glenwood Beach and generally its moist, springy base, is rich in a very much mixed flora, justified by the great diversity of soils, moisture, exposure, and drainage. A few of the more important species are here enumerated, with a few notes on some remarkable examples of plant distribution, somewhat outside the area limited by the Glenwood Beach.

The moist slope is rich in shade-loving sedges and grasses. Mention may be made of species of manna grass (Leersia, Glyceria) and wild rye (Elymus); sedges (Carex lupulina, lupuliformis, hystricina); wild garlic (Allium canadense) and wild leek (A. tricoccum); Solomon's seal (Polygonatum commutatum); greenbriers (Smilax hispida, ecirrhata, and herbacea); and nearly all the willows (Salix), except the true marsh S. candida and S. pedicellaris and the dune S. syrticola, some ten species in all. Very fine examples of balsam poplar (Populus balsamifera) are occasionally seen. The canoe birch (Betula alba papyrifera) is common, also, bloodroot (Sanguinaria), Hepatica, wild ginger (Asarum reflexum), Anemone canadensis, blackberry (Rubus allegheniensis), cranes bill (Geranium maculatum), jewel weeds (Impatiens biflora and palida), the hairy-wood, common blue, and two white violets (Viola sororia, papilionacea, blanda, and pallens), marsh betony (Pedicularis lanceolata), red honeysuckle (Lonicera dioica), elder (Sambucus racemosa), golden-rods (Solidago), and asters (Aster).

Just west of the beach crest are two large areas, one of beard's tongue (Pentstemon digitalis), the other of orange hawkweed (Hieracium aurantiacum),—the latter very numerous and brilliant, but far removed from the recorded range. Evidently it has been established for a long time, as some acres of open woodland are now overrun. Tartarian honeysuckle (Lonicera tartarica) is common in the same wood. Mountain ash (Pyrus americana) is occasional. The thought is suggested that the old-time nursery of Mr. Douglass, which was located near here, may have been a center of distribution for some of the above species....

NOTE: Since the above was written a large portion of the southern end of the moor has been taken over for manufacturing purposes, the land graded, ridges, marshes, and sloughs destroyed, and plants by the wholesale exterminated. This commercial invasion has paused for the present at least, some two miles to the south of Dead River outlet. Furthermore, a very large tract of land east and south-east of Beach Station on the C. & N. W. Ry. is now a dairy pasture, and the wonderful meadow flora of the pastured area is exterminated. Gentians that existed by the thousands ten years ago are with difficulty found by the dozen or even less. It would appear that this favored plant refuge is soon to be a memory. A further rumor is that the whole tract is to be made into a model suburban residence area. This is so alarming that determined effort is being made by plant lovers, in Waukegan and elsewhere, to have the moorland set aside as a plant and bird refuge. The success of this undertaking is not as yet assured.

Thus, one by one, the fine plant havens-of-refuge are "yielding up the ghost" to ruthless, if necessary, urban extensions of activity. Doubtless, destruction awaits much of our native flora that is unfortunately within commuting distance of Chicago, a fate to be greatly deplored but not to be prevented.

〜〜〜 HERMAN SILAS PEPOON (1860–1941) HAS ALWAYS STRUCK ME as an unusual person, if for no other reason than he abandoned a career in medicine to teach science at a Chicago public school, Lake View High School. (When, do you suppose, has that happened last?) No doubt this gave him greater freedom to botanize, but he apparently truly valued his role as a teacher to be a "molder of character."

Pepoon was born in Warren, Illinois, and received his undergraduate degree at the University of Illinois and his medical training at Hahnemann Medical School in Chicago. From 1892 until his retirement in 1930, he taught at Lake View; over the course of his tenure there, he had over ten thousand students. Each of these, according to V. O. Graham, was touched by their distinguished pedagogue: "his buoyant spirit changed work from drudgery to joyous effort."

Besides the classroom, Pepoon influenced many through his books *Studies of Plant Life* (1900), *Representative Plants* (1912), and his still-wonderful *An Annotated Flora of the Chicago Area* (1927). This last book remains a classic, capturing like no other the region as it was in his day. While many such works are dry compendia of records, his discussion of places and specific plants still inspire modern botanists. Floyd Swink and Gcrould Wilhelm, for example, in their monumental *Plants of the Chicago Region* (1994), quote Pepoon's treatment of the prairie Indian plantain (*Cacalia plantaginea*): "Wet prairies west of Chicago, very abundant, covering many acres of the marsh grasslands with its white blanket of bloom." They then comment: "Pepoon's statement, though no doubt true then, now reads like a myth."

Pepoon was keenly aware that what he described was vanishing fast. Perhaps the most poignant passage in his *Annotated Flora* is found in his account of Goose Lake, a trove of botanical treasures nestled in the Indiana Dunes that was to be destroyed many decades later by the construction of Burns Harbor and associated steel mills: "Well do I remember the pilgrimage with Professor Hill, Professor Umbach and Dr. Moffatt to gloat on the *one* specimen of ebony fern that grew on a shaded bluff, near the east shore of Goose Lake, and also the dessert of [yellow-eyed grass], [pipewort], margined fern, a half-dozen species of Rynchospora, [three-way sedge], and five species of [hair cap moss] which Professor Hill, with justifiable pride, exhibited. The days are gone, the men are largely passed on, the flowers have disappeared, and into our hearts a feeling of sadness comes to realize that never again can these things be."

Sources: V. O. Graham, "Herman Silas Pepoon," *Chicago Naturalist* 5, no. 1 (1941); *LakeReView* [Lake View High School paper], January 15, 1930.

Dune Boy: The Early Years of a Naturalist
Edwin Way Teale

THE DEATH OF A TREE

◌◌◌ For a great tree death comes as a gradual transformation. Its vitality ebbs slowly. Even when life has abandoned it entirely it remains a majestic thing. On some hilltop a dead tree may dominate the landscape for miles around. Alone among living things it retains its character and dignity after death. Plants wither; animals disintegrate. But a dead tree may be as arresting, as filled with personality, in death as it is in life. Even in its final moments, when the massive trunk lies prone and it has moldered into a ridge covered with mosses and fungi, it arrives at a fitting and a noble end. It enriches and refreshes the earth. And later, as part of other green and growing things, it rises again.

The death of the great white oak which gave our Indiana homestead its name and which played such an important part in our daily lives was so gentle a transition that we never knew just when it ceased to be a living organism.

It had stood there, toward the sunset from the farmhouse, rooted in that same spot for 200 years or more. How many generations of red squirrels had rattled up and down its gray-black bark! How many generations of robins had sung from its upper branches! How many humans, from how many lands, had paused beneath its shade!

The passing of this venerable giant made a profound impression upon my young mind. Just what caused its death was then a mystery. Looking back, I believe the deep drainage ditches, which had been cut through the dune-country marshes a few years before, had lowered the water-table just sufficiently to affect the roots of the old oak. Millions of delicate root-tips were injured. As they began to wither, the whole vast underground system of

---

nourishment broke down and the tree was no longer able to send sap to the upper branches.

Like a river flowing into a desert, the life stream of the tree dwindled and disappeared before it reached the topmost twigs. They died first. The leaf at the tip of each twig, the last to unfold, was the first to wither and fall. Then, little by little, the twig itself became dead and dry. This process of dissolution, in the manner of a movie run backward, reversed the development of growth. Just as, cell by cell, the twig had grown outward toward the tip, so now death spread, cell by cell, backward from the tip.

Sadly we watched the blight work from twig to branch, from smaller branch to larger branch, until the whole top of the tree was dead and bare. For years those dry, barkless upper branches remained intact. Their wood became gray and polished by the winds. When thunderstorms rolled over the farm from the northwest the dead branches shone like silver against the black and swollen sky. Robins and veeries sang from these lofty perches, gilded by the sunset long after the purple of advancing dusk filled the spaces below.

Then, one by one, their resiliency gone, the topmost limbs crashed to earth, carried away by the fury of storm-winds. In fragments and patches, bark from the upper trunk littered the ground below. The protecting skin of the tree was broken. In through the gaps poured a host of microscopic enemies, the organisms of decay.

Ghostly white fungus penetrated into the sap-wood. It worked its way downward along the unused tubes, those vertical channels through which had flowed the life-blood of the oak. The continued flow of this sap might have kept out the fungus. But sap rises only to branches clothed with leaves. As each limb became blighted and leafless, the sap-level dropped to the next living branch below. And close on the heels of this descending fluid followed the fungus. From branch to branch its silent, deadly descent continued.

Soft and flabby, so unsubstantial it can be crushed without apparent pressure between a thumb and forefinger, this pale fungus is yet able to penetrate through the hardest of woods. This amazing and paradoxical feat is accomplished by means of digestive enzymes which the fungus secretes and which dissolve the wood as strong acids might do. These fungus-enzymes, science has learned, are virtually the same as those produced by the single-celled protozoa which live in the bodies of the termites and enable those insects to digest the cellulose in wood.

Advancing in the form of thin white threads, which branch again and again, the fungus works its way from side to side as well as downward through the trunk of a dying tree. Beyond the reach of our eyes the fungus kept spreading within the body of the old oak, branching into a kind of vast, interlacing root-system of its own, pale and ghostly.

Behind the fungus, along the dead upper trunk, yellow-hammers drummed on the dry wood. I saw them, with their chisel-bills, hewing out nesting holes which, in turn, admitted new organisms of decay. In effect, the dissolution of a great tree is like the slow turning of an immense wheel of life. Each stage of its decline and decay brings a whole new, interdependent population of dwellers and their parasites.

Even while the lower branches of the oak were still green, insect wreckers were already at work above them. First to arrive were the bark beetles. In the earliest stages their fare was the tender inner layer of the bark, the living bond between the trunk and its covering. As death spread downward in the oak, as freezing and storms loosened the bark, the beetles descended, foot by foot. Some of them left behind elaborate patterns, branching mazes of tunnels that took on the appearance of fantastic "thousand-leggers" engraved on wood.

During the winter when I was twelve years old a gale of abnormal force swept the Great Lakes region. Gusts reached almost hurricane proportions. Weakened by the work of the fungus, bacteria, woodpeckers, and beetles, the whole top of the tree snapped off some seventy feet from the ground. After that the progress of its dissolution was rapid.

Finally the last of the lower leaves disappeared. The green badge of life returned no more. On summer days the sound of the wind sweeping through the old oak had a winter shrillness. No more was there the rustling of a multitude of leaves above our hammock; no more was there the "plump!" of falling acorns. Leaves and acorns, life and progress, were at an end.

In the days that followed, as the bark loosened to the base, the wheel of life, which had its hub in the now-dead oak, grew larger.

I saw carpenter ants hurrying this way and that over the lower tree-trunk. Ichneumon flies, trailing deadly, drill-like ovipositors, hovered above the bark in search of buried larvae on which to lay their eggs. Carpenter bees, their black abdomens glistening like patent leather, bit their way into the dry wood of the dead branches. Click beetles and sow-bugs and small spiders found security beneath fragments of the loosened bark. And around the base of the tree swift-legged carabid beetles hunted their insect prey under cover of darkness.

Yellowish brown, the wood-flour of the powder-post beetles began to sift about the foot of the oak. It, in turn, attracted the larvae of the Darkling beetles. Thus, link by link, the chain of life expanded. To the expert eye the condition of the wood, the bark, the ground about the base of the oak—all told of the action of the inter-related forms of life attracted by the death and decay of a tree.

But below all this activity, beyond the power of human sight to detect, other changes were taking place. The underground root system, comprising almost as much wood as was visible in the tree rising above-ground, was also altering.

Fungus, entering the damaged root-tips or working downward from the infected trunk, followed the sap channels and hastened decay. The great main roots, spreading out as far as the widest branches of the tree itself, altered rapidly. Their fibers grew brittle; their old pliancy disappeared; their bark split and loosened. The breakdown of the upper tree found its counterpart, within the darkness of the earth, in the dissolution of the lower roots.

I remember well the day the great oak came down. I was fourteen at the time. Gramp had measured distances and planned his cutting operations in advance. He chopped away for fully half an hour before he had a V-shaped bite cut exactly in position to bring the trunk crashing in the place desired. Hours filled with the whine of the cross-cut saw followed.

Then came the great moment. A few last, quick strokes. A slow, deliberate swaying. The crack of parting fibers. Then a long "sw-o-sh!" that rose in pitch as the towering trunk arced downward at increasing speed. There followed a vast tumult of crashing, crackling sound; the dance of splintered branches; a haze of dead, swirling grass. Then a slow settling of small objects and silence. All was over. Lone oak was gone.

Gram, I remember, brushed away what she remarked was dust in her eyes with a corner of her apron and went inside. She had known and loved that one great tree since she had come to the farm as a bride of sixteen. She had seen it under all conditions and through eyes colored by many moods. Her children had grown up under its shadow and I, a grandchild, had known its shade. Its passing was like the passing of an old, old friend. For all of us there seemed an empty space in our sky in the days that followed.

Gramp and I set to work, attacking the fallen giant. Great piles of cord-wood, mounds of broken branches for kindling, grew around the prostrate trunk as the weeks went by. Eventually only the huge, circular table of the low stump remained—reddish brown and slowly dissolving into dust.

For two winters wood from the old oak fed the kitchen range and the dining-room stove. It had a clean, well-seasoned smell. And it burned with a clear and leaping flame, continuing—unlike the quickly consumed poplar and elm—for an admirable length of time. Like the old tree itself, the fibers of these sticks had character and endurance to the very end.

〰〰〰 EDWIN WAY TEALE (1899–1980) WAS UNDOUBTEDLY ONE OF the twentieth century's most celebrated nature writers. Born in Joliet to a railroad mechanic and a teacher, he spent his summers on his grandparents' farm near Furnessville in the Indiana Dunes. His book *Dune Boy* recounts those halcyon days, when a nascent interest in nature bloomed into a lifelong calling. While not yet in his teens, he had decided that writing about nature would be his career and that changing his middle name from Alfred to Way would make him sound more the part. (I wonder if such a change would help a nature writer in his fifties.)

Teale received degrees in English literature from Earlham College in Rich-mond, Indiana, and Columbia University, where he wrote his master's the-sis on the treatment of nature in the writings of Wordsworth and Scott. He worked in various editorial jobs, including a thirteen-year stint as a features writer for *Popular Science*. Throughout his life, Teale manifested a liking for gadgets and tinkering. During one of his early summers with his indulgent grandparents, he constructed a glider that, according to his father, actually became airborne for a few seconds. More significantly, he developed exper-tise in photography and focused many hours (and lenses) on insects. A col-lection of these pictures, along with an engaging text, became his first book on natural history, *Grassroots Jungles* (1937). The work was well received, and after several more successful books on insects and/or photography, he quit his day job at the age of forty-two to pursue freelance writing full-time. He was to author eighteen books and edit eleven others over the next forty years, be-coming, in the words of biographer Anne Keen, "a kind of twentieth-century Thoreau." His four-volume exploration of North America's seasons—*North with Spring* (1951), *Autumn Across America* (1956), *Journey into Summer* (1961), and *Wandering through Winter* (1965)—are perhaps his masterpieces, with

Wandering through Winter receiving a Pulitzer Prize. For the last twenty-one years of his life, he and his wife lived on a 130-acre estate in Hampton, Connecticut.

Source: Anne T. Keen, "Edwin Way Teale," *American National Biography* (New York: Oxford University Press, 1999), vol. 21, pp. 419–20.

Land Animals

Pigeons!

Darius Cook

CRB CRB CRB A gentleman from Berrien informs us that about 3 miles and a half from that village, the pigeons have taken possession of the woods, about 5 miles square, where they are nesting, and that there is from 10 to 75 nests on each tree. Large branches of trees are broken by them and the ground is strewed with eggs. On approaching the spot, one would imagine that he was near the Falls of Niagara, so incessant and loud is their thunder.

Notes on the Ornithology of Wisconsin

Philo R. Hoy

CRB CRB CRB With few exceptions, the facts contained in the following brief Notes were obtained from personal observations made within 15 miles of Racine, Wisconsin. This city is situated on the western shore of Lake Michigan, at the extreme southern point of the heavy timbered district where the great prairies approach near the lake from the west, and is [in a] remarkably favorable position for ornithological investigations. It would appear that this is a grand point, a kind of rendezvous, that birds make during their migrations. Here, within the last seven years, I have noticed 283 species of birds, about one-twentieth of all known to naturalists, many of which, considered rare in other sections, are found here in the greatest abundance. It will be seen that a striking peculiarity of the ornithological fauna of this section is that southern birds go further north in summer, while the northern species go further south in winter than they do east of the great lakes. [In the original article, birds are

Darius Cook, "Pigeons!" *Niles (MI) Republican*, May 6, 1843.

Philo R. Hoy, "Notes on the Ornithology of Wisconsin," *Proceedings of the Philadelphia Academy of Sciences* (March 2, 1853): 804–13.

identified only by the scientific names then in vogue. I have replaced those with currently used common names.]

Golden eagle: I have a fine specimen shot near Racine, Dec. 1853. It is a fact worthy of note that this noble eagle, in the absence of rocky cliffs for its eyrie, does occasionally nest on trees. One instance occurred between Racine and Milwaukee in 1851. The nest was fixed in the triple forks of a large oak.

Washington's eagle [One of J. J. Audubon's bigger blunders was describing the immature bald eagle as a species separate from the adult, an issue addressed by Hoy.]: I procured in 1850 a living bird that had been slightly wounded, which answered to Audubon's description of this *doubtful* species. I kept it in an ample cage upwards of two years, but before its death it underwent changes in plumage which led me to believe that, had it lived, it would have proved to be the white-headed species. I put several species of hawks and owls into the same apartment, several of which the eagle killed and devoured without ceremony. When a fowl was introduced, he pounced upon it and without attempting to kill, proceeded to pluck it with the greatest unconcern, notwithstanding its piteous screams and struggles. It is my opinion that the Bird of Washington will prove to be only an unusually marked and fine immature white-headed eagle.

Merlin: This active little falcon is numerous, especially in spring and fall during the migration of the smaller birds. A few nest with us, many more in the pine forests of the northern part of the State. Those that nest in this vicinity regularly morning and evening visit the lake shore, in quest of bank swallows, which they seize with great dexterity while on the wing.

Northern Goshawk: This daring and powerful hawk I found in all seasons; the old birds only remained during the winter, the young retiring further south. The young are ... more bold and daring, much more destructive to the poultry yards than the more sly and cautious old ones—a peculiarity not, however, confined to this species.

Swallow-tailed Kite: This kite was numerous within ten miles of Racine where they have nested up to the year 1849, since which time they have abandoned the region. I have not seen one since 1850. They nested on tall elm trees about the 10th of June, and left us about the first of September.

Long-eared Owl: More numerous in the vicinity than any other owl. The young leave the nest about the middle of June.

Great-horned Owl: One of our most numerous species. I once put a remarkably large and fine owl of this species into the same cage with the "Washington Eagle," previously mentioned, which soon resulted in a contest. The moment a bird was given to the owl, the eagle demanded it in his usual peremptory manner, which was promptly resisted with so much spirit and determination that for a time I was in doubt as to the result; but finally the eagle had to stand aside, and witness the owl devour the coveted morsel. After several similar contests, it was mutually settled that possession gave an undisputed right, the owl not being disposed to act on the offensive. I had a fine red-shouldered hawk in the same aviary which the owl killed and ate the second night.

Common Nighthawk: Numerous. They leave us by the 15th of September. On the 10th of this month, 1850, for two hours before dark, these birds formed one continuous flock, moving south. They reminded me, by their vast numbers, of passenger pigeons, more than night hawks. Next day not one was to be seen.

Wood Thrush: Common. Wishing to add to my collection a pair of this species, together with their nest and eggs, I shot the female and was about to secure the nest, when the male, which had been watching me in the vicinity, commenced singing; and as I approached the spot glided off still further from the nest, all the time pouring forth the most mellow and plaintive strains I have ever before heard uttered by this most melodious of songbirds. After I had been enticed to a considerable distance, he returned to the vicinity of the nest; three or four times I followed this bird in the same manner before I succeeded in shooting him. This movement and effect of his tender song, so far enlisted my sympathies that I regretted exceedingly my cruelty in destroying his nest and mate. It is common for birds to resort to various stratagems for the purpose of attracting intruders from their nests, but this is the only instance with which I am acquainted where the charms of their music was employed for this object.

Chestnut-sided Warbler: This beautiful little warbler is exceedingly abundant. It prefers localities with a dense underbrush, especially hazel, thinly covered with trees. In such situations it is not uncommon to hear the songs of a dozen males at the same time. They construct a nest of blades of grass and thin strips of bark intermingled with caterpillars' web, fixed in a low bush (generally hazel), seldom more than two or three feet from the ground.... If the nest be approached when the female is in it, she will drop to the ground and hobble along with one wing dragging, uttering at the same time a *peeping* note of distress. I once caught a young bird of this species that had just left the nest; the parent birds, in their alarm for its safety, approached so near to me that I caught the male in my hand. I let them both go, upon which the joy of the old bird appeared to be greater for the escape of the young fledgling than for his own release.

Parula Warbler: Common. The beautiful pensile nest of this bird has never, to my knowledge, been described. Audubon undoubtedly erred in attributing the nest described by him to this species. That presented by me to the Collection of the Academy is formed by interlacing and sewing together, with a few blades of grass, the pendant lichen (*Usnea barbata*) which grew upon a dead horizontal branch of an oak fifty or sixty feet from the ground. A hole, just large enough for the bird to enter, is left in the angle immediately under the branch, which forms a complete roof for the nest; it is finished with a slight lining of hair. The whole forms a beautiful basket of moss, which is admirably adapted to the purpose intended, so effectually concealed, so light and airy, that it would be almost impossible to suggest an improvement, and certainly one of the most interesting specimens of ornithological architecture.

➤➤➤➤ PHILO ROMAYNE HOY (1816–1892) HAS LONG BEEN A HERO of mine. He was the first serious bird student in this region, providing information on what the avifauna was like when the prairie-forest ecotone was still intact. Hoy wrote of a time when local breeding species included golden eagles, merlins, swallow-tailed kites, and goshawks—birds that very likely will never nest anywhere near here again. And he wrote with a feeling that is evident in this selection.

While birds were his special love, virtually all things, both alive and inanimate, were subjects of engagement. As partial evidence of Hoy's eclectic zoological interests, his name has been attached to the amphipod *Diporeia hoyi*, the fish *Coregonus hoyi* (bloater), and the pygmy shrew (*Sorex hoyi*).

Dr. Hoy obviously had a wonderful curiosity. He examined the stomach contents of the deepwater ciscoes brought in by fishermen to make the first-ever collections of Lake Michigan's benthic fauna. He found an organism called the opossum shrimp (*Mysis relicta*), which at that time was known only from oceans. This mystery, along with Hoy's prodding, led the Chicago Academy of Sciences and the Wisconsin Academy of Sciences, Arts and Letters to undertake a study of Lake Michigan's depths in the spring of 1871. Unfortunately, the specimens were destroyed when the CAS burned to the ground during the Chicago fire a few months later.

In 1922 Hoy's daughter, Mrs. William Miller, gave a retrospective on her father, which was later published in a local newspaper. Dr. Hoy was born and educated in Ohio, establishing a medical practice in Fair Haven for several years before taking his family west to Racine in 1846. While traveling the pitted roads of the region to reach patients, he carried not only medicines, splints, and other accoutrements typical of a physician's work, but also "a gun, his butterfly net and sometimes a fish net, his pocket lens, a botany box, a collecting bottle for insects and, best of all, his alert all observing eyes."

One time he asked a group to look through a microscope to observe a mosquito mounted on the slide. As he declaimed on the marvelous stinging mechanism of the insect, he was interrupted by a minister: "After all, doctor, isn't it a pretty small thing for a man to be spending time on." Hoy replied, "Sir, a thing not too small for the creator to make is not too small for me to study."

Source: Mrs. William Henry Miller, "Dr. Philo Hoy: Pioneer Scientist: Review of His Life," *Racine Journal News*, May 2, 1922.

The Quadrupeds of Illinois, Injurious and Beneficial to the Farmer
Robert Kennicott

COMMON GREY RABBIT OR HARE (*LEPUS SYLVATICUS*, BACHMAN)

᭱ ᭱ ᭱ … The grey rabbit exists nearly throughout the eastern half of the Union, and south of the Gulf of Mexico, but it is not found far northward. It is abundant at least as far west as Iowa and Minnesota. This well-known species is fond of dry, level ground, rather thinly wooded, and interspersed with thickets and open spots. In Northern Illinois, where the prairies are traversed by streams, bordered by trees, or dotted with groves, the grey rabbit is very abundant, particularly in the groves and edges of the larger woodlands where clumps of hazels and briers are numerous. This hare is properly an inhabitant of the woods, and though sometimes abounding for several miles on the prairies, it is not so much at home there as in the neighborhood of trees, among which it finds better shelter from its innumerable enemies. Many, which spend the summer on the prairies, are believed to return to the woods in winter. As the wild prairies become settled, the rabbits are observed to live further from the forests, seeking shelter about fences and stacks. In the hilly and heavily-timbered regions of Southern Illinois, this species is less abundant.

The "form" of the grey rabbit is in some concealed position, by the side of a log, in a small brush heap, or at the root of a bunch of briers and weeds, if in the

Robert Kennicott, "The Quadrupeds of Illinois, Injurious and Beneficial to the Farmer," *Transactions of Illinois State Agricultural Society* (1857): 78–84.

woods; or it is frequently situated in the grass at the edge of the prairie, or of the sloughs that run into groves and outskirts of the woods; and here, as well as on the prairie, it is where the overhanging grass shelters and conceals it. The form, in fact, is only a particular spot to which the rabbit retires to spend the day, and is merely a slight depression, sometimes with a few grasses and leaves drawn together, little or no art being ever used in its construction; though, as before mentioned, it is usually in a position somewhat concealed. The same individual sometimes has several forms and, in winter, one is frequently chosen in a more sheltered position than that used in summer. But in winter, it does not always occupy a form, often being found in hollow trees, whether fallen or upright, as well as in holes in the ground. It usually retires, however, to these situations only in severely cold and stormy weather, or for refuge when pursued.

Though holes in the ground are often occupied, they are not dug by the rabbit, but are the deserted burrows of some other animals. It is true that the female scratches shallow holes in which to bring forth her young, in open fields, but it is rarely indeed that rabbits of this species dig burrows. I am credibly informed of a few instances in which they have been known to dig holes for themselves in hillsides; but these may be considered as departures from their natural habits. This rabbit is not pugnacious, several even taking refuge in the same hole; but though they exist in astonishing abundance in particular localities, they are not naturally gregarious.

The grey rabbit is exceedingly timid, and rarely or never makes the slightest resistance when attacked by other than its own kind. Its only attempts to escape its enemies are by speed and stratagem. When pursued, an old male exhibits as much cunning as a fox—doubling, turning aside, and permitting the dog to pass, and then running on the back track; going through water, which it dislikes; and frequently springing upon a log and sitting motionless, while the dog, in plain sight, beats around within a few feet of the spot. Usually, when one of these animals is started by dogs, it runs a short distance, and, unless closely pursued, turns aside and stops. The dog generally passes it, when it at once returns to the neighborhood of its form; or, if unable to do so, directly, an old one will frequently manage, by repeated doubling, to elude its pursuer, and reach its form again by a circuitous route. Should it be closely followed by a fleet dog, it will make for a burrow, or a hollow tree, which has an opening at the ground into a cavity extending some distance above, up which it forces itself by bracing against the sides. Young rabbits are not so apt to double and attempt to turn back to their forms, but often run immediately to a tree; and an old one will sometimes take to a hole without much doubling, especially if it has before been chased and found refuge in the same retreat.

When seized, the grey rabbit never makes any attempt to bite. It utters a clear, sharp, wailing cry, like *que-a-a-a*, which is its only note, and is never heard except in distress. At other times, this animal appears to be voiceless except that in fighting, or playing together, the males produce a low, purring sound, scarcely above the breath. They also make a noise by stamping upon the ground with the hind-feet.

Like its congeners, this species has a very acute sense of hearing, and, when running, it stops and listens to any extraordinary sound. Though it has not good "bottom," its speed for a certain distance is great, enabling it to out-run almost any dog. It always travels by leaps, its powerful hind-legs and the immense muscles of the back enabling it to take long bounds, sometimes of 10 or 15 feet, in which it is but little aided by the weak fore-legs. It never appears to run or "trot," and, when it walks at all, as in eating, it rests the hind-feet upon the ground, only moving a short distance on a walk, and more generally hopping along by jumps of about a foot. This, like all other hares, is nocturnal, or, perhaps, more properly, crepuscular, moving about for food and amuse-ment chiefly by twilight, or on moonlight nights. It is frequently seen standing, however, on open ground in the sunshine, especially in spring and summer.

The position of the rabbit's feet, in running, is not always understood. I well remember my astonishment when, upon examining their tracks the first time, I found, as I thought, that they always ran backwards. For, the slight tracks of the fore-feet are really situated behind the larger and more widely separated prints of the long hind ones. As this animal springs, the fore-feet strike the sur-face near one another, while the hind-feet are spread apart and brought to the ground some distance in advance, outside of them; as these strike, the fore-feet, which have touched the surface but lightly, are lifted, and the spring is again made with the hind legs alone. In making the longest leaps, the forefeet strike in a line, one behind the other, and at some distance in the rear of the hind ones, as if they had been again raised before the latter had touched the surface.

Rabbits are very active, moving about at all times, except in very cold and stormy weather, when they keep close in their retreats, sometimes not leaving them for a day or two, and not unfrequently lying in their forms in the tall grass completely buried under the snow. Wherever two or three of these animals occupy a neighborhood, long well-worn paths may be found beaten in a single night, after a light snow in mild weather. Particular paths are used even when there is no snow, the same track being travelled repeatedly by one or more individuals.

The food of the grey rabbit is grass and other herbage, the tender shoots of briars, and various shrubs, as well as the buds, twigs, and sometimes, perhaps, the bark of trees. I have never observed that it gnaws hard-shelled nuts, like those of the hickory, though it is said to eat chestnuts; nor does it generally, if ever, dig through the snow for food. It does not hold food in the paws when eating, like many rodents, nor does it usually sit erect upon the tarsi. The domesticated rabbit, in eating a twig, holds it in its lips, and continues, without laying it down or ceasing to masticate rapidly, to cut off pieces from the end with the incisors, until the whole is devoured. This species doubtless eats in the same manner.

Rabbits are sometimes quite injurious in gardens, by devouring young plants of beans, cabbages, lettuces, and all kinds of vegetables; and where very abundant, they occasionally damage harvest fields, though they do not appear to feed very generally upon ripened grain. But their most serious injury is the destruction of fruit-trees, by cutting off the shoots and, perhaps, sometimes gnawing the bark. Their damage to fruit-growers in this way is at times very great, and leads to bitter complaints. When the ground is covered with snow, they enter gardens and nurseries and bite off and devour small shrubs and fruit-trees; or, if the snow be of sufficient height to enable them to reach the branches of orchard-trees, these, too, are eaten and their tops sadly disfigured. The branches are taken off with the rabbit's incisors so smoothly as to leave the appearance of their having been cut with a knife, and more than one orchardist has wrathfully sought the persons who "stole scions." Rabbits are said to kill fruit-trees by gnawing the bark from the trunks, and in this manner to have utterly ruined large and valuable orchards. Fortunately, however, this reported bark-gnawing appears to be generally, if not always, done only when the rabbits cannot reach the buds and branches upon which they prefer to feed, eating the entire branch. In hunting these quadrupeds, every winter, and working every summer, for ten years, in a very large nursery of fruit-trees, where they were numerous, I have never seen a tree from which bark had been gnawed by them, though thousands were severely "pruned," the rabbits, in deep snows, appearing to feed entirely upon the twigs and buds of the young apple trees. From the larger limbs they cut off the buds, of which they are fond; and in the woods, in winter, they can be tracked to living forest trees, recently felled, to which they repair to feed upon the buds. They also feed in winter upon the buds and young shoots of briars, sumach, hazel, thorn, oak, hickory, basswood, poplar, and other shrubs and trees.

It is highly probable that, injurious as rabbits are considered, by gnawing bark, the mischief charged to them is often, if not generally, done by meadow mice alone. It must be remembered that in deep snows the arvicolæ can readily climb some distance up the trunks of the trees, and I have frequently observed them to gnaw bark at a height of two feet or more from the ground. If these animals do gnaw the bark of fruit trees, as reported, it must be when they cannot reach the limbs or obtain any other food. A gentleman living on a prairie farm in Northern Illinois, informs me that, though many rabbits frequent his orchard throughout the year, he has never had a single tree barked by them; and in such a situation they might certainly be expected to gnaw bark, if ever. Though I am inclined to believe that they do not injure the farmer by bark-gnawing to the extent usually supposed, yet I by no means wish to defend them from the just charge of committing great havoc in nurseries and gardens by biting off young plants; but would rather suggest that the true criminals—the meadow mice—be destroyed, as the best means of checking the evil.

The grey rabbit is very prolific, producing young three or four times a year, and usually from four to six at a birth. In open ground the female scratches a shallow hollow, in which to bring forth her young. In this she forms a nest of soft leaves and grasses, well lined with fur from her own body; and when she is absent, the young are always completely covered and concealed in this nest, which they leave at an early age, and separate from the mother as soon as able to take care of themselves.

It is pleasant to observe that an animal usually so timid and unresisting will fight bravely for its young. A naturalist tells me that he once saw a grey rabbit attack a large black snake, which was holding one of her young in its coils. She fought by springing over the snake, and striking back with her hind-feet, which is the usual mode of defence of this species. Her blows were delivered with force and precision, and so rapidly that the snake was struck nearly every time, despite his attempts to evade them. As she passed, the snake aimed at her with his fangs, but though he often scratched off a mouthful of hair, he was plainly getting the worst of the battle, when the naturalist interfered. Another instance is related in which a rabbit was observed to pursue a hawk in the act of carrying off her young.

The grey rabbit is not only preyed upon by various carnivorous mammals, but by many rapacious birds found here, as well as by the larger snakes. The musteline mammals or animals of the weasel family, are the most to be dreaded. They search out the retreats of these animals, and, as most of them can enter wherever the latter pass, they readily follow, and kill them unresisted. I suspect the little brown weasel (*Putorius cigognanii*) subsists largely upon them in winter in this region, as does the larger white weasel (*Putorius noveboracensis*) which is also said to be their worst enemy at the East. I have repeatedly observed the track of the common mink for a great distance, as it wound about logs and brush-heaps, often entering hollow trees and burrows, sometimes following a rabbit's track, till finally I have come to where an unhappy victim has been pulled down from a tree in which it had in vain sought refuge. In Northern Illinois, numerous cats, which have escaped from domestication, and live in the woods like wild animals, frequently prey upon them. Among the birds, the great horned owl is noted as a successful rabbit-catcher. The white owl occasionally seizes one, in winter, as it sits on its form on the prairie; and the red-tailed buzzard, or "hen-hawk," as it is called, frequently swoops upon one of them in summer. Their young are destroyed in great numbers, as they fall an easy prey to any animal which finds them, when too small to escape by flight; and a large proportion of the whole number produced are probably thus doomed before the period of maturity. Many rabbits are infested by the larvae of a large gadfly, (oestrus,) and are hence said by hunters to have the "wolf." In their fur live astonishing numbers of a peculiar flea, apparently differing from the common species.

In cultivated districts, where many of the natural enemies appointed to check their increase are destroyed, the rabbits frequently multiply to such an extent as to render their extermination a matter of importance. Then they are easily trapped or snared, and may readily be poisoned by arsenic or strychnine, placed in a bait of apple, turnip, or other vegetable; but the most effectual mode is to encourage the hunting of them.

As before stated, the grey rabbit often has his form situated in the tall grass, at the edge of the prairie, or in sloughs running into the woods. By walking along between these and the trees, where there is generally a space clear of cover, while a dog beats the grass beyond, one may get a shot at them, as they will almost always make straight for the woods. The rabbit will generally "lay" to the dog, giving him a fair chance for a "point," so that one may come up and take a shot as he goes off in a direct line, if that is preferred to a cross shot. If he cannot be brought down at first, the dog, by following on the track, will start him the second time, when he may be shot as he comes back, unless the dog should compel him to retreat into a tree. The finest shot should be used in shooting rabbits, for they are very easily killed, and generally drop at a slight wound. I have, more than once, shot one, however, without injuring a bone, when he would run half a mile, and then fall dead without a struggle. A more primitive mode of hunting them, I believe, is practised by boys, which is to go armed with a small sharpened pole and some matches, accompanied by a dog to chase them into hollow trees. Sometimes the hole in the tree is such that one can reach the animal with his hand, or pull him down with a short hooked stick: but when he is out of reach, and the boy without an axe for cutting into the hollow of the tree, a stick is introduced to "poke him out", and shortly after, in an agony of fright and pain, he rushes down the hollow, and the boy quickly grasps the legs of the captive. When the game cannot be brought down with the stick, leaves are collected and fired at the entrance of the hollow, and in a short time, the suffocating animal unavoidably descends.

When chased on the prairie, if there are no stacks of hay nor grain under which to find refuge, the rabbits take to the long, heavy sedge grass (carex) in the sloughs, where, by doubling and shifting about, they generally elude the pursuit of the dogs. They are also snared in great numbers upon their path-ways. The following is a very simple but successful method of capturing them: A small thickly branched tree is felled across the path-way, with the limbs so arranged as to leave but a single narrow passage; an elastic sapling is then bent down over this, and tied by a cord to the fallen tree, or to a hooked peg, driven into the ground, at the side of the opening; this is not tied to the peg by a common square "hard knot," but only with what is called a "single bow-knot," so that the pulling of the end of the cord frees the whole. In order to prevent the strain given by the bent tree from pulling it out, an enlargement is first formed by knotting the cord just within the point at which it passes from the bent tree, under the part of the cord passing around the peg, so that, although this protuberance does not permit the cord to be drawn through from above, neither does it interfere with the loosening of the knot by drawing out the bow, if the other or lower end be pulled. This lower loose end of the cord is formed into a noose a little larger than a rabbit's head, and placed open in the path, so that the animal, in attempting to pass, readily puts his head through, but in his endeavor to force through his shoulders not only tightens the noose around his neck, but pulls out the bow; thus loosening the knot, when the bent sapling, being freed from its attachment, springs up and breaks the rabbit's neck, or suspends him until he is strangled. A very smooth, tightly-twisted cord should be used; the noose is sometimes formed of brass wire, which keeps its position and slips easily, and is not liable to be cut by the animal before entering, like the cord. A little practice is necessary in learning to arrange the knot, so that, when loosened, the noose will not be drawn up on the wrong side and entangled; and the arrangement of the whole will be better understood after a few experiments. In consequence of the rabbit's well known habit of travelling in accustomed paths, which may be discovered even in summer, it is not necessary to use any bait, though pieces of apple, parsnip, or cabbage placed in the path on each side of the snare might more fully insure success; and the snare may also be set at the entrance of a little pen, or hollow tree, in which is placed a bait. I learn from a gentleman of Pembina that, on the Red River of the North, the Indians subsist, in hard winters, when game is scarce, almost wholly upon hares caught in this way. Grouse, quails, and many other animals can also be successfully snared; and it is said that even the moose and the deer have been caught in snares constructed on a larger scale. Hares may also be caught in steel traps and "dead-falls"; and, in fact, they will enter almost any kind of trap.

The grey rabbit frequently takes up its abode about farm-yards, and I have often observed individuals living all winter under stacks and buildings situated within a few rods of dwellings, making nightly sallies into the garden, greatly to the injury of the plants, many of which they destroyed. In one instance, within my observation, a mink did good service, and amply paid for the two or three fowls he consumed, by ridding a farm-yard of several rabbits which had thus taken up their quarters under the barn and hay stacks, and were making sad havoc among some choice plants in the flower garden. As long as the mink remained, no rabbits were observed on the premises, though before and after his visit their tracks were seen in every direction, despite the presence of two dogs accustomed to hunting. Where rabbits are troublesome, and no fowls are kept, the presence of minks and weasels is desirable, especially of the weasels, the good offices of which in the destruction of rats and mice, both in the field and farm-yard, often save a single farmer more than the value of all the fowls destroyed for years in a large neighborhood; yet there are few who ever willingly spare the life of a weasel. Indeed, it is frequently killed while in the very act of hunting the far greater enemies of agriculture. A gardener once expressed to me his satisfaction at having slain several garter-snakes and a green snake, which had caused great alarm and discomfort about his home. They infested his rose-bushes; but the good gardener knew not that they resorted thither to destroy the green slugs of which he had so long complained, and that the snakes themselves were harmless to man. So, too, whoever kills weasels on his farm, at a distance from the poultry, might find it profitable to consider what they feed upon.

Grey rabbits sometimes form a considerable item of human food, and are sold in our city markets, in winter, in large numbers at a price of 10 or 12-1/2 cents each. The flesh is rather dry, and without much flavor, and is generally not deemed eatable in summer.

〰〰〰 ROBERT KENNICOTT (1835–1866) DIED MUCH TOO YOUNG, but even in his short life he was a key actor in the early efforts to study the natural history of this region. Born to supportive and prominent parents, Kennicott manifested his strong interest in zoology as a child. Like a number of other naturalists in this anthology, he was sickly as a youngster and thus allowed uncommon freedom to engage in his own pursuits, which for Robert was either searching the family homestead ("The Grove," now a park in Glenview, Illinois) and its environs for specimens or reading and writing. At the age of seventeen, he moved to northern Ohio to study nature-related subjects under the celebrated Dr. Jared P. Kirtland, for whom at least one snake and one rare warbler are named. (It was in fact Robert Kennicott who so honored his former teacher with respect to the reptile, an individual taken at the Grove.) He also started writing and sending specimens to an even more renowned scientist, Spencer Fullerton Baird, of the Smithsonian Institution. (We can see the strong role of Robert's father, John, in getting his gifted son connected with all the right people.)

In between his extensive collecting trips to such destinations as Washington, D.C., southern Illinois, and arctic Canada, Kennicott kept busy at home. He wrote the first annotated faunal list for northeast Illinois, a report that appeared in 1855 in the *Transactions of the Illinois State Agricultural Society*, a journal edited by his father. Two years later he published an account on the state's mammals for the same publication.

Kennicott also studied medicine in Chicago, but he is best known for his work on behalf of Northwestern University and the Chicago Academy of Sciences. Northwestern then held more natural history material than any other institution in the West. Kennicott enlarged it through his own collections and acted as curator. Later he shifted his loyalties to the Chicago Academy of Sciences, which in his view should become the "Smithsonian of the West." (Unfortunately, as mentioned earlier, it burned to the ground in 1871.)

In 1865 Kennicott embarked on his second arctic expedition, this time as leader. Their mission was to construct a telegraph line from San Francisco through Russian America, where a line would eventually cross the Bering Strait to Russia and beyond. Never endowed with robust health, Kennicott died on the banks of the Yukon River a year later. While the country lost one of its finest naturalists, the efforts that he and his colleagues expended in exploring this vast northern land made it easier for Congress to ratify "Seward's Folly."

Source: Jack White, "Ecological and Cultural Significance of the MacArthur Foundation's Property at the Grove" (1995), unpublished report by Ecological Services, Urbana, Illinois, to the MacArthur Foundation.

Recollections of Bird-Life in Pioneer Days
Halvor L. Skavlem

Some of the most lasting and vivid impressions of my boyhood,—I may well say childhood days—relate to and recall pictures of bird-life in Southern Wisconsin, somewhat more than half a century ago.

We hark back to the time of the ponderous slow moving, breaking team, consisting of five to seven yoke of oxen, hitched to a long cable of heavy log-chains attached to a crudely but strongly built "Breaker," with a beam like a young saw-log and a mould-board made of iron bars that turned over furrows two feet or more in width.

Those great unwieldy breaking teams, consisting of 10 to 14 large oxen, are yet distinctly outlined on memory's page, and reminiscently, I see them crawling like some huge Brobdignagian Caterpillar around and around the doomed "land" — "land," in breaking parlance, being that piece of the wild selected for cultivation,—leaving a black trail behind, that, day by day, increased in width, bringing certain ruin and destruction,—absolute annihilation,—to the plant habitants who had held undisputed possession for untold centuries.

The mild-eyed, slow-moving ox teams were not only instruments to the destruction of the centuries-old flower-parks of the wilderness, but with them came tragedies in bird-life, resultant from the inevitable changes from nature's rules of the wild, to man's artificial sway. Often in preparing or

Halvor L. Skavlem, "Recollections of Bird-Life in Pioneer Days," *By the Wayside* 13, no. 10 (June 1912): 57–59.

planning for the breaking of a new piece of land, the same was guarded from the prairie fires of the fall and early spring, so that it could be "fired" at the time of breaking. This would commence the latter part of May and continue on through June and July, covering the nesting season of the numerous species of bird-life, that had for untold generations made this beautiful park region of the Rock River Valley, their summer home.

It was in the early fifties that I, then a little tow-headed tot, chased butterflies and gathered armsfull of prairie flowers, at the same time "spotting" bird nests of many and various kinds, on a piece of land destined to be civilized by the big plow that very season.

I distinctly remember the large eggs of the "Prairie Snipe" and the still larger ones of the "Crooked-bill" or "Big-Snipe." The former I later learned to know as BARTRAMIA LONGICAUDA [Upland Sandpiper], and the latter, long after they had entirely disappeared, I found had the book name of NUMENIOUS LONGIROSTRA, or LONG BILLED CURLEW. These snipes were so numerous at this particular season, that a bird student might have been misled to the conclusion that they were nesting in colonies. But, undoubtedly, the true explanation was that this protected piece of prairie with its dead grass unburned, was the ideal condition for the ground-nesting prairie birds.

The snipe were not the only birds that appeared in unusual numbers, but all bird-life seemed to regard this particular piece of land as a perfect paradise for a summer home.

Bob-White would mount the top of a dead sumach and call to his mate,— "Wheat—most—ripe" "Wheat—most—ripe," while she sat patiently brooding the nest-full of snow-white eggs in the thick bunch of dead grass nearby.

Near the little knoll at the farther side of the prairie, where earlier in the season the Prairie Chicken clan held their camp-meeting when many a lively scrap between the gallants of the company was settled to the entire satisfaction of the coy hens who would always give expression of their approval with a timid "ye-es,—ye-es—yes, yes, yes,—ye-es," these same matronly hens were now quietly tending their domestic duties, silently slipping off and on their well-filled nests even so cunningly hidden under the tufts of dead grass. Some of the nests were already far advanced towards that stage when the peeping egg should announce the arrival of the covey of young chicks; indeed, some of the most enterprising ones had already added their quota to the bird census of the season.

The patches of hazelbrush that looked like tiny islands of green set in a field spangled with the many colored gems of Painted-cups, Pinks and Blazing stars, were densely populated with a variety of bush-loving birds. Conspicuous among these were the Brown-thrashers and Cat-birds, who opened the morning services at day-break with bird melody rivaling the overrated Avian Opera of the old world.

Evening vespers were softly chanted by the Robin and the "Vesper-bird"; "Cheewinks" rustled in the dead leaves that mulched the hazel-groves, while untold and unknown varieties of just little "ground-birds" and "bush-tits" animated every nook and corner of this bird paradise, during the long June days away back in the early Fifties of the last century.

This is but a repetition of the annual picture of this favored locality,— during the preceeding years, decades and centuries,—when nature's rules were supreme, before the Paleface's Art and greed and their Chief Manito, Mammon had invaded the sacred precincts of this part of the natural world.

A slow-moving monster comes creeping up the trail over the picture of this pleasant June day. It is the great breaking team slowly and solemnly approaching the new-made home of the pioneer settler. The patient-looking oxen are unyoked and the driver with his great long whip playing a snapping tune that sounds like a scattering volley of pistol shots, "herds the cattle" with many a "haw" and "gee" to a nearby part of the common, where there is good "feed" and restful shade until they are "rounded-up" the next morning to continue their work of breaking the wilderness.

The time has now come to "fire the land." All conditions are favorable for a good "burn";—a clear warm afternoon, a gentle breeze away from the homestead;—the dry grass under the flower spangled green and dead leaves that mulch the hazelbrush will burn like powder.

All hands now set to work starting the fire,—pulling up great bundles of dry grass they ignite the outer end of the bundle and then run along the edge of the "land" scattering the ignited grass as they go, down one side and up the other. The little boy is all excitement helping pa with little bundles of dead grass because he too must act his part in the new order of things; and soon the land is all encircled with flame and great clouds of vapor-like smoke roll upwards and onwards signaling distant neighbors that they are burning "breaking-land" where new fields are being born.

But what of our bird friends, the old habitants of the land, Boh-White and his interesting family, the Prairie Snipe and their big eggs or their curious, odd-looking long-billed babies, the Brown-thrashers, Cat-bird, Bobolink and Lark, that filled the morning air with their songs of happiness and swelled with bird pride in anticipation of happy little families? What of the hundreds of happy bird homes that the morning sun brightened and warmed? All,—all are gone. A black, scorched and desolate scar profusely sprinkled with wrecks of nests, scorched eggs and charred bodies of little baby birds, disfigure the face of Mother Earth. Oh, could I but command the language of "Christopher North" or John Muir in word painting, I would BURN this horrible bird-tragedy into the brains of my readers,—young and old,—so they would never consent to the burning of grass or bush during the nesting season.

I doubt if anyone of the human agents of this pathetic bird-tragedy gave a single thought to the bird victims of their fire, or even noticed a single distressed and bewildered mother bird hovering over the smoking ruin of her family home.

It was not until the next day that the little boy realized the loss of his flowery play-ground and the many bird-nests that he had "spotted" with boyish ingenuity. He started for the "Big Snipe" nest but where was it? All his marks were gone, some of the large green plants were still standing, but scorched, blackened and wilted, DEAD, all DEAD. Here comes the big snipe, with silent but graceful motion she sails a circle around the distracted child, then utters her harsh call, indicating both anger and distress. Soon her fellow sufferers respond from all points of the compass and the air is full of the big long-billed birds angrily screaming and scolding, now and then making threatening dives at the thoroughly scared and crying lad. Grandpa comes to the rescue, and to soothe the troubled child he tells him he may pick all the eggs he wants. With his little homemade cap for basket, he starts his collection with the baked eggs of the big snipe and,—though his little bare feet are sorely pricked by the sharp stubs of the burned grass,—he soon fills his cap with eggs,—baked and burned,—large and small,—spotted, speckled and white. Grandpa now directs the way to the house and in his eagerness to show his treasure the boy starts on the run, stubs his toe and falls. Memory fails to tell what became of the eggs and cap, but I distinctly remember that Grandpa wore a blue peaked knit cap, doubled over on the side with tassel dangling from the tip end,—you can see a picture of it in Ross Brown's "Land of Thor."

〰〰〰 HALVOR L. SKAVLEM (1846–1938) WAS BORN AND GREW UP IN Newark, Rock County, Wisconsin (near Beloit), one county west of Walworth. His parents were pioneer farmers, and he was the product of a "common school education" that did not extend even to the end of high school; but like other early naturalists, he had the brains, curiosity, and drive that enabled him to attain a high level of achievement. At the age of fifteen, he walked to Iowa, where he began teaching at Fort Dodge. He had the money somehow to buy property that held both a strip of timber and a house that became the school. So in addition to his $45-a-month salary, the school board paid him on a monthly basis $10 for rent and $75 for firewood.

In 1873 he returned to Newark Township, where he married Gunnil Olmstead and acquired a farm. While working his land, he became fascinated by Indian relics and the people who left them. The University of Wisconsin recognized his expertise by hiring him to conduct surveys of native mounds. Skavlem was particularly noted for his ability, said to be unique among white men, to make arrowheads that were indistinguishable from originals.

Some years later he became county sheriff and moved back to Janesville. On two occasions he stood down mobs, promising that although they might lynch the prisoner or blow up the dam, he would kill some of them in the attempt. His public service also included membership on the boards of the county and the city public library (where he was librarian for three years).

Running through his other activities was a deep interest in natural history. Skavlem amassed one of the largest private bird collections in the state, as well as an extensive herbarium of local plants. (The former was donated to the Janesville Public Library and the latter to the University of Wisconsin.) Preparatory to the World's Columbian Exposition of 1893, the Smithsonian Institution hired him to collect mushrooms for the museum's display at the fair. He also published several articles and possessed at the time of his death a library of fifteen hundred volumes across a wide range of disciplines.

Source: "Halvor L. Skavlem Dead at 98; World Known Authority on Indian Lore and Botany," *Janesville* [WI] *Daily Gazette,* January 5, 1938.

Birds of Northeastern Illinois
Edward W. Nelson

᭑᭞ ᭑᭞ ᭑᭞ Alice's Thrush [Gray-cheeked Thrush]: Very abundant migrant; frequently open woods and the borders of adjacent fields. May 1 to 20,

Edward W. Nelson, "Birds of Northeastern Illinois," *Bulletin of the Essex Institute* 8 (1876): 90–155.

September 1 to October 5. I have rarely heard this species sing except during damp, gloomy days in spring, when trees and bushes were dripping with a fine misty rain. On such occasions, I have often been greeted by the clear metallic notes of this thrush rising clear and strong, filling the air with a sweet, indescribable melody, and then dying away in measured cadence until the last notes are scarcely distinguishable. As the first strain ends, the sound is re-echoed by hidden musicians on every hand, until every tree seemed to give forth the weird music.

Brown Thrasher: Common summer resident. Arrives April 20, nests the middle of May, and departs in September. That the nest of the species is often placed in trees and bushes for protection against some apparent danger I have no doubt, but in many cases this site is chosen from a mere whim of the bird. I have found in one such "scrub oak" grove, on a sandy ridge, some half-dozen nests for several seasons in succession, and each year about one-half the nests were in the trees, and the remainder were built at the bases of saplings or bushes, yet I could find no apparent cause for the location of the nests in the trees. The young were in each case reared with equal safety.

Golden Eagle: Not very common during winter. Arrives in November and departs early in spring. Formerly nested throughout the state. In December 1874, while hunting prairie chickens in a field a few miles south of Chicago, my friend, Mr. T. Morris, was suddenly attacked with great fury by a pair of these birds, they darting so close that had he been prepared he could easily have touched the first one with his gun. As it arose to renew the attack he fired a small charge of number six shot, and brought it down dead. The second then darted at him, and so rapidly that he did not fire until it had turned and was soaring up, but so near that the charge passed through the primaries in a body disabling but not injuring the bird, which was then captured alive. The cause of this attack was explained by the proximity to a carcass upon which these birds had been feeding. The craw of the dead eagle contained a large quantity of carrion, as I learned upon skinning it.

Western Piping Plover: Very common summer resident along the lake shore, breeding on the flat, pebbly beach between the sand dunes and shore. Arrives the middle of April and proceeds at once to breeding. . . .

Some thirty pairs were breeding along the beach at [Waukegan] within a space of two miles, and I afterwards found the birds as numerous at several points along the shore. Every effort was made to discover their nests without success, although the birds were continually circling about or standing at a short distance uttering an occasional note of alarm. The first day of July, the year previous, Dr. Velie obtained young but a very few days old at this same locality, showing that there is considerable variation in the time of breeding. This was also shown by specimens obtained the last of May, and which I think were later arrivals than those found breeding in April having the ova just approaching maturity.

Wilson's Snipe: Abundant during the migrations and not a very rare summer resident....

Morning and evening and throughout cloudy days in the early part of the breeding season the male has a curious habit of mounting high overhead, then descending obliquely for some distance, and as it turns upward strikes rapidly with its wings producing a loud whistling sound with each stroke. This maneuver is repeated again and again, and appears to be performed for the same purpose as is the "booming" of the night-hawk. Besides this sound the Wilson's Snipe has a peculiar, sharp cry during this season, which is uttered when the bird is disturbed. I first became acquainted with this note in May 1876, when while walking along a marshy strip of land, I was surprised to hear a loud ka-ka-ka-ka-ka, uttered with great force and in a rather loud, harsh tone. Turning quickly I was still more astonished to find the author to be one of these birds. It was flying restlessly from post to post along a fence and showed the greatest uneasiness at my presence, the notes being repeated at short intervals. Although its nest was probably near, I could not discover it.

Field Plover [Upland Sandpiper]: Very common summer resident. Arrives early in April and departs in September. Frequents in greatest abundance the borders of marshes and half wild prairies. Quite difficult to approach when it first arrives, but during the breeding season becomes perfectly reckless, and hovers over head or follows through the grass within a few yards until it has escorted the intruder well off its domain. The presence of a dog in the vicinity of its nesting place is the signal for a general onslaught by all the birds of the vicinity, which hover over the dog, and with loud cries endeavor to drive it away. Being but little appreciated as game it is seldom hunted in this vicinity.

Esquimaux Curlew: Rather common during the migrations. Arrives a little later than the [Long-billed Curlew] and passes north with short delay. Returns the last of September and in October. Frequents wet prairies with the Golden Plover.

Night Heron [Black-crowned Night-Heron] Common: Owing to its fre-
quenting the almost impenetrable wild rice swamps this species would be
overlooked on a transient visit to their haunts. The first of July 1874, I saw a
few young of the year in Calumet Marshes, but it was not until June 1876 that I
learned anything regarding their habits in this state. The middle of this month, in
company with my friend Mr. T.H. Douglas, I visited Grass Lake, Lake County,
Illinois, some miles west of Waukegan. This "lake" is simply a widening of the
Fox River, which flows through its center, producing a shallow body of water
a mile wide and about three miles long. A large portion of the lake is covered
with a dense growth of wild rice. While collecting near a large patch of this
we were surprised to see a number of night herons arise from the interior of
the patch and commence circling about uttering hoarse cries. Upon examining the place we were still more surprised to find that the birds were breeding
in this apparently improbable location. During this and the following day we
examined, within an area of two acres, at least fifty nests of this species. They
were all placed in the midst of a particularly dense bunch of rice, the stiff last
year's stalks of which, converging slightly near the roots, formed a convenient
base for their support. The nests were all well-built structures, composed of
innumerable small pieces of dead rice stalks, varying from two to ten inches
in length. Some of the nests were quite mathematically built, the material being arranged so that the usual cylindrical form would become either a decided
pentagonal or hexagonal figure. The nests averaged from twelve to fifteen
inches in diameter at the top and from ten to thirty inches in depth. So firmly
were they built that I several times stood upon a large nest to take a more extended view, and did it but little damage. A few contained fresh eggs, and a
few had young from nine to ten days old, but the majority contained eggs with
half grown embryos. The parents exhibited great solicitude when we were in
the vicinity, but were so cautious that we succeeded in shooting but two.

Short-tailed Tern [Black Tern]: The following notes upon the breeding
habits of this species comprise my observations during the last two seasons,
during which time I have examined between two and three hundred nests. In
nearly every instance the eggs are deposited in a well-built nest formed of the
surrounding material. In prairie sloughs the nests are generally located well
out from shore, in from one to two and one-half feet of water, and in the midst
of the fine wiry grass growing in such places. In such situations the nests are
formed of a mass of the surrounding grass, consisting of both living blades
and the dead straws floating in the water. These are heaped into a conical

mass, upon the apex of which, resting but an inch or two above the surface of the water, the eggs were placed. As would be supposed these structures were often quite bulky. In one instance I collected all the eggs deposited in a small prairie slough, and upon visiting the place about a week later found the birds had built smaller nests in shallow water and deposited a second set of eggs. These were removed, and upon a third visit I found many of the birds were nesting upon the masses of dead weeds or upon old muskrat houses. The sets taken from the above nests averaged as follows: first, three eggs; second, two eggs; third, one egg. When the nests are built upon a small lake, where the water is too deep for their nest to rest upon the bottom, they generally build a slight nest of grass stems upon a floating bog, mass of dead reeds, or old muskrat houses, but a well built nest will be found in nine cases out of ten. Early in May, when farmers are ploughing near a place frequented by these terns, they often follow behind the plough and pick up the earth worms and larvae exposed.

An unfledged young one, which I once took home, became very familiar in a few hours, and would come, upon being called by a squeaking noise, and take a fly from my fingers. It was also quite expert at capturing flies upon the floor, but it was some time before it learned to distinguish between a fragment of dirt or a nail head and the insect. Although but little over a week old it could run rapidly from place to place and appeared quite contented with the change of quarters, and but for an unfortunate accident which caused its death would, I think, have been easily raised.

〜〜〜 EDWARD WILLIAM NELSON (1855–1934) WAS BORN IN AMOS-keag, New Hampshire. He lived in Manchester, New Hampshire, with his family until the Civil War took his parents—his father was killed and his mother became an army nurse in Baltimore. He and his brothers then went to live with their grandparents in the Adirondack Mountains of New York. It was here that Nelson developed his early interest in wildlife.

After the war his mother moved to Chicago, where she established a successful dressmaking business. The boys joined her in 1868, and Nelson began exploring the lakefront, the Calumet area, and other biologically rich locations in pursuit of specimens. Birds, fish, and insects were all worthy of his attention. His insect collection was particularly large, and he might have become an entomologist had it not been for the Chicago Fire of October 1871. Forced to leave their home, the Nelsons joined the streams of people fleeing the fire. Edward had tied the boxes that housed his insects in a package that he could hold in one hand. He laid his treasure down on the street for just a moment, but when he reached to retrieve the boxes, they had already been pilfered.

Nelson's only formal schooling was from 1872 to 1875, when he attended the Cook County Normal School, although he later received honorary master's and doctoral degrees from Yale and George Washington, respectively. Afterward he taught at Dalton, Illinois. (I wonder if anyone associated with the town's schools today realizes what a distinguished scientist once toiled there so long ago.) But he soon concluded that fieldwork was more to his liking than pedagogy, so he left Chicago in 1876 for Washington, D.C., where he hoped to find employment with the Smithsonian Institution. Despite his relatively short tenure in the Chicago area, Nelson produced two papers on Illinois birds (one on northeastern Illinois and the other on southern Illinois) and two on Illinois fish (one dealing with the entire state and the other on "Chicago and vicinity").

By the time Nelson embarked for Washington, D.C., he had already amassed a large group of prominent personal contacts, including E. D. Cope (paleontology), Robert Ridgway (ornithology), and Spencer Fullerton Baird (director of the Smithsonian). He failed to get an appointment to the Smithsonian, but Ridgway and Baird did get him a position as a weather observer on an expedition to Alaska. While in the arctic, he made important contributions to the fields of biology, ethnography, and geography. (His penchant for gathering tribal art and other common items prompted the native people to call him "the man who collected good for nothing things.") According to biographer Margaret Lantis, "in his time he was unique among Alaska field workers in the fullness and accuracy of his observation, and probably has not yet been surpassed in effective range of interest."

Nelson returned from Alaska in 1881, and the remaining fifty-three years of his life were so varied they are difficult to summarize. While writing up his Alaskan research, he developed tuberculosis and eventually moved to Arizona, where he recuperated under the care of his mother. To the extent that his strength allowed, he stayed busy, establishing a cattle ranch with his brother and serving as county clerk of Apache County. In 1890 he joined the U.S. Biological Survey as a special field agent. He was with the survey for forty years, serving as chief from 1916 to 1927. During his tenure with the agency, he conducted field studies in California, Guatemala, and every state in Mexico. He also published extensively (including his landmark "The Rabbits of North America"), negotiated international bird treaties, sponsored national legislation to protect birds, and established the U.S. Reindeer Experiment Station in Fairbanks. According to a compilation made in 1935, 107 taxa of plants, insects, reptiles, amphibians, fish, mollusks, birds, and mammals were named after him, as well as a lagoon, island, and mountain range. The recently split Nelson's sharp-tailed sparrow also bears his name.

Sources: Margaret Lantis, "Edward William Nelson," *Anthropological Papers of the University of Alaska* 3, no. 1 (1954): 5–16; Edward A. Goldman, "Edward William Nelson—Naturalist," *The Auk* 52, no. 2 (April 1935): 135–48.

Letter [Least Bittern]
X.Y.

᭤᭤᭤ One morning as we were returning from a pleasant and successful hunt at Hyde Lake (connected to Wolf Lake, just east of the Calumet River) my companion threw up his gun to shoot a small bittern which was clinging to a stout stalk of rice. Noticing the constrained and unnatural position of the bird upon the rice stalk, I threw up his gun, at the same time calling his attention to the circumstances. As we approached he did not fly but fluttered upon his swinging perch, held as it were by some unseen agency. Running the boat up to him, we found that he had struck the rice stalk with his sharp spear-like bill, splitting it, and his head passing on through, the halves of the stalk had sprang close up to his neck and hung him. He was grasping the stalk with both feet and in this way had saved his precious neck. [The author rescued the bird, which was so weak it needed a drink of water before it flew off.]

Tragedy in the Home of the Owls
Edward B. Clark

᭤᭤᭤ One night last week a thunder storm broke over the southern part of the city.... Then came a gust of wind that shrieked and tore with all the fury of a miniature cyclone.

X.Y., "Letter [Least Bittern]," *American Field* (May 8, 1886): 436.

Edward B. Clark, "Tragedy in the Home of the Owls," *Chicago Tribune*, November 10, 1896.

On Lake Ave. is an old homestead that dates back to the 1850s. On it are some of the oaks that thickly covered portions of Hyde Park before people preferred lawns to trees. Two of these oaks were landmarks. They were old when the Fort Dearborn massacre took place. One by one their branches had decayed and dropped off. Then the owners hand-trained ivy up the gnarled and stubborn trunks and they stood side by side in all the dignity of a borrowed green old age.

Further back than the memory of the oldest inhabitants one of these oaks had been the home of screech owls. The upper third of the massive trunk was hollow and in the hollow lived a family of these curious birds. Perhaps they lived there the year round, but at any rate every spring a pair of owls put in an appearance. If they were not the same pair no one could tell the difference. They may have been successive generations of the original pair that were there when the old homestead was built, but one owl differed not from another in oddity of appearance or melodiousness of screech. The first warm days of April every year would find two funny little balls of brown feathers at the holes in the old tree. At dusk some one would be startled by a swish of wings and a vague shape that flitted among the oaks and maples and balms of Gilead. Later in the evening, a plaintive screech would go up at frequent intervals from the shadow of the big willows and some one would say: "Spring has come, sure; there are the owls."

And if further proof were needed, next day the old place would be alive with red-headed woodpeckers, which drummed ferociously on the old trunk and screamed discordant protests from the other old oak. There has been war to the death between the owls and the woodpeckers over that oak for nobody knows how many years.

Then every summer about the middle of June, some evening just at dusk the two old owls could be found sitting on a limb of a maple. They are about as big as a boy's mitten and dark brown. They wear an air of conscious pride and are dignified beyond all telling. Then, if one looks about carefully, he will find not far away, the cause of their pride and dignity. Big balls of fluffy light brown feathers will be seen perched on a limb in the midst of some thick bunch of foliage. There are usually three young owls and never less than two.

When the miniature cyclone came shrieking out of the west that night a tongue of wind went straight through the old homestead on Lake Avenue. And the owl tree was directly in its path. The wind littered the lawn with broken twigs, carried away the top of a sturdy young oak, and then swooped down on the old owl tree.

The giant's hour had come. Its massive trunk was torn from the ground and hurled prostrate across the driveway. It fell with the force of a steam hammer, and hollowed by age, burst into a thousand pieces.

In the morning a pitying hand pulled aside the fragments and there in the hollow which for generations had been the home of the screech owls, were three little balls of light brown feathers flattened and crushed as if the wind had wreaked a permanent spite. The same hand drew aside the other fragments in search of the old owls. But nothing more than a patch or two of dark brown feathers could be found. But with the feeling that some one or something was watching him, the searcher looked up and there, in the balm of Gilead, huddled close in to the trunk, sat the two old owls side by side, wing touching wing. They were watching the searcher with wide blinking eyes.

And the searcher felt as if he had come unwittingly upon a tragedy.

Birds of Lakeside and Prairie
Edward B. Clark

WHERE THE BLACK TERN BUILDS

᭰ ᭰ ᭰ The little village of Worth lies just beyond the smoke of factory-filled Chicago. It is on the marshes of Worth that the black terns build their nests; it is in the thorn thickets that hedge the pastures that the loggerhead shrikes make their homes; the rails, the redwings, and the wrens haunt the reedy swamps; and the hawks and the crows live in the heavy timber. Outside of a race-track and the many birds that flock in its fields, Worth has few attractions to offer. The race-track draws thousands of people daily for a short season, but the birds' visitors are few. In no other place, perhaps, so near the great city, could the black terns nest in peace. Certain it is that Worth is the only place readily accessible to the city bird-student where these "soft-breasted birds of the sea" may be found during the season of courtship and house-keeping. Black terns are abundant in the shop windows and upon the hats of thoughtless women. The shop birds and the bonnet birds are wired and twisted into positions of grotesque ugliness. There never was a line of beauty in the stuffed bird of a milliner. Would that woman could see it! The

Edward B. Clark, *Birds of Lakeside and Prairie* (Chicago: A. W. Mumford, 1902), 87–93.

black terns of Worth are living; the sweep of their wings is as graceful as are the curving blades of the swamp flags. There is a price upon the head of the black tern because the milliner covets the bird that it may be used as a means for a second temptation of woman. Neither the black tern of Worth nor the Wilson's tern nesting in northern Wisconsin can long survive the demands of fashion for which the word cruel is far too feeble an adjective.

I wandered one late May day through the music-filled fields of Worth. My destination was the Phillips farm, which lies about a mile from the depot. The orioles were whistling wherever a treetop offered a swaying perch. The meadows were literally filled with singing bobolinks. I passed a little country school-house; the children were singing the opening song of the morning. On the ridge-pole above them was perched a black-throated bunting, who was adding his mite of music to swell the chorus. A little farther on I made the acquaintance, that morning, of the grasshopper sparrow. It is a tiny field-loving bird, with a song which much resembles the sound made by the insect for which it is named. One of the sparrows took perch on a slender weed which its weight was not sufficient to bend, and there gave me a sample of its vocal power, though, perhaps, I might better say vocal weakness. It will not do, however, to despise the grasshopper sparrow's song, for some day when greed has caused the killing of all the larger birds we may turn for enjoyment to this humble little feathered rustic.

On either side of the Phillips farmhouse there is an orchard, while hedges that do duty as fences extend in all directions. On that May morning at the end of the porch there were four wild rose bushes in full bloom. A syringa, with its burden of white blossoms, flanked the line of roses. In the syringa bush a catbird was singing, and strangely enough, he forgot to throw into the midst of his melody the harsh note that so often mars his performance. I stood for a minute enjoying the bloom of the roses and the song of the bird. The singer left the discordant element out of its song, to be sure, but discord came in the shape of an English sparrow, who viciously attacked the catbird who had been presumptuous enough to lift its voice in a British sparrow's presence. The American fought faithfully, but it was no match for the heavy-beaked alien. I drove the sparrow away. A few minutes afterward I found its big bulky home in a cherry tree. I tore the nest down and destroyed the eggs. Cruel? Not a bit of it. Cruel to one kind of bird, perhaps, but kindness to an hundred others. Go thou and do likewise.

At the end of a little lane that leads pastureward from the house is an Osage orange, half tree and half shrub. It is the sole surviving corner-piece of two hedges of bygone days. In this growth was a nest of the loggerhead shrike. This bird spends its winters in the South, but comes to this latitude to breed, replacing here the great northern shrike which comes from the far North in the winter and scurries back Arctic-ward at the first suggestion of spring. The loggerhead lives on small birds, small snakes, and large insects. Being a predatory creature, it supposedly should be possessed of some courage, and yet here was a loggerhead shrike that had five dependent young ones in its nest, and still did not dare to come within a field's width of its home while trespassing man was about. A robin or a jay would have been at the post of danger, and if it could have done nothing else, would have roundly berated the intruder. The loggerhead sat on the faraway fence-post and was apparently perfectly unconcerned while effort was made to peek into its nest. Some friends who had joined me undertook to take a snap-shot of the shrike's home and young. The nest was so well fortified with twigs and branches, each of which carried a score of thorns, that the photographing process was beset with difficulties. To the right of the nest, pierced through the neck and hanging from a thorn was the half-eaten body of a small snake, placed there by the shrike perhaps to provide the larder against any future scarcity of living game. As soon as we had left the vicinity of the nest the shrike went back to its young and doubtless gave them each a bit of snake steak to make them forget their fright.

The Worth marsh, which stretches away for acres from the foot of the orchard, is a fruitful field for the study of bird-life. When we had opened the old-fashioned gate at the lane's end, we could see a glistening patch of clear water far beyond the rushes' tops. The dark forms of birds were wheeling about above its surface and their cries were borne down to us by the breeze. We skirted the marsh and approached the open water, and there through our glasses had a perfect view of the darting birds. They were dark, almost black, but there was a gloss to their feathers which the sun's rays let us see from time to time as the birds kept up their changeful flight. They were black terns that had left the waters of the larger lakes to come to this place of sedges to rear their young. The red-winged blackbirds nest by hundreds in the reeds of this great swamp. At the time of our visit the nesting season was at its height. As we walked into the swamp regardless of mud and water, the male redwings met us and hovered over our heads. They asked us more vigorously than politely to turn back. The redwing is protected by law in the state of Illinois, but in nearly all the other states he is put beyond the statute's pale. The bird unquestionably has a weakness for grain, but the good that he does in insect-eating fairly balances the evil of his life. That he is a beauty in his black blouse with its shoulder knots of scarlet and gold, none will gainsay. Can't we give a kernel or two of corn ungrudgingly to a creature that adds something of living beauty to the dreary wastes of swamp-land?

The long-billed marsh wrens are abundant in the Worth country. These birds have the curious habit of building several nests before they make up their minds which one to occupy. The scientists have been hard at work for years trying to find a reason for this bit of wren freakiness. The scientists are still at work, for no one yet knows the reason save the wren, and the wren won't tell. We flushed from the edge of the marsh that morning a Bartram's sandpiper. This bird is, I believe, the largest of the sandpiper kind. It makes its summer home at Worth, and occasionally has for a neighbor its plover cousin, the lesser yellow-legs. When splashing through the water to get a better look at the sandpiper who had taken to some high ground, I found floating the broken egg of a king rail. The egg told the story of a nest built too low, of heavy night rains, and a flooded abode. King rails are interesting creatures, notwithstanding the fact that it is to be doubted if they have any brain. They are big, blundering, stupid birds who get themselves into all sorts of predicaments, out of which, of themselves, they can find no means of extrication. A friend of mine once found a king rail standing in the middle of the sidewalk near the corner of Schiller Street and the Lake Shore Drive in Chicago. The bird paid but little heed to passers, but seemed to lack the wit to get away from such uncongenial surroundings as stone pavements and brick walls. The men in a North Clark Street barber-shop in the same city were astounded one day to see an ungainly bird make its way through the open door to the center of the shop, where it calmly surveyed the surroundings. Another king rail took possession of a bedroom in the second story of a Chicago residence. The bird absolutely refused to allow itself to be "shooed" out of the window through which it had come. It showed no fear of human beings, and allowed itself to be picked up without resistance. When it was put through the window it took to flight readily enough, but the chances are that before it had traveled far it managed to get into some other fix.

There is something of the savage left in us all. I am free to confess that I like to see birds fight. I don't mean that I wish them to fight, but if they must fight I like to see the fracas. In a tree in a field back from the Worth swamp was a scarlet tanager. It was sitting there peacefully enough, and apparently enjoying the view, when a bluejay dropped down from above and went at it beak and claw. I fully expected to see the tanager turn tail and flee before the face of his assailant, but it surprised me and won my admiration by doing nothing of the kind. It gave the bluejay blow for blow. The combatants half flew, half fell to the ground, clawing, pecking, scratching, and screaming. There was a bewildering brilliancy of moving color. There was another witness to this fight besides the human beings who were looking at it with all the interest ever centered on a ring contest. The bluejay's mate was in the treetop, but made no effort to take a claw in the affair until she thought that her spouse was getting the worst of it. Then she came down hurtling, and joining forces with her mate, soon convinced the tanager that it had enough. The jays did not follow the defeated bird, who made off like a scarlet streak to the shelter of the woods.

On our way back to the farm-house we saw a hawk quartering the marsh in search of prey. It was doubtless a marsh harrier, though it looked much like a duck hawk. I have elsewhere spoken of my admiration for the hawk family. The duck hawk is a true falcon. He is the epicure of the feathered race. He disdains mice and barnyard fowl, and lives largely upon game. His delight is in the chase, and the rapidity of his flight is as the passage of light. He over-takes the teal or the mallard, and seizing his quarry in midair, bears it away for a feast. The utter fearlessness of this wandering falcon was shown not long ago at Calumet Lake. Some duck hunters had built a blind, behind which they crouched in their boats. Two ducks came into the decoys. Both men fired a barrel each, and both missed. At that instant, like a bolt from the sky, a falcon descended and struck down one of the ducks within twenty yards of the blind. Instantly the hidden hunters fired the second barrels of their guns at hawk and duck and both birds fell to the water. The men put out from behind their blinds to pick up the birds. The duck was dead; the hawk, still living, though wounded unto death, remained with its talons sunk deep into the feathers of its quarry, and facing the oncomers with blazing eyes stood ready to give them battle. They killed the falcon with the stroke of an oar. The hand of man is ever against the hawk. When the last duck hawk is dead there will have passed a creature with more of the essence of true courage in its being than exists in the carcasses of a dozen of the cowards who have brought extinction to its race.

➤➤➤➤ COLONEL EDWARD BRAYTON CLARK (1860–1941) WAS BORN in Utica, New York, and graduated from the U.S. Military Academy at West Point. He helped raise an Illinois regiment for the Spanish-American War, and his participation in World War I led to his being decorated by France as a Chevalier of the Legion of Honor. Clark's taste for adventure manifested itself not only in being a soldier, but also as a newspaper correspondent for the *Boston Herald, Chicago Tribune,* and *Chicago Post* (where he served as their Washington, D.C., correspondent). He covered the Sioux insurrection in 1890–91 and the Garza uprising in Texas a year later.

But natural history and conservation were also important aspects of Clark's work. Birds often appeared as the subjects of his newspaper stories, as well as at least one book and numerous magazine articles. He was one of the organizers of the Illinois Audubon Society and later served the group as vice president.

Clark's one book, *Birds of Lakeside and Prairie,* may be the earliest writing I have come across that is devoted to *looking* at birds. Up until maybe the 1930s or so, when someone went out in search of animals (regardless of taxa), they wanted to return with something that they found, either dead or alive. Even the early conservationists like Charles Eifrig and Benjamin Gault collected birds, if only to document the rarities. But Clark's approach is much more in the vein of the modern birder who is generally satisfied with observations or photos.

I have selected two of his writings: an uncredited article Clark wrote for the *Chicago Tribune* and a chapter from *Birds of Lakeside and Prairie,* dealing with a marsh in Worth, Cook County, which was destroyed when the Cal-Sag Channel was built in the early 1920s.

Sources: Christine Brooks, Worth, Illinois, Historical Museum; "Col. Edward Brayton Clark," *Chicago Tribune,* September 23, 1941; "Clark, Edward Brayton," *Who Was Who in America,* vol. 5, 1969–1973 (Chicago: Marquis Who's Who, 1973).

Nature Sketches in Temperate America
Joseph Lane Hancock

THE TREE-TOAD

⌇⌇⌇ Scarcely a week passed during spring and summer that I did not come across the tree-toad in my walks afield. In April its song was heard at twilight, emanating from the lowlands at the border of ponds, ditches, and

Joseph Lane Hancock, *Nature Sketches in Temperate America* (Chicago: A. C. McClurg & Co., 1911), 72–82, 182–86.

wet meadows, or from the shrubs and trees nearby. During summer, one of its favorite places of retreat in the daytime was in the shade of the wooden rafters under the roof of a neighbor's porch. Again, the little pockets on the trunks of beech trees in the woods were especially attractive places for them. In September, the openings into the old apple trees often afforded choice locations in which they took up their residence, staying in the cavernous interior until frosts. . . .

In this species the stripes on the legs are much less variable than those on other parts of the body. . . . The color of this individual was almost white with grayish mottlings. Against the background of the bark it presented an excellent example of protective resemblance. This effect was even more striking before the little subject suddenly turned crosswise on the bark as I was about to open the camera shutter. It is only occasionally that one hears the sound of the tree-toad in July. In the middle of the day I heard one while it was on the bark of a walnut tree. It uttered a slight clattering note, similar to the notes made by the red-headed woodpecker, and could scarcely be distinguished from it.

One of these toads, which I had taken indoors, fed eagerly on grasshoppers and other insects, taken from my fingers. On one occasion it caught sight of a full-grown female walking-stick insect which it seized about midway along its body. Then the toad doubled the insect up, bringing the two halves together, and slowly but surely began to swallow it. To facilitate the operation the toad used its front feet, much as we would our hands, guiding and pushing the morsel down its throat. In view of the great length of the walking-stick's body, it was surprising that such a small animal could swallow this insect whole without at least first removing the slender legs; but the tree-toad was fully equal to the occasion. The result of the meal was that the ungainly walking-stick produced a noticeable angular protuberance at the side of the toad's body, which, however, was only temporary as it only lasted for an hour or thereabouts after it was swallowed.

The appetite of the tree-toad for insects is prodigious, especially after its day's fasting. One evening I offered my toad guest one of the familiar sawfly larvæ, which is a large, light colored, bulky grub that feeds on the heart-shaped willow leaves. After the larva, which was at first curled up from fright, began to gain confidence and straightened out in its attempt to crawl, the toad became much interested in its actions. Then, after watching it intently a little longer the toad finally sprang toward it, catching hold of the forward extremity, and hastily swallowed a fourth of its body. The powerful larva now protested by suddenly curling its body, at the same time becoming perfectly rigid. The toad was put at a great disadvantage by this manœuvre and was obliged to disgorge the part swallowed, but did not do so without a struggle to retain its hold. The larva then remained perfectly quiet, seeming to be conscious of danger, while the toad in the meantime made no further attempt to swallow it again, though standing over it and watching quite a while for a final move on the part of the larva.

The performance of suddenly curling the body is quite characteristic of this and some other larvæ, and here I had convincing proof that this habit could be useful in the preservation of the individual against the attacks of toads, and has quite a biological significance. How birds would act toward this larva would be quite interesting to determine. The boldness and heroic courage of this little tree-toad in attempting to swallow so large a larva was admirable. But what must be said of the courage of this same little batrachian which was displayed a few minutes later! I happened to have at hand a plump, half-grown larva of a large moth, *Telea polyphemus*. This green larva had been feeding on the leaves of hazel and was much larger than that of the saw-fly species. On seeing this larva the toad at once seized it and in a few moments had it half-way down, but after struggling some time with it the toad again disgorged its bulky prey. I later supplied the toad with its regular diet of beetles and small moths. Occasionally it ate cabbage butterflies and small dragon-flies which I caught in a net during the day and reserved for its supper.

The tree-toad varies much in color, depending on the background environment, humidity, and degree of light. On certain occasions it changes from yellowish white to a deep gray or brown, after moving to a new position in a darker place. Similarly, sometimes it may change from white to a beautiful green, or green and gray, if moved to a background of green leaves. This color transformation requires about an hour's time. The male and female are very similar in size and color, the former being distinguished principally by his darker throat and larger tympanum, or ear. This species can be distinguished from other tree-toads by the orange-yellow or brownish concealed coloring of the leg surfaces where they fold in contact. The eggs are laid in the spring of the year in small clusters in the water of ponds and marshes, and are attached to the stems of plants. They hatch into tadpoles within two or three days after they are laid. Dickerson states in her interesting "Frog-Book" that about seven weeks are required for the tadpoles to complete their metamorphosis. The little green frogs then leave the water for their land excursions.

I believe from observation that the tree-toads, when common in a locality, exercise an important influence or control over insect life, especially in beech forests and swamps, and act as a selective factor.

THE HABITS OF THE WALKING-STICK

Among the most curious members of the Orthoptera, or grasshopper family, are the walking-sticks. This name is applied to these insects because of the resemblance their long cylindrical bodies bear to sticks or twigs of trees. I have found them in greatest abundance among the undergrowth and herbage in the mixed beech forests. Here, between the first of June and the last of August, the various green stages of the young insects are frequently met with, while after this period the large, mature gray and brown forms scatter out among the open woods. At this period, moreover, the adults are often found on the trunks of the fruit trees in orchards. The males are much smaller than the females, and not infrequently the former retain their green color in adult life.

There is but one generation of these variable insects each year, the numerous eggs being dropped one at a time on the ground by the females in September. These eggs usually fall among the leaves in the forests. They are shining black and have a bright stripe on one side which gives them a near resemblance to a small seed of some leguminous plant. Being laid in the fall, the eggs remain on the ground during the winter and hatch in the following spring. The first young appear, as I have above intimated, about the first week in June, but the exact time of incubating and hatching varies to a considerable extent. On June 22, 1906, I found the average size of the young to be three-eighths of an inch in length.

I never fully appreciated the value of the green color to young walking-sticks, and conversely, the use of the gray and brown colors to the adults, until one day on examining the foliage of a sapling oak I happened to be in a position to look down upon a cluster of its rich green leaves. Here I observed a young, half-grown walking-stick astride one of the leaves. His body was directly in line with the middle of the leaf, with his head directed toward the stem. When I first discovered him his forelegs were, as usual, closed together alongside the slender antennæ which projected forward. The leaves of this oak were deep green, with the light pale green veins contrasting somewhat conspicuously. The position of the insect was such that he stood over the veins, the legs being arranged almost parallel with them. From this view of his body, he was so closely in accord with the veins that he was almost invisible. I wondered at the time if this position was a mere coincidence or was a common behavior. It is obvious that in this attitude and at a little distance the young walking-stick may stand on the upper surface of the leaves and defy the sharp-eyed birds or other vertebrates to discover his attenuate form.

While I found this resemblance of the walking-stick to the central and radiating veins of the oak and other leaves to be remarkably perfect, there is another point that I noticed in this connection worthy of consideration. I found that when these Orthoptera become unduly alarmed their usual impulse seems to be to seek the under surface of the leaf. After some study I also found that he is here even more protected than when standing on the upper surface, not only on account of the shadows cast on his body, but from the fact that the assimilation is made far more perfect by reason of the median and lateral veins of the leaves being more strongly and roundly in relief here than they are shown above.... The resemblance of the walking-stick to the venation of the oak can be demonstrated in quite a striking manner, if the main median vein with two pairs of laterals are cut out carefully with a pair of scissors. The result is a fair *facsimile* of the young walking-stick.

Of course, in the position which the young walking-stick naturally assumes in foraging, it does not always accurately lie over the veins of the leaf, for the veins slightly alternate. I have found, however, by observations, that if the insect stands at any angle on the leaf, and if he is viewed from below, his resemblance to the veins is sufficient to protect him. The nearer the position of the body corresponds to the central vein structure the greater the safety from attacks. The oak is doubtless the most often chosen as the natural habitat of the common walking-stick, but I have often found the young at home on a variety of herbage, shrubs, and small trees, where they live on the leaves. This in no way mitigates the special adaptation of his body referred to, for I have found that his attenuated form resembles almost as well the leaf veins of various other plants....

The adult walking-sticks often exhibit a most singular performance when they are attacked or frightened. Either they walk off with an ungainly stride in their attempt to escape, or they may fall to the ground. In the latter case they often "play possum" by lying perfectly rigid on the ground among the *debris*, feigning death perfectly. On such occasions they will sometimes allow themselves to be handled and still retain the stiffened attitude for minutes, or even hours afterwards, without regaining their feet. One of these feigning insects which I brought indoors was left on a table at ten o'clock in the forenoon and it stayed in the same position on its back until half-past three in the afternoon of the same day, or a period of six and a half hours. In the meantime it passed faecal matter three times, showing that internal functions were actively carried on during this apparently intense nervous suspense....

In the above word sketch, I have stated that the eggs of these insects bear a close resemblance to the seed of plants, and secondly, that the young show remarkable adaptation to a life on the leaves, their attenuate bodies being deceptively like the venation of the leaves. Finally, I will now call attention to the fact that the adults are protected in their resemblance to twigs even to the minutest detail. In the fall the sexes are found together in large numbers, in pairs, often on the bark of the forest and fruit trees. Again, I have found them on such shrubs as the sumach; here they are all but invisible as they hold on the main stems. When I discover them sometimes by accident, I often marvel that they are ever seen at all by human eyes, so close do their bodies imitate the twigs. Even the leaf scars are wonderfully represented. And why, may I ask, do these insects choose as a place of refuge the very spot on the sumach where there are dead branches near the main stem?

Here is certainly a group of insects for the naturalist to work on. The problems presented are certainly accessible by experimental research, and if they are coupled with observations in the great biological experiment grounds of nature, there is still room for interesting discoveries. Such adaptations as are shown in the form and color of the walking-sticks are the most striking of any in our temperate fauna. They doubtless offer the most powerful argument in favor of natural selection....

THE EVENING PRIMROSE SPIDER TRAP

What a poetical mood nature was in when she evolved the evening primrose with its dainty yellow blossoms, that forecast the night by opening just before sundown! If it were not for the opening of these flowers toward sunset, there would not be accommodation for the night-flying moths, which depend upon such flowers for their honey. And were it not for these flowers, what would become of the yellow and rose-colored moth which frequents the half-open blossoms in daytime for the protection it there enjoys by its harmonious coloring? Again, how little do these insect guests of the evening primrose suspect that within the delicate flower petals an invisible animated trap sometimes awaits their coming, which in a twinkling and without warning literally catches them in the jaws of death. Evidence of this secret trap may not be found in every evening primrose. In fact, I examined many of the flowers of these plants before discovering the remarkable secret trap referred to.

Those who have not witnessed it certainly have something interesting in store for them. In the wild sandy meadow where the primrose plants are abundant, I have seen a dead moth, a fly, or perhaps a bee lying on the open flower, or below it on the ground. When examining more closely into the cause of this insect carnage, I have been utterly taken by surprise on lifting one of these insects to find it in the grasp of an almost invisible, yellowish white spider. I had passed the dead insects on the flowers many times before I had seen the spider. What wonder is it, then, that I had not seen her before.

She has elongated front legs, and a peculiar inclination to walk sideways or backwards with equal facility, and through this resemblance to a crab's gait she is sometimes called the crab-spider, *Misumena vatia*. Having become familiar with her appearance, her attainments are more readily understood. In the early part of September, one was detected in a flower of the evening primrose, in the attitude ready to seize her prey. Her appearance was of such a curious character that I have endeavored to depict her in my drawings. Here she rested with her small, atrophied, third pair of legs touching the stamen-cross on each side, while the first and second pairs of legs were widely spread apart, ready for instant action when the time came. I watched her at intervals, but saw only an occasional insect flying in the vicinity, as the flower did not happen to be in the most favorable location. She stayed here at her station all night and into the next day without capturing a victim.

When the bright sunlight came out and the blossoms began to fade, she changed her position. It was interesting to see her hanging her legs out over the edge of the petals. Later, I saw her slyly walk out of the now fast-wilting flower. Then she climbed up the stem to the summit, where the new yellow petals were just peeping out at the top. She stopped there, placing her yellowish abdomen as near as possible to the slightly exposed portions of the flower bud which bore the same color. At the top of the plant there were six buds in various processes of development, but the spider directly selected the very one which showed promise of opening when evening dawned. All day she sat patiently waiting, as if conscious that the new flower would, when spreading its petals, attract food within reach. When the petals finally unfolded she walked from the underneath surface to the upper part of the flower near the stamens, at the centre, and stealthily arranged her body into a living trap.

After this period of long waiting the reward finally came. First, a light spotted flower beetle flew near and made an aerial descent on the flower. But evidently it was not fitting food, for I was surprised to see it ignored. A moment later, however, she fully awakened when a fly came to the flower and she immediately seized it, using her long legs in the operation of clasping it. After bringing the fly to her mandibles she let go and continued her meal by holding it with her mouth parts, at the same time again spreading her legs wide apart.

Another observation, which directs our attention to the instincts of this spider, was made one late afternoon. I saw another spider of this species on the blosssoms of one of the primrose plants. The following early morning she had changed over to another flower on the same plant, having been driven out by the collapsing petals, as previously described. The following evening the spider was in the newly opened flower, but on the third morning she had left, leaving a dead bee as the remnant of her repast.

But why did she leave? What unerring instinct was exercised in telling her that the succession of flowers was at an end? For it afterwards developed that the flowers had ceased blossoming. In a certain light, I caught sight of a number of her spun threads connecting the tops of the different flowers. Here lay tell-tale evidence, for the spider had the habit of spinning out and leaving a web behind her. These threads marked perfectly the course of her movements in my absence.

A little study of the various strands revealed the fact that she had examined critically the tops of the buds and had discovered in some way the failing nutrition of the flowers. She had shown evidence of deliberation and had gone to and fro in her travels over the bud tops, as was indicated by the various threads....

➤➤➤ DR. JOSEPH LANE HANCOCK (1864–1919) WAS BORN, RAISED, educated (Northwestern University), practiced medicine, and pursued his strong interests in natural history in and around Chicago. An all-around naturalist, he became the foremost authority on the North American *Tettigidea*, a genus of pygmy grasshopper (*Tetrigidea*). Donald Culross Peattie wrote in one of his 1937 columns for the *Chicago Daily News* that "if you own Shelford's *Animal Communities in Temperate America*, E. R. Downing's A *Naturalist in the Great Lakes Region*, and Hancock's *Nature Sketches in Temperate America*, then you have the three best general nature books for this countryside." Hancock's work is certainly the least well known of these among modern readers, which is unfortunate since it is both very readable and retains a freshness that people would still find appealing today. The book was written to provide local illustrations of Darwinian evolution and natural selection. Augmenting the text are numerous photographs and drawings that are the product of the author's hand, further manifesting Hancock's wide range of talents. Finally, a majority of the observations he records in his book were made at his summer home in Lakeside, Berrien County.

Source: http://iltrails.org/cook/chicagobios.htm.

Moths of the Limberlost
Gene Stratton-Porter

THE ROBIN MOTH (CECROPIA)
🐛🐛🐛 When only a little child, wandering alone among the fruits and flowers of our country garden, on a dead peach limb beside the fence I found it—my first Cecropia....

It clung to the rough bark, slowly opening and closing large wings of gray velvet down, margined with bands made of shades of gray, tan, and black; banded with a broad stripe of red terra cotta colour with an inside margin of white, widest on the back pair. Both pairs of wings were decorated with half-moons of white, outlined in black and strongly flushed with terra cotta; the front pair near the outer margin had oval markings of blue-black, shaded with gray, outlined with half circles of white, and secondary circles of black. When the wings were raised I could see a face of terra cotta, with small eyes, a broad band of white across the forehead, and an abdomen of terra cotta banded with snowy white above, and spotted with white beneath. Its legs were hairy, and the antennæ antlered like small branching ferns. Of course I

Gene Stratton-Porter, *Moths of the Limberlost* (Garden City, NY: Doubleday, Page, and Co., 1912), 81, 87–92, 99–100, 130–34.

thought it was a butterfly, and for a time was too filled with wonder to move. Then creeping close, the next time the wings were raised above its body, with the nerveless touch of a robust child I captured it.

I was ten miles from home, but I had spent all my life until the last year on that farm, and I knew and loved every foot of it. To leave it for a city home and the confinement of school almost had broken my heart, but it really was time for me to be having some formal education. It had been the greatest possible treat to be allowed to return to the country for a week, but now my one idea was to go home with my treasure. None of my people had seen a sight like that. If they had, they would have told me.

Borrowing a two-gallon stone jar from the tenant's wife, I searched the garden for flowers sufficiently rare for lining. Nothing so pleased me as some gorgeous deep red peony blooms. Never having been allowed to break the flowers when that was my mother's home, I did not think of doing it because she was not there to know. I knelt and gathered all the fallen petals that were fresh, and then spreading my apron on the ground, jarred the plant, not harder than a light wind might, and all that fell in this manner it seemed right to take. The selection was very pleasing for the yellow glaze of the jar; the rich red of the petals, and the gray velvet of my prize made a picture over which I stood trembling in delight. The moth was promptly christened the Half-luna, because my father had taught me that luna was the moon, and the half moons on the wings were its most prominent markings.

The tenant's wife wanted me to put it in a pasteboard box, but I stubbornly insisted on having the jar, why, I do not know, but I suppose it was because my father's word was gospel to me, and he had said that the best place to keep my specimens was the cellar window, and I must have thought the jar the nearest equivalent to the cellar. The Half-luna did not mind in the least, but went on lazily opening and closing its wings, yet making no attempt to fly. If I had known what it was, or anything of its condition, I would have understood that it had emerged from the cocoon that morning, and never had flown, but was establishing circulation preparatory to taking wing. Being only a small, very ignorant girl, the greatest thing I knew for sure was what I loved.

Tying my sunbonnet over the top of the jar, I stationed myself on the horse block at the front gate. Every passing team was hailed with lifted hand, just as I had seen my father do, and in as perfect an imitation of his voice as a scared little girl making her first venture alone in the big world could muster, I asked, "Which way, Friend?"

For several long, hot hours people went to every point of the compass, but at last a bony young farmer, with a fat wife, and a fatter baby, in a big wagon, were going to my city, and they said I might ride. With quaking heart I handed up my jar, and climbed in, covering all those ten miles in the June sunshine, on a board laid across the wagon bed, tightly clasping the two-gallon jar in my aching arms. The farmer's wife was quite concerned about me. She asked if I had butter, and I said; "Yes, the kind that flies."

I slipped the bonnet enough to let them peep. She did not seem to think much of it, but the farmer laughed until his tanned face was red as an Indian's. His wife insisted on me putting down the jar, and offered to set her foot on it so that it would not "jounce" much, but I did not propose to risk it "jounc-ing" at all, and clung to it persistently. Then she offered to tie her apron over the top of the jar if I would put my bonnet on my head, but I was afraid to attempt the exchange for fear my butterfly would try to escape, and I might crush it, a thing I almost never had allowed to happen.

The farmer's wife stuck her elbow into his ribs, and said, "How's that for the queerest spec'men ye ever see?" The farmer answered, "I never saw nothin' like it before." Then she said, "Aw pshaw! I didn't mean in the jar!" Then they both laughed. I thought they were amused at me, but I had no intention of risking an injury to my Half-luna....

As we neared the city I heard the farmer's wife tell him that he must take me to my home. He said he would not do any such a thing, but she said he must. She explained that she knew me, and it would not be decent to put me down where they were going, and leave me to walk home and carry that heavy jar. So the farmer took me to our gate. I thanked him as politely as I knew how, and kissed his wife and the fat baby in payment for their kindness, for I was very grateful. I was so tired I scarcely could set down the jar and straighten my cramped arms when I had the opportunity.

I had expected my family to be delighted over my treasure, but they ex-hibited an astonishing indifference, and were far more concerned over the state of my blistered face. I would not hear of putting my Half-luna on the basement screen as they suggested, but enthroned it in state on the best lace curtains at a parlour window, covered the sill with leaves and flowers, and went to bed happy. The following morning my sisters said a curtain was ru-ined, and when they removed it to attempt restoration, the general concen-sus of opinion seemed to be that something was a nuisance, I could not tell whether it was I, or the Half-luna. On coming to the parlour a little later, lad-ened with leaves and flowers, my treasure was gone. The cook was sure it had flown from the door over some one's head, and she said very tersely that it was a burning shame, and if such carelessness as that ever occurred again she would quit her job. Such is the confidence of a child that I accepted my loss as an inevitable accident, and tried to be brave to comfort her, although my heart was almost broken. Of course they freed my moth. They never would have dared but that the little mother's couch stood all day empty now, and her chair unused beside it. My disappointment was so deep and far-reaching it made me ill; then they scolded me, and said I had half killed myself carrying that heavy jar in the hot sunshine, although the pain from which I suffered was neither in my arms nor sunburned face....

In connection with Cecropia there came to me the most delightful experience of my life. One perfect night during the middle of May, all the world white with tree bloom, touched to radiance with brilliant moonlight, intoxicating with countless blending perfumes, I placed a female Cecropia on the screen of my sleeping-room door and retired. The lot on which the Cabin stands is sloping, so that, although the front foundations are low, my door is at least five feet above the ground, and opens on a circular porch, from which steps lead down between two apple trees, at that time sheeted in bloom. Past midnight I was awakened by soft touches on the screen, faint pullings at the wire. I went to the door and found the porch, orchard, and night-sky alive with Cecropias holding high carnival. I had not supposed there were so many in all this world. From every direction they came floating like birds down the moonbeams. I carefully removed the female from the door to a window close beside, and stepped on the porch. No doubt I was permeated with the odour of the moth. As I advanced to the top step, that lay even with the middle branches of the apple trees, the exquisite big creatures came swarming around me. I could feel them on my hair, my shoulders, and see them settling on my gown and outstretched hands. Far as I could penetrate the night-sky more were coming. They settled on the bloom-laden branches, on the porch pillars, on me indiscriminately. I stepped inside the door with one on each hand and five clinging to my gown. This experience, I am sure, suggested Mrs. Comstock's moth hunting in the Limberlost. Then I went back to the veranda and revelled with the moths until dawn drove them to shelter. One magnificent specimen, birdlike above all the others, I followed across the orchard and yard to a grape arbour, where I picked him from the under side of a leaf after he had settled for the coming day. Repeatedly I counted close to a hundred, and then they would so confuse me by flight I could not be sure

I was not numbering the same one twice. With eight males, some of them fine large moths, one superb, from which to choose, my female mated with an insistent, frowsy little scrub lacking two feet and having torn and ragged wings. I needed no surer proof that she had very dim vision.

~~~~ GENE STRATTON-PORTER (1863–1924) ENJOYED GLORIOUS success as a writer in the early years of the twentieth century. Her several novels (including A Girl of the Limberlost [1909]) sold over 8 million copies and were translated into at least thirteen languages including Afrikaans, Arabic, and Korean. Quoting her publishers, Arthur Shumaker writes that "during the last seventeen years of her life her books sold at the rate of 1,700 copies a day, including Sundays, and that if placed end to end, they would form a line of 1,110 miles."

Born Geneva Stratton on a farm in Wabash County, Indiana, she was the youngest of twelve children. Being the lowest in seniority posed some difficulties, but also conferred one great privilege over her siblings: no one objected to her roaming freely in the fields and nearby woods, especially after her mother, Mary, was stricken with typhoid, from which she never fully recovered. An energetic and curious child, Gene became captivated by the local birds, whose family ways she learned intimately.

One of her earliest memories concerned the finding of a woodpecker shot by a brother as it fed on cherries. She tried to revive it (at one point filling its mouth with gooseberries), but eventually her father was forced to introduce her to the concept of death. She mused about the unfortunate bird and offered an agreement to her father: "If you will make the boys stop shooting woodpeckers, I will not eat another cherry. The birds may have all of mine." Her mother said her sacrifice was unnecessary as there was enough fruit for the family and the birds, and she put a stop to the killing.

Her father, Mark Stratton, a farmer and Methodist minister, was an accomplished orator who used to entertain the family with his recitations. Gene would devise and polish her own routines from a makeshift stage in the orchard. And before she could even write, she composed stories and poems in her head that she implored her mother to "set down."

A national depression forced Mark Stratton to retire from farming, and in 1875 he moved the family to Wabash. Gene did not adjust well to city living. Visits to the farm helped (it was on one of these that she collected her first cecropia moth), but she resented both her teachers who had no concern for her interests and her fellow students who had "finer shoes, and prettier dresses and hair ribbons" than she did.

In 1881 Gene began attending the annual Chautauquas held at Sylvan Lake, near Rome City in Noble County (third county east of LaPorte). These inspirational convocations combined entertainment, religion, and culture in an effort to broaden the intellectual and moral horizons of America's still largely rural population. But for Stratton, the major importance of the trips was that she became enthralled with the varied landscapes and avifauna around Sylvan Lake, and that she caught the eye of another attendee, Charles Dorwin Porter from Geneva, Adams County.

Porter owned two drugstores and aspired to become a banker. He was too shy to introduce himself to Stratton in person, so some months later he wrote her a letter (having received her address from a cousin who knew one of Stratton's brothers-in-law). They eventually met, and although she was not immediately smitten, their correspondence continued. In these epistles she referred to him as Mr. Porter (a custom she continued even in wedlock) and expressed her belief that marriage was primarily for the husband, as it provided few if any benefits for the wife. Despite this lack of enthusiasm on her part, his ardor grew, she relented, and the wedding took place in April 1886.

Charles and his new bride moved to Decatur, halfway between his stores in Geneva and Fort Wayne. To spend more time with Gene, he soon gave up the Fort Wayne store. Then Gene discovered she was pregnant, and in August 1887 she gave birth to her only child, Jeanette. Never really happy in Decatur, she prevailed upon Charles to move to the more rural Geneva.

To augment the funds available to her, Gene began writing and decided to focus on the subjects she loved most, birds. Her first submission was rejected because the editor wanted photographs, and she refused to rely on standard shots of long-dead specimens. Gene would eventually become an accomplished photographer, utilizing the most sophisticated equipment available. As her biographer writes, "Gene grew adept at rigging her photographic equipment from trees with the aid of a wagonload of ropes and tall ladders. She used different lures to attract birds to the camera, such as beefsteak wired to limbs and bushes."

Gene, knew, of course, she needed access to wild birds, and in resolving that challenge she discovered the wonderful place to which she has become inextricably connected. Less than a mile away sprawled thirteen thousand acres of wet woods and open marsh known as the Limberlost. Over the years Gene would explore this largely roadless wetland as she sought wildflowers, moths, and the birds whose nesting habits she photographed in detail and with great care.

By 1913, however, this wilderness had shrunk dramatically due to draining and logging. Gene and her family then relocated to Sylvan Lake, where she created her estate named Wildflower Woods. She lived and wrote here for ten more years, before moving to Southern California, where she died in a car accident.

Gene's career as a writer took off with an article that appeared in *Recreation* in 1900. Entitled "A New Experience in Millinery," it is her firsthand account of buying a hat and then asking the clerk for a scissors, which she used to remove the birds. The magazine then began running a column by her called "Camera Notes." She was to contribute to many magazines, most particularly *Outing* and *McCall's*. Although her fame and fortune rested on novels, she was most pleased with her nonfiction books, among which are *Birds of the Bible, Moths of the Limberlost*, and *What I Have Done with Birds*.

While Gene's writings were enthusiastically embraced by the public, she was generally panned by critics, who found the novels excessively sentimental and cloyingly sweet. One of the most disparaging assessments comes from Arthur Shumaker, who was not merely content with savaging the writing but took on her readers as well:

> The truth is that her typical writings have always appealed essentially to the simple-minded—children, the old, and those who have been buffeted by the waves of life until they have nearly sunk or have been shipwrecked. Mrs. Porter received large quantities of letters of appreciation from the poverty-stricken, the sick, the inhabitants of reform schools and prisons, stating how much her books had helped them to face reality. These unfortunate people received a therapeutic effect from their imaginative return to nature and the idyllic simple life, releasing them from the world of men, where they had experienced disaster.

In the face of such reviews, it would take outright courage to admit liking the works of Stratton-Porter. She herself countered by rejecting what she claimed was the growing trend among authors and critics to "teach that no book is *true to life* unless it is true to the *worst in life*." (To get a really balanced view of life, I suppose, spend half of your time reading Stratton-Porter and the other half Theodore Dreiser.) But there were those not yet addled or incarcerated who did admire her books. The paleontologist R. R. Rowley was so impressed by *What I Have Done with Birds* that he named a trilobite after her that is now called *Australosutura strattonporteri*. William Lyon Phelps of Yale said she was "a public institution, like Yellowstone Park."

Another critic, F. T. Cooper, wrote that her prose could make even the housebound "hear the birds and smell the flowers and watch the shifting seasons and alternating sunshine and rain." Surely, even if her stories were devoid of other attributes, the gift of making nature live through her words would be reason enough to respect the writings of this singular woman.

Sources: James Cook and Richard Kurkewicz, paleontologists and members of the Western Trilobite Association, e-mails to author, January 2006; Fredric Taber Cooper, "The Popularity of Gene Stratton-Porter," *The Bookman* (August 1918): 671–72; Judith Reick Long, *Gene Stratton-Porter: Novelist and Naturalist* (Indianapolis: Indiana Historical Society, 1990); Barbara Stedman, ed., *Our Land, Our Literature*, http://www.bsu.edu/web/landandlit, Ball State University, Muncie, Indiana; Arthur W. Shumaker, *A History of Indiana Literature* (Indianapolis: Indiana Historical Society, 1962).

A Day among the Waterfowl and Its Sequel

G. Eifrig

⌀⌀ ⌀⌀ ⌀⌀ Several years ago a friend and the writer decided to take a walk from Arlington Heights to Elk Grove and back, to see what birds we would meet with. The day fixed, May 30th, 1914, came and we started at five in the morning. It was one of those rare, perfect days, and we commenced seeing things from the very threshold of the house. We saw fifty-six species of birds that day, or rather till about three in the afternoon. Beside the common birds, such as Robins, Bluebirds, Thrashers, House Wrens, Bobolinks, Meadowlarks, Mourning Doves, Redwings, Grackles. etc., we saw Henslow's Sparrows, Grasshopper Sparrows, Indigo-birds, Dickcissels, Migrant Shrikes, Traill's Flycatchers, Cerulean and Blackpoll Warblers, and many more. After awhile we came to a large "slough," i.e. a swamp with much open water, shallow on the sides but three to four feet and more deep in the centre. While we were hearing and seeing King Rails, Spotted Sand-

G. Eifrig, "A Day among the Waterfowl and Its Sequel," *Audubon Bulletin* (Winter 1918–19): 44–46.

pipers, Killdeer, Bitterns, Redwings, Red-shouldered and Marsh Hawks, a Short-eared Owl, Prairie Chickens and a multitude of other sights and sounds—nature was at her best that day—I suddenly spied several Yellow-headed Blackbirds flying into the marsh. That made me excited, as they are becoming rare in our parts, and, throwing aside all impediments, I walked, with clothing on, after them into the water, not any too warm then. I wanted to see the nests of the Yellow heads. When the water reached up to the hips, I suddenly saw a circular, compact mass of old cat-tail before me, and looking more closely saw seven large whitish eggs on it, finely sprinkled with black. It was the nest of a Coot or Mudhen. Nearby was another one. Then a Black Tern or Sea Swallow arose from the marsh nearby, circled around us, uttering cries of apparent displeasure at our trespassing on her watery domain. Keeping the yellow heads in view I pushed on. A rod or so farther on, I saw another floating mass of cat-tail before me, heaped up with freshly put on material. Knowing the tricks of the Pied-billed Grebe or Dipper, I recognized it as one of her nests, and lifting off the moist cat-tail from the top saw the seven buffy eggs of this well-known denizen of open swamps. The bird had heard our approach, had hastily put on the concealing vegetation, making the nest look like many of the other masses of similar material to be seen on all sides, and had slunk off. Nearby was another with two eggs, camouflaged the same way. By this time I had lost my Yellow heads, and could not locate the nests, especially since the water was getting deeper, and cooler, and wetter, as it seemed to me. There were nests of the gaudily dressed Redwing on all sides, also the queer globular ones of the Prairie Marsh Wren, who protested vigorously against our intrusion. So we turned back toward shore, but not before we had seen a Blue-winged Teal, with its white crescent on each cheek, which no doubt had a nest in there somewhere. I was told that in some years quite a number of wild ducks had nested there. Coming to shore, we noticed that our wet clothes did not improve our appearance, or our feeling of comfort, but we did not get time to mind it, because new things turned up right along such as Little Green Herons, Upland Plover, Black-crowned Night Herons, and, to cap these experiences with a climax, we flushed a Prairie Hen from her nest of thirteen eggs. It was in a clump of alfalfa, about ten feet from a road where many autos passed daily. Some of our clothing had dried on us, and we kept on talking about our experiences in the water. Such is the wonderfully rich and interesting life of the marsh. There, life is fairly piled up, flora and fauna display themselves more lavishly than in most places on dry land.

Now comes the sequel. August 19th of the following year, we went over the same route, to the same place. But imagine my surprise when instead of seeing the graceful *scirpus Lacustris*, wild rice, cat-tails, etc., and hearing the cries and calls of marsh and water birds, we saw cows peacefully browsing, looking at us with mildly inquiring eyes—not even a trace of marsh or water flora and fauna remaining. The beautiful slough had been drained by the owner of the farm, who had no eye for the beauties of wild life, but only for the coin, the milk from the kine, and later the crops the newly won field would bring him.

The moral of the tale is plain. If we want to retain the native birds of such areas, and enjoy in the open the varied form, voice and color, we must set aside preserves before it is too late. Nor should we then think only of the birds of the forest, grove and prairie, but also of the water birds. They are suffering more and more by the increasing cultivation of the soil, which is necessary to feed us humans. They have to leave us, not of their choice, but because they are compelled by being deprived of their nesting sites and conditions.

Recently the writer read Mr. E. W. Nelson's "Birds of Northeastern Illinois" written in 1876. I was astounded at the difference in conditions as regards the bird life of Chicago then and now. Of the pretty and interesting Wilson's Phalarope he says: "Very common summer resident in this vicinity." It is almost absent now. He gives many shore-birds and wild ducks and others as breeding in the Calumet and Sag regions and in other places, that are now hardly seen here, let alone being found breeding here. Their haunts, so full of rich and interesting life, have now largely been turned into fields and factories, some chemical factories with their evil smelling effluvia and smoke. It may be necessary, but how one hates to see such changes!

Who of the readers of this Bulletin, owning a large or small slough or marsh, perhaps undrainable and therefore useless to him of her, will communicate with the Illinois Audubon Society, with a view of making this a swamp bird-sanctuary? Should we not preserve some of these natural monuments and remainders as well as remainders of what this region once was, and not force them to leave us forever? Let us do it, and do it soon.

~~~~ CHARLES WILLIAM GUSTAVE EIFRIG (1871–1949) WAS BORN in Germany but moved with his family to Freedom, Pennsylvania, when he was seven. Following the completion of his studies at Concordia College (Fort Wayne, Indiana) and Concordia Theological Seminary (St. Louis),

Eifrig served as pastor at three congregations over the next fourteen years. (Eifrig received an honorary doctorate from Valparaiso University.) In part, I suspect, because of his deep interest in nature acquired as a child, he changed career paths by becoming a professor at Concordia Teachers College at Addison and then River Forest, Illinois. During his lengthy tenure there (1909–42), he wrote numerous ornithological articles in such journals as the *Auk*, the *Wilson Bulletin*, and the *Audubon Bulletin*, as well as two textbooks for elementary students, one on mammals and one on fish, reptiles, and amphibians. He also served as president of the Illinois Audubon Society. As a field ornithologist, one of his most remarkable finds was ten to fifteen Bachman's sparrows in breeding condition and seemingly on territory in River Forest, Cook County, well north of any possible nesting location than they have been found since. (And the species is now absent from the entire state of Illinois.)

More than even through his writings, his greatest influence, perhaps, was to imbue his hundreds of students with an appreciation of nature that they would have otherwise lacked. In his obituary of Eifrig, Dr. Alfred Lewy wrote that "he did much to overcome the suspicion of science among the Lutherans.... As his pupils go out into the world, preaching the love of nature as well as their theological doctrines, we hope to see Dr. Eifrig's influence spread in ever widening circles."

Source: Alfred Lewy, "Professor C. W. G. Eifrig," *Audubon Bulletin*, no. 77 (March 1951): 8.

Memoirs of a Naturalist
Herbert L. Stoddard Sr.

～～～ By early 1913 I was settled in Chicago and hard at work in the new N. W. Harris Public School Extension of the Field Museum. The Harris Extension was responsible for preparing natural history exhibits for the Chicago schools. S. C. Simms, an anthropologist, was curator of the extension at the time and later served as director of the museum.

I had joined the staff of the Field Museum with the understanding that I was to make my own field studies and collect the material for the displays.

Herbert L. Stoddard Sr., *Memoirs of a Naturalist* (Norman: University of Oklahoma Press, 1969): 111–12, 114–20, 122–28. Courtesy of the University of Oklahoma Press.

Simms gave me a free hand in every respect during the seven years I remained with the Harris Extension. I carried on most of my field collecting in Illinois, Indiana, and Wisconsin, remaining in the field as long as I felt necessary. The curator was interested in results—and also evidently aware of my independent spirit. He let me have the freedom I needed to achieve them.…

I soon found that there were splendid bird-collecting regions in and near Chicago. The cattail marshes, low prairies, and savannas around Wild* and Hyde lakes, the unique Indiana Dunes at the south end of Lake Michigan, the rolling prairies and prairie sloughs and the diversified farmlands of Calumet Sag, west of the city—all these regions were outstanding bird ranges. Other regions were the "Blind Slough" and the "Feeder," near Chicago Ridge, where grebes, rails, gallinules, bitterns, blackbirds, black terns, and other species nested in profusion; Willow Springs, where rose-breasted grosbeaks nested; and Worth and Addison, the best nearby ranges for longspurs. Kouts, Indiana, near the Kankakee River, south of Chicago, was a noted stop for migratory warblers and waterfowl. This region had been explored by Frank Woodruff, of the Museum of the Chicago Academy of Sciences, and other ornithologists.

I soon became familiar with all these localities. Armed with the old Parker 10-gauge with its .45–90 auxiliary barrel, a variety of ammunition ranging from "squib" auxiliary loads for small birds to the heaviest 10-gauge loads for large species, I was well equipped to collect anything from hummingbirds to eagles. Almost daily during migration seasons I would leave the city on an "owl" car about four o'clock in the morning. By nine o'clock, when most of the other employees were arriving at work, I would be back at my workbench skinning the specimens I had obtained.

I spent most weekends collecting in the Indiana Dunes, returning late Sunday nights loaded with specimens and accessory material. I spent every possible minute on some phase of my work, still emulating the work habits of Carl Akeley, who had left the Field Museum a short time earlier to join the staff of the American Museum in New York. The entertainments of Chicago held little charm for me, and my drab room near the Illinois Central tracks even less.

Simms introduced me to an Austrian taxidermist named Charles Brandler, who had worked with Akeley on large animals in the Milwaukee shop and

---

*Stoddard clearly meant Wolf Lake. Editor.

had accompanied him to Chicago as an assistant. Brandler had mastered and used the revolutionary taxidermic methods developed by Akeley and had become an expert large-mammal taxidermist in his own right. Brandler had also worked for a time in the Milwaukee Museum. Two of his excellent groups there were of the African wild ass and the little Veracruz deer.

Brandler was a picturesque character with black hair, beard, and long mustache. Besides being a skilled taxidermist, he was also an enthusiastic hunter, trapper, and field collector. Because of his friendship for Simms and his desire to help the Harris Extension off to a good start, he befriended me in many ways. He helped me find a room only a half-mile from the museum.

Brandler also took me into partnership with him in muskrat trapping in Jackson Park. The muskrats were undermining the banks of the ornamental lagoons, and he had obtained permission from the superintendent of the South Parks Commission to trap the animals, on the strict condition that the traps were set after dark and removed before daylight. Brandler and I ran trap lines during the early spring months. Not infrequently we would have a dozen or more animals out of the traps, skinned, and stretched by the time most residents of the city were sitting down to breakfast.

I thoroughly enjoyed these nocturnal activities, which were also profitable, for we received about two dollars each for prime skins. During the first season we obtained enough matched skins for a fur coat Simms presented to his wife—mainly for the novelty of having a coat made from furs trapped within the Chicago city limits. Billy O'Brien, the museum's fur dresser, spent many spare-time hours tanning the skins.

In those years the Field Museum was situated in the Fine Arts Building, which had been built in Jackson Park for the World's Fair of 1893. While working at my bench in the museum, I had only to raise my eyes to feast them on the scene outside the window—the wide expanse of tree-studded lawns and banks of shrubbery in the park, which stretched far into the distance. Off to one side I could see the sparkling whitecaps of Lake Michigan. Jackson Park, with its miles of fine drives, golf courses, yacht harbor, lagoons, and wooded islands, was a man-made oasis in a desert of brick and mortar.

The museum building itself, though solidly built, looked somewhat shabby from the outside. It was a stucco-veneered edifice, and for years large patches of stucco had been falling off. It would have looked much worse but for the genius of Carl Akeley, who during his years at the museum had applied himself to solving the problem. The worst parts of the exterior had been gone over with liquid cement applied with a "cement gun" Akeley invented especially for the job. The gun, only one of Akeley's outstanding inventions, was to be invaluable in World War I, when it would be used in France to line trenches and to make cement boats.

In the spring of 1913 I began making notes on bird migrations in Jackson Park. Swarms of warblers, thrushes, and other birds would pour into brushy parts of the park during the half-light of daybreak. Many a morning I listed fifteen to eighteen species of gaily plumaged warblers and a variety of other birds, all within a couple of hours. The lagoons attracted ducks, gulls, terns, grebes, and other waterfowl, especially when high winds roughened the surface of Lake Michigan. Wooded Island, just south of the museum, was the best place in the park for birding. At the peak of a migratory wave the shrubbery and trees were filled with nocturnal migrants. Between 1913 and 1920 I made many long bird lists there.

There were similar oases attractive to migrating birds on the Illinois-Indiana line, right in the midst of a factory district, and there were still others, at intervals of several miles, all the way to the south end of Lake Michigan, in Indiana. One refuge that I often visited was the one mentioned earlier, lying between Hyde Lake and Wolf Lake. There a strip of shrubbery and stunted trees occupied a ridge lying between prairie and savanna. The only mark man had made on the region was a single-track railroad running through the strip. Luxuriant growth covered the ground among the shrubs. The lakes had low areas of grassy shore, extensive areas of cattail marsh, and prairie-like stretches of short-grass country. The wide variety of vegetation attracted many varieties of birds.

This region quickly became my favorite local collecting ground. The area was encircled by high steel mills, chemical works, and factories. The air was filled with chemical fumes, and resident birds were almost black from the smoke-soiled air and vegetation. There were large colonies of yellow-headed and red-wing blackbirds, as well as black terns and other water birds. King, Virginia, and sora rails nested in abundance there, and the region was the only known local nesting ground of the tiny, rare black rail. It was here that I collected a specimen of this bird, which was given the name *Creciscus jamaicensis stoddardi* by Henry K. Coale, a prominent Chicago realtor and amateur ornithologist, who became a close friend and collecting companion. This supposed subspecies, which bore my name for over a decade, is no longer on the A.O.U. *Check-list of North American Birds*.

At that time Wilson's phalarope nested sparingly on the shore of Wolf Lake, on the Indiana side of the line, and on one occasion I collected the downy young of a nesting pair. Marsh hawks, American and least bitterns, Traill's flycatchers, yellow warblers, and prairie horned larks were among the birds that nested in abundance there. At daybreak on May 21, 1915, I saw migrant birds pour down into the shrubbery and ground vegetation from the sky in a spectacular manner. That morning I saw over thirty Connecticut warblers (more than I have ever seen at any one time in all my subsequent years) and managed to collect seven specimens.

On another trip I saw a female king rail calmly incubating her eggs in a marshy spot within eight feet of the railroad track. Train after train would thunder by, but throughout all the noise, vibration, and whirling of dust and cinders, the mother bird never stirred from her nest.

The dense, low-growing thickets around Hyde Lake were the haunts of hobos, and I often found shells of king rail and Florida gallinule eggs the hobos had eaten. There were similar thickets near Lake Calumet, which I reached by walking from the terminus of the Stony Island streetcar. It was in the savannas near this lake that E. W. Nelson, the pioneer Illinois ornithologist, had found Nelson's sharp-tailed sparrow "undoubtedly breeding" on June 12, 1875. Later Frank Woodruff found them nesting in the grass bordering the lake. However, though I searched many times, I was never able to find any. I did occasionally collect Leconte's sparrow during migrations. Jacksnipe and pectoral sandpipers were abundant on the low, short-grass prairies, while common, Forster's and Caspian terns came to the lake in large numbers when spring storms were roiling the waters of Lake Michigan. Gulls, ducks, and grebes could be expected on Lake Calumet in the spring.

Traill's flycatchers could be found in the thickets of dense, chest-high shrubbery scattered throughout the prairies east of the lake. On one occasion, while I was collecting group material for the little flycatchers and yellow warblers, I had an amusing experience with some hobos encamped there. On that trip I had not taken the Parker since I had expected to be loaded down with group material on return journey. But I did have a shoulder satchel containing among other things a little pistol-like combination gun known as the Marble Game Getter. This gun had a collapsible stock and a .22-caliber barrel above and a .44-caliber barrel below, in which I used shells loaded with fine shot when collecting small birds.

While I was poking around in the shrubbery looking for flycatcher nests, three burly hobos began to close in on me, evidently intending to search me for money or valuables. There were no other human beings in sight, and I probably looked like easy game. They approached, well spread out to catch me no matter which direction I ran. When they were about thirty or forty paces away, I pulled out the little handgun and casually picked off a flycatcher perched on a bush about fifty feet away. Of course, the hobos had no idea that the .44 barrel was loaded with dust shot. To them I was a *gunman*, able to pick off a tiny bird with a solid bullet. The expressions on their faces as they backed away and disappeared into the bushes were unforgettable....

Colin Sanborn, who was to become curator of mammalogy of the museum, was active in bird collecting and study during my Chicago years. He was a protégé of Henry K. Coale, mentioned earlier. Coale had a cabin in the woods near Highland Park, north of Chicago. In the spring, at the height of the warbler migrations, Sanborn and I would join him there for a few days' collecting. We three spent many happy hours together around the skinning table at "Coale Cabin," as we put up our specimens and discussed the finds of the day....

Coale told Sanborn and me about collecting birds near 40th Street, in what is now near downtown Chicago. As for *my* old collecting grounds at 108th Street, today they are deep within the city. There was one muddy morass not far north of Lake Calumet, near the Stony Island car tracks, where Brandler and I collected some fine shore birds in the years 1913 to 1919. Any bird that fell near the center of the area was almost impossible to retrieve from the chest-deep mud. Before I left Chicago in 1920, the area had been subdivided for residential lots and named, of all things, "Southlawn Highlands"!

In addition to local collecting trips and those to Prairie du Sac, I also made weekend trips to the Indiana Dunes. At that time there were only one or two fishermen's shacks in that entire, splendid stretch of dune and shore line extending more than twenty miles in a horseshoe curve around the south end of Lake Michigan. A few naturalists from Chicago frequented this wonderland during the summer months, but in the winter I seldom crossed a human trail in the snow. Red foxes, western fox squirrels, red squirrels, great horned owls, and ruffed grouse were abundant throughout the region.

The dunes were highly diversified. Traveling dunes were common near the shore line; in my time a great one was moving in and threatening to engulf parts of Michigan City. In some places old, stabilized dunes covered with diversified forest and ground cover extended for miles inland; in other places such formations were narrow. There were deep hollows, well protected from the wind, where northern scrub pine, ground juniper, and deciduous undergrowth thrived. There the red squirrels abounded. Ponds, cattail marshes, and grassy savannas lay between the dunes paralleling the lake shore. Poison sumac covered both floating and stabilized bogs near Mineral Springs. There were picturesque growths of white pine and great stretches covered with scrub oak. Wet savanna and broken country of abandoned farms, together with some cultivated regions, could be found further inland. In short, the region was a paradise for the naturalist.

Then a botanist in the University of Chicago wrote a glowing article about the wonders of the dunes for *The National Geographic*. Within a few short years the wildness and splendid isolation of the region had given way to clusters of cottages. Graphophones blared in the hollows where the red squirrels had chattered and scolded. Tin cans and broken bottles littered the savannas. Beach-grass plains and beach-plum thickets became lovers' rendezvous. Spotted sandpipers and piping plovers' eggs were destroyed by swimmers. The whole region was engulfed by humanity. But until 1920 the dune region was my "happy hunting grounds," and I prefer to remember it as it was during those years.

June 26, 1915, was a momentous day in my life. On that day I forsook bachelorhood and married Ada Wechselberg, a member of the family with whom I had boarded during my years in Milwaukee. After our marriage in Milwaukee, Ada and I went to Chicago. Leon Pray and his wife, who were away on vacation, loaned us their flat for our first month of married life. By the time the Prays returned to Chicago, we had found an apartment of our own on University Avenue, where we remained for the rest of our stay in Chicago.

After our marriage I continued the weekend field trips. I sometimes feel guilty when I recall the days and weeks my wife was alone while I was on extended trips. Fortunately, though city born and bred, my wife always understood my need for these retreats from urban living and bore her lonely hours without complaint. It is likely that the life of a naturalist's wife is not an easy one.

Leon Walters and I frequently made trips to the dunes together. We would travel as far as Millers, Dune Park, or Mineral Springs on the South Bend Electric Train. Then each of us would go his own way after agreeing on a meeting place for the return journey or a campsite for the night. We had several ten-cent frying pans hidden in hollow logs here and there throughout the dunes. We usually carried a small roll of breakfast bacon in our packs and supplemented our diet with fried fox squirrels, Blanding's turtles, cottontail rabbits, and even the roughed-out bodies of surplus specimens. It was my good fortune to obtain the first four records of the Arctic three-toed woodpecker for Indiana in that region, the first one at Dune Park on March 11, 1917.

For some time the Department of Zoology of the museum had been disregarding local field work, apparently considering the nearby regions worked out. However, Osgood and Cory began to take notice when I brought in a long-tailed jaeger (on September 21, 1915), and a roseate tern (on August 14, 1916), both taken at the lower end of Lake Michigan. Simms and I had no intention of using specimens of special scientific value in the Harris Extension displays, but we took pleasure in obtaining such rare skins. Moreover, we took a perverse delight in teasing the staff members in the Zoology Department. At times a verbal sparring match ensued between Simms and Cory when Cory demanded for his department some specimen that I had brought in. These matches often had to be settled by the museum's director.

Simms dearly loved a good scrap, and Cory and Simms had a particularly amusing "go around" when Coale described a black rail (taken at Hyde Lake on May 30, 1916) as a new species, confining the typical form to Jamaica, where it was first taken. It was to Cory unthinkable that a coveted specimen should repose in the humble Harris Extension collection. Finally I recommended to Simms that we turn it over to the Zoology Department, after they had tried by every means to get it from us. Simms enjoyed all this maneuvering, though Cory was serious about it and thought we were.

A great fence that enclosed the steel mills at Gary shut off the beaches to the west, but there was a little-frequented stretch from the fence to Millers, about five miles east, that I found exceptionally productive of shore birds, probably because the ruddy glow of the furnaces attracted huge numbers of insects to the region. Black-bellied plovers were particularly abundant in both spring and fall. An adult blue goose (the second record for Indiana) was collected there on October 21, 1916. When taken, the bird was with about forty blue geese and six or seven snow geese. Several other flocks were transversing the area on that wild, windy day. It was my only experience with these spectacular birds in the Chicago region. Another second record for the state was a buff-breasted sandpiper secured nearby on August 30, 1916.

There were some old dunes heavily grown up to juniper cedar and wild "frost" grapes, just behind the beaches. I remember seeing sizable flocks of evening grosbeaks among the cedars and northern scrub pines.

The shore birds were more abundant on this stretch than elsewhere. Several species, which had been all but exterminated during the period of intense slaughter for sport and market, were just beginning to increase in numbers again, and we were always on the lookout for them. I am glad to say that the increase has continued. Several of the larger species, which were reduced to low numbers between 1900 and 1920, subsequently reappeared in quantity from Key West to the wilds of Canada. The prohibition against spring shooting of migratory birds from Canada, effected by the Migratory Bird Treaty, and later the removal of most of the birds from the list of game that can be shot at any time have been effective measures in saving them. Shore birds that are still very rare, or that have passed from the scene, like the Eskimo curlew, are those that were not protected from guns in their journeys outside the United States.

I had several companions on the local bird-hunting trips. Among them were W. D. Richardson and his wife, of Chicago. One day in 1913, while I was traveling on an electric car bound for the dunes, Richardson introduced himself and his wife, explaining that he was a bird photographer, that he wanted to get pictures of a great horned owl at the nest, and that he had heard I had several under observation in the dunes. At first I was not happy about having strange companions along, and I vowed to myself that I would put them to the test. I struck off across the dunes at a grueling pace, heading toward the first owl nest, in a broken-off oak several miles away. To my surprise, the Richardsons easily kept up. By the time we had arrived at the tree, climbed up to the nest, and photographed the two downy young, the pair of naturalists had completely won me over. Thus began one of the most cherished friendships of my life.

〜〜〜 HERBERT STODDARD (1889–1968) WAS A NATURALIST IN THE grand sense of the word, which in my mind signifies a deep knowledge of the widest range of natural history topics from animals to plants, and the processes that affected both. Born in Rockford, Stoddard learned the ways of the natural world independently of any formal training, for he quit school in 1905 before entering high school.

Stoddard worked at a variety of jobs until he fell in with a well-connected local taxidermist. One time the two of them were visiting the Ringling Brothers Circus facilities in Baraboo, Wisconsin, when a four-ton hippopotamus died. The circus decided to donate the specimen to the Milwaukee Public Museum, which then sent its chief taxidermist to disassemble the animal. Stoddard was assigned to assist in the process. The carcass had been moved to an unheated building that provided little shelter from the biting cold: "When we threw a steaming hippo steak at the wall it would stick and freeze to the siding." Despite the low temperatures, when it came time to remove the skeleton, "evil smelling steam arose from the interior of the carcass throughout our work."

At the urging of the highly impressed taxidermist, the museum offered Stoddard a job, which he accepted. He stayed at Milwaukee from 1910 to 1913, when he joined the staff of the Field Museum. His main task was in preparing specimens for the Harris Extension, dioramas designed for exhibition at area schools. During his seven years in Chicago, Stoddard made numerous trips throughout the region, particularly the Indiana Dunes.

Upon leaving the Field Museum, Stoddard returned to the museum in Milwaukee, where he stayed until 1924. His career took a very different turn when he agreed to lead a federal study on the management and life history of the bobwhite. This resulted in his classic treatise, *The Bobwhite Quail: Its Habits, Preservation, and Increase.* Most of his time thereafter was spent near Thomasville, Georgia, where he continued his studies on game birds and the role of fire in maintaining southern pine forests. His memoir lists a total of sixty-five publications that he either authored or co-authored.

Source: Herbert L. Stoddard Sr., *Memoirs of a Naturalist* (Norman: University of Oklahoma Press, 1969). Courtesy of the University of Oklahoma Press.

Bird Notes from Trailside Museum
Mary Block

⟳⟳⟳ A saw-whet owl found good hunting behind the big Public Service Company gas tank near Maywood. A ridiculously tame, innocent looking ball of fluff, he sat on a loop of grape vine in a thicket. But while no one was looking, it hunted efficiently. One observer saw it eating a meadow mouse; it could hardly be criticized for that but from directly beneath its perch there was retrieved a male towhee, dead, and with both wings pulled out. Later in the day, I found just behind its perch another towhee, freshly killed, and minus a tail. When they were skinned each showed that he had been pierced though the chest from the back by sharp talons. The next Sunday, April 8, it was seen calmly sitting in its grapevine loop, tearing to pieces a junco. Some boys reported to me later in the day that when they passed the little fellow, it was just beginning on a fresh junco, and that nearby they had found a decapitated cardinal, and a similarly headless robin.

[Notes on a Woodcock]
Stephen S. Gregory Jr.

⟳⟳⟳ There is a small patch of woods in Northfield Township, Cook County, Illinois, which seems to afford attractive sanctuary for many of the birds of the district. It is only about seven acres in extent and rather sparse, but with a fair proportion of underbrush. The ground slopes away towards the north branch of the Chicago River and the Skokie flats. At the woods the elevation does not amount to much, but it is perhaps the highest point in the neighborhood. To the northeast corner of this patch of woodland I came by direction on Sunday the 10th of April, 1927, to enjoy a glimpse of a woodcock on its nest. I was told where the nest was in a general way and stood within a few feet of the bird for some moments knowing that it was close by, yet unable to detect its presence. Nothing appeared to my earnest scrutiny but a mottled expanse of dry leaves and branches as yet unadorned with foli-

Mary Block, "Bird Notes from Trailside Museum," *Audubon Bulletin*, nos. 15–16 (1924–25): 63.

Stephen S. Gregory Jr. file, Chicago Academy of Sciences, date unknown. Courtesy of the Chicago Academy of Sciences—Notebaert Nature Museum.

age. Unexpectedly, my patience was rewarded as the outline of the bird in every detail flashed clearly into my vision. And it was right on the spot where I had looked so hard a moment before with unseeing eyes. It was then mid-afternoon and although the nest was several yards in from the edge of the woods and not reached by the direct rays of the sun, sufficient light filtered through to lead me to believe that I might make some photographic record.

I am quite inexperienced in photographing birds and have no special apparatus. My camera is an old model Graflex 3-1/4 x 4-1/4, fitted with a Zeiss Tessar lens, working at f 4.5, and of course, a focal plane shutter. I was a little uncertain as to how the bird would accept my advances and therefore started in a rather conservative manner, taking my first picture at a distance of several feet, with an exposure of one-tenth second and diaphragm open to f 4.5. I varied my exposure from this position by changing the stop two or three times, and then, as my subject did not seem to be disturbed, took courage and moved eventually to the closest point which the bellows of the camera would accommodate. At this point, as nearly as I could estimate it, my lens was about eighteen inches from the bird. I was not very well satisfied with the woodcock's position. It lay facing in an easterly direction, with its head down and bill well buried in the leaves. That long bill is such a distinctive feature of the woodcock that I was quite anxious to have a better view of it and finally endeavored, as carefully as I could, to move some of the dry leaves in such a way as to offer a better field for the lens. But these leaves were really dry and they crackled ominously. I suppose that is the reason that the bird jumped just at that point and flew a few rods, lighting at the edge of the woods, where it stopped to preen its feathers. I photographed the nest with the four eggs and then left, returning for a moment, half an hour later, to find the bird still absent.

I had no opportunity to visit this interesting nest again until one week later, the 17th of April. In the meantime heavy flooding rains made me foolishly apprehensive for the safety and welfare of the nest. This time I found the bird at substantially the same time of day, lying facing in a general northwesterly direction. There was quite a breeze blowing through the woods and it rustled and whirled the dead leaves about sufficiently to hamper me somewhat in my effort to get pictures. One leaf eddied around in the breeze and finally moved over and lodged itself on the side of the bird's head, completely covering its eye. I admired the patience with which it retained its position, never flinching, giving no sign of discomfort or displeasure, other than the flickering of its eyelid. I had better luck, or perhaps was a little more careful or skillful, in arranging the leaves on this occasion. I found that by moving my hand very slowly, I could, with entire success, remove the one that was by the bird's head and others which interfered too much with my view. The light was sufficiently evenly distributed so that it was possible for me to move around the bird and take it from three different angles before retiring, without putting it off the nest.

On my third visit, on the 24th of April, the bird had taken still a third position on the nest, this time facing in a northeasterly direction. I started my operations as usual at some little distance and worked slowly up to the minimum range and this time did more scene shifting than on either of the previous visits. I felt that it was likely to be my last view of the bird on the nest and was anxious to make the most of the chance. I think the last three pictures in themselves illustrate quite well what good fortune I had in eliminating interfering foliage. Again I marveled at the steadiness of that little bird, for I could detect no response to motion with my hands at fairly close range, no muscular reaction, except the winking of the eye. This, of course, was not true when I reached the point of removing leaves which were either touching or practically touching the bird itself. It did flinch a little then, and much to my delight, raised up its head. After each re-arrangement, I made more exposures, of course, before daring to proceed much further, but I finally reached the point where it seemed to me if I could take out from under the bill of the bird one leaf which curled about it at its tip, I could get a view of the whole bill, which so far had been denied me. This seemed worth the chance as nothing more could be accomplished without eliminating this last obstacle. I suppose, if I had more experience with woodcocks, I should have known that with care I could do about as I pleased with the bird on its nest and perhaps I should not have been so gratified when I found that it remained after my last tampering with the leaves, and even permitted me to stroke its feathers when I was through photographing. It did exhibit some sign of apprehension at this last invasion, but did not leave the nest and I left it there to find on the succeeding Sunday that the little ones had been safely hatched and taken away.

Chicago, June 10, 1927.

❦❦❦ STEPHEN S. GREGORY JR. (1888–1964) WAS A LIFELONG resident of the Chicago area, leaving only to obtain his AB at Yale University and master's of engineering at the University of Wisconsin. It does not appear that he practiced engineering, but rather went into the field of financial planning. He founded and then served as president of the investment advising firm Gregory, De Long, and Holt, Inc. Gregory was a particularly valued member of the American Ornithologists' Union (AOU), the Chicago Academy of Sciences (CAS), and the American Society of Mammalogists (ASM): these groups, of course, boasted lots of naturalists among their ranks (of which he was certainly one), but many fewer authorities on investing. He was quite generous in lending his expertise in this arena to these organizations.

Gregory's interest in birds began early in his youth. As a teenager he contributed sightings to the U.S. Biological Survey, meriting this letter from W. W. Cooke in 1907: "Thank you most heartily for your extra good notes on the arrival of birds this spring, We are glad to enroll you on the list of our observers."

He came of birding age in a time when collecting was still in vogue. His personal collection began with a clapper rail that he shot while on a trip to Mississippi. He evidently obtained his federal permit after the deed so that he could at least keep the bird legally. (He eventually was licensed in Mississippi, Illinois, and Michigan.) Over the years, his holdings grew to over four thousand specimens, most obtained through the purchase of other collections, among which were those of such local ornithologists as Benjamin Gault and Ruthven Deane. After his death, the greater part of the collection—including four passenger pigeons, five ivory-billed woodpeckers, and six Carolina parakeets—reached the CAS, where they were housed in new cabinets donated by Gregory's friends. Gregory's wife donated the remaining portion of the birds, mostly taken in the Huron Mountains of Michigan, to the Museum of Zoology at the University of Michigan.

For people long dead it is of course difficult to assess their personal attributes, but documents in Steve Gregory's file at the CAS make him out to be exceptionally likable. For one thing, he apparently went out of his way to help young ornithologists. At AOU meetings, when the biggies would be socializing in someone's room, he made it a point to call less-established people to join them. He also retained his modesty, not always present among the wealthy and accomplished, to mention just two groups that randomly come to mind. Gregory had sent some specimens to his friend Donald Dickey, a curator at the Southwest Museum in Southern California. On April 3, 1921, Dickey responded to whatever Gregory had written with this letter: "No apologies are needed for the 'taxidermy.' You can make as good a study skin as I now.... It is inevitable that men ... who have done thousands of skins should acquire the technique ... in a way we cannot aspire to without more practice than one who has other things to do in life can aspire to."

The piece on the woodcock used in the anthology is from an unpublished paper in the same CAS file of Gregory material. There is a note from the archivist expressing some reservations that Gregory wrote the piece, but without stating any reasons. However, I see no reason to doubt his authorship, given that the location was near his home in Northbrook (Cook County) and that Gregory did develop an interest in photography.

Source: Alfred M. Bailey, "Obituaries: Stephen S. Gregory, Jr.," *The Auk* 82 (1965): 683–84.

Bird Life of a Roadside Marsh

Alfred M. Bailey

∽∽∽ The Illinois prairies are dotted with small marshes of a few acres in extent, and many of these swampy places border highways. In the early spring they resemble small lakes, but as the vegetation grows, the water disappears from view, and an area of luxuriant vegetation is all that remains, to the casual observer, to set the area aside from its surroundings.

I had an occasion to observe the bird life of one of these little marshes two years ago while photographing birds, and was amazed at the wealth of marsh forms nesting so close to the main highway. Along Roosevelt Road, due north of Hinsdale, was a marsh of three or four acres with water of approximately two feet in depth. I made my first visit on May 5 and found the dead tules so dense that they hid the water. At the west end, however, there was a sheet of open water, and along the border, among the dead growths I found three nests of the Pied-billed Grebe, one of which contained eight eggs, while in the other two were seven eggs each. These nests were floating masses of vegetation, and the eggs were concealed beneath decaying marsh growths. I photographed the grebes during the next three weeks, and in that time I never saw an adult bird—except from my photographic blind. The old ones were so shy, they always slipped from their nests before I was near enough to see them.

In the rushes, near the open water, were several nests of Red-winged Blackbirds, and in half an hour's search I found six nests of the American Coot, with sets of eggs ranging from two to eight in number. An American

Alfred M. Bailey, "Bird Life of a Roadside Marsh," *Audubon Bulletin*, no. 21 (1931): 38–39.

Bittern was observed, standing motionless among the reeds, and I found it was upon its nesting platform and that there was one egg.

The rails started building their nests later in May, and several unfinished ones were observed May 16, but I was unable to determine the species, as there were no eggs. A few days later I found a nest of the Virginia Rail with two eggs (which was found abandoned on the next visit), and on May 29 I found another of the same species with nine eggs. Thirty feet away in a similar location, in the tules a few inches above the water, was a nest of the Sora Rail, which had been destroyed, the eggs having been kicked into the water. Muskrats are abundant in this marsh, and it is probable that one of these animals used the nest as a resting platform. This same date I found a nest of the King Rail with six eggs, a bulky platform in a dense stand of marsh vegetation.

The Least Bittern and the Long-billed Marsh Wren seem to nest later than their neighbors, for I located two nests of the former and six of the wrens during the middle of July.

These roadside marshes then, inconspicuous areas of low lands, are of great importance in that they furnish breeding places for many forms of bird life, and efforts to drain the swampy places should be discouraged.

During the season in this little pond which I learned to call "my marsh," as I visited it so many times, I observed no fewer than nine nests of the Pied-billed Grebe, six of the American Coot, two of the Virginia Rail, one of the Sora Rail, two of the King Rail, two of the Least Bittern, one of the American Bittern, twenty-five of the Red-winged Blackbird, and eight of the Long-billed Marsh Wren.

Times change. Two years have passed, and "my marsh" is no more. Civilization is ahead, and hurrying automobiles have gained fifteen seconds of time by the straightening of Roosevelt Road. A concrete highway now bisects the low area which was once the nesting grounds of more than one hundred pairs of marsh birds; the marsh has been drained except for comparatively small places, and "my" birds have had to move elsewhere.

✦✦✦ IN HIS OBITUARY OF ALFRED M. BAILEY (1894–1978), ALLAN Phillips wrote that with Bailey's passing "ornithology [lost] one of its veteran explorers and pioneer popularizers, a leading early photographer and cinematographer." Bailey, born, raised, and educated (University of Iowa) in Iowa, spent most of his career as a museum director, an ideal role for one possessed of the eclectic talents mentioned by Phillips. Initially, he served as curator of birds and mammals at the Louisiana State Museum, before joining the staff of the Field Museum in 1926. By the next year, he was already at the helm of the Chicago Academy of Sciences (1927–36), which he left for the directorship of the Denver Museum of Natural History, a position he was to hold for thirty-three years.

Bailey refused to allow his many responsibilities as director to keep him deskbound, for he believed that "field work is the lifeblood of natural history museums." He pursued birds and other natural history subjects in various places around the world, including Ethiopia, arctic Alaska, and Mexico. (While in Durango, Mexico, he collected a sparrow that was almost ruined in the process. He considered tossing it, but decided to prepare what was left as a specimen. Good thing, for the bird proved to be of a new genus and species that was named for him, *Xenospiza baileyi*.) Of his more than two hundred publications, the best known is probably the two-volume *Birds of Colorado* (coauthored with R. J. Niedrach). Two different schools, Norwich University and the University of Denver, granted him honorary doctorates.

Source: Allan R. Phillips, "In Memoriam: Alfred M. Bailey," *The Auk* 98, no. 1 (January 1981): 173–75.

~~~~~~

American Egrets in the Lake Region

W. J. Beecher

☙☙☙ It seems quite impossible for me to recall any field experiences of the past summer without recalling also the prolonged drouth that so profoundly influenced them. Even now in fancy I can see the nacreous rising of clouds on every horizon, falsely promising a rain that never came. Again I see the dazzling white of highways turning upward, ever upward to sharp contrast with the hazy blue; and the hot blast from the prairies carries on its wings the taste of dust! Rain? it could not rain!

The season is at its height: every day the extraordinary heat seems to increase, the sunlight to grow more brilliant; from Chicago comes a steady stream of resorters and cottagers, seeking the relief that only the lakes can afford. And sometimes even the lakes seem a part of the abnormal fantasy;

W. J. Beecher, "American Egrets in the Lake Region," *Audubon Bulletin*, nos. 24–25 (1934–35): 8–10.

for scanning the slough a half-mile out, one often imagines he can distinguish the immaculate figures of long-legged, long-necked tropic birds....

... Just off the southwest corner of Long Lake there is a dip in the long swell of the morainic upland,—one of those frequent small areas where the flora of the gravelly hillsides gives way to the lush, green vegetation of the slough,—a depression which in normal years is the reservoir for quite a water supply. Formerly, before the coming of the railroad, a narrow channel wended through the marsh, connecting it with the lake; perhaps it was the cutting off of this feeder by the embankment that cause the pond to dwindle to almost nothing last summer. Looking from the railroad elevation on the north, one sees a lagoon singularly lacking in the cover necessary for marsh birds. The eye encounters almost no cat-tails or tules; there is only a narrow margin of saw grass about its border. Eastward, just at the foot of the slope is a wall of osiers that formerly marked the highest reach of the water; on the western margin, where the green swell of the goat pasture rolls away from the boggy shore, a quaint white summer cottage snuggles, its back to the woods. Deep-rooted in the muddy brink lie the rotting hulks of two sun-bleached duck boats; the water has long since shrunken from these, leaving a blackened field of lily pads, fantastically curled and seared as by a glowing iron.

Just over the eastern rise appears the roof of a pretentious residence, and across the tracks within easy stone's throw a large roadhouse squats; and somehow one gets the impression that this ancient little tarn is the only place hereabouts that remains untainted by the encroachment of Civilization.... But never while it maintained a normal level did the place distinguish itself as a haven for birds....

Frank Amos brought the news: he had seen them from the train—fully half a hundred of the large white herons. It was six o'clock in the evening, but Long Lake was only four miles away; I wheeled out my much-abused road racer and left at once.

The seasoned old hunter was right; singly and in pairs egrets began to pass overhead as I neared my destination, all flying east. It could not be far now; I cut across a narrow field, threw the bike over a rail fence and began to follow the tracks. A few minutes later I came abreast of a small woodlot, and suddenly in the hollow beyond I saw it! There was a brief impression of the dried-up pond, scarcely a hundred yards across,—of a sea of puddles alive with killdeers and yellowlegs. But that which immediately arrested all attention was the jagged white line of egrets that graced a long flat where the last of the open water lay.

The barren insignificance of the place was at once evident and I wondered what possible inducement could bring these wary birds to assemble there. Then I understood; halfway down the embankment and all but lost in the weed growth, a rambling barbed wire fence struggled. Fences meant privacy, and privacy—in this case I thought—meant protection. Apparently this was a family affair with the egret assemblage the object of interest; there were half a dozen great blue herons and a score of night herons and bitterns loitering about on the flats,—and curiously enough the egrets held solemnly aloof from food-seeking while all the others searched diligently for whatever the pond offered. I was surprised at my inability to pick a single little blue heron out of the flock. While I was absorbed in the senatorial calm of the honored guests, a thickset little German who lived across the tracks joined me; his name, he said, was Krenschen.

We stood there while the sun sank low,—and the old fellow, waxing genial, began to recount a history of the place;—how they had fought the closing of the pond by the railroad, and how they had lost. No, he had never seen the white herons before this year,—all through August they had been using the pond, coming in greatest numbers at evening. Never in his memory had the place attracted so much outside attention. Locally it was famous only for the size of its turtles. Krenschen wanted to know where they came from,—why they were here. I told him they lived mostly in the Gulf states, and brought their families north for a kind of vacation in late summer; they were due to return soon.

From every direction now the egrets began to arrive, seemingly converging from all the lakes; thirty feet over the pond each one set his wings in exactly the same way, the black legs dropped forward, arresting progress,—and he floated easily to rest in the midst of that ever-increasing throng on the flat. Counting the outstretched necks there were seventy-four birds.

As the glowing sun drops slowly beyond the verge of the world, the long dark shadows stalk out of the woods to the westward. Slowly they creep over the little hollow and its murky pond,—slowly up the eastern slope. Across the hollow of heaven is one mad striving of color as the deepening blue of the eastern sky begins to devour the riotous glowing of the west; and high up in the delicate blending of violet and gold, a whistling flock of blackbirds pursue their way toward some favored roost.

Twilight: my friend Krenschen has gone to town; over the pond a sudden hush has fallen. Into the brief interval the chirping of a solitary cricket drifts; a frog twangs gutturally. Then—a soft rustling of wings from the mud flat—a series of low, rasping croaks, and half the great army is in the air; another susurration like the soughing of wind in the aspens, and the last of the egrets are a-wing. Slowly, majestically they mount over the pond ... there is the unforgettable picture of their white figures against the deepening blue of the east—reflecting the last delicate bloom of the afterglow. They wheel, break ranks and strain away over the woods, their spectral figures beating silently into that transparent dusk. Entranced, I stand gazing after this fading vision, asking myself if what the eye sees can be true or is merely an apparition,—some spirit of the tropics come to haunt our boreal sloughs like a will-o-the-wisp....

...Rain? it could not rain! And yet it did,—torrentially and incessantly with the passing of August,—filling every gully and ravine, washing out crops, turning fields to lakes. I stood on the embankment again on a grey September morning; not an egret was in sight; only the wild plaintive cry of the killdeers and yellowlegs came to my ears. I found Krenschen feeding his chickens....

"Where are the herons today?" I called, though I knew the worst had happened. "The herons? They are gone!"—his blue eyes were wide with the wonder of it. "I haf not seen one since the other night."

That was all. I never saw the egrets at "the pond" again though I returned many times during the ensuing days. True, I continued to see them in small groups on the lakes, and received an authentic report of a subsequent gathering on Duck Lake a mile to the south; there were seventy in that flock—it must have been the same one that used to frequent the pond. But so far as I know an egret was never seen there again.

Perhaps the sight of such a congregation of these beautiful birds is not so remarkable after all; but the swift recovery of the egrets from their near-extinction by plume hunters a few years ago—the recovery of their trust in man—is very significant. To me it means that the extermination of the Passenger Pigeon will never be quite forgotten. We know now that, for the future, we shall reap as we have sown; it is at least reassuring to know that we are striving to plant well.

~~~ WILLIAM JOHN BEECHER (1914–2002) IS BEST KNOWN AS the director of the Chicago Academy of Sciences (CAS), a position he held from 1958 to 1983. He once told a friend that he would never wed because he was already married to the academy. In 1983 this relationship that he cherished so highly was rent asunder when he was forced to resign. I do not believe that he ever fully recovered from that heartbreaking event.

But what a time he had during that quarter of a century. Dr. Beecher was undoubtedly Chicago's best known ornithologist: he was the expert the newspapers went to if a bird-related item arose. He was also very shrewd in working with the media in promoting conservation. The most obvious example was the day he exhibited the thousand or so dead birds that he had collected over the course of one morning from the base of the John Hancock Building. He called in the press and sounded the first local alarm on the avoidable tragedy of building-induced bird mortality. His was the first voice in a choir that has since grown to include many, including Mayor Daley. (The fact is he was both a tireless and one of the most effective contestants in many of Illinois's critical conservation battles, including those to protect Goose Lake Prairie in Grundy County and Volo Bog in Lake County.)

Except for the three years he served in the southwest Pacific during World War II (and even while at war, he managed to send 288 specimens to the Field Museum), Beecher spent his entire life in this region. His interest in matters avian was kindled when Brother Anselm of St. Patrick Academy gave the class an assignment on birds; Beecher selected the red-headed woodpecker and was hooked. After high school he entered the University of Chicago, where he spent over ten years pursuing the three degrees he would earn there, including a doctorate in 1954. His master's research examined the distribution of birds in the wetlands of Fox Lake (Lake County, Illinois) and was widely lauded when published as *Nesting Birds and the Vegetation Substrate* (1942). He switched gears for his dissertation, focusing on the "anatomy of jaw musculature and skull structure."

Documenting and preserving a region's biodiversity requires first and foremost people willing and able to do it. It is a strength of the Chicago area to have numerous institutions, both private and public, that provide employment to such experts. Bill Beecher's history provides a good example: prior to his term at the CAS, he worked at the Field Museum from 1937 to 1942 and the Cook County Forest Preserve District's Little Red School House Nature Center from 1954 to 1957.

When he left CAS, Beecher devoted extensive energy marketing the seven-power binocular glasses that he invented. Of little value to birders, they did prove helpful to patients with certain eye disorders. Whatever profit he made on the glasses Beecher gave to conservation causes, particularly the protection of tropical forests.

Bill Beecher is one of the only two authors in this anthology whom I personally knew. During my butterfly-collecting phase of the early 1960s, my father made an appointment for us to meet him at his office in CAS. After first explaining that his real area of expertise was birds, Beecher gave us a tour of the museum, whose habitat dioramas of local settings were exquisite. His innovative use of "paintings, shadows, and color transparencies" instilled a realism that still haunts the memories of us old-timers privileged to have experienced the displays before they were destroyed when the academy left the building. For me, the most vividly recalled of the exhibits was one depicting cougars denning on a slope of a Palos moraine. But that tour didn't merely bestow memories—Dr. Beecher outfitted me with a butterfly net that I still use today, forty years later.

Source: Peter E. Lowther and Mary Hennen, "In Memoriam: William J. Beecher, 1914–2002," *The Auk* 120, no. 1 (2003): 199–201.

~~~~~~

Green Herons
Donald Boyd

☙☙☙ The green heron is the most common bird of this family with us. At least one or two individuals may be found about every creek, pond, or marsh. This does not mean that its nest may necessarily be found near where the bird is seen. As a rule these birds travel some distance from their nesting places to feed. Usually they follow a well defined, direct route between nesting and feeding places, and since their only journeys are always to and from these places it is not a difficult matter to find one or the other by following their air line trail. This matter is simplified as to the nesting site since they are frequently found in small colonies. The different members of the colony generally have their individual feeding grounds. Many journeys are made between these places when there are four or five half grown young in the nest. As one approaches the heronry the converging lines of the birds on their homeward route gives the exact location of the place very readily.

The nest is usually a very poorly constructed affair. Sticks and twigs thrown together at random apparently without lining, making a flat platform is the only effort, architecturally. Often the oval, pale, blue eggs may be seen

Donald Boyd, "Green Herons," unpublished essay (undated), Calumet Archives, Indiana University Northwest (Gary).

through the twigs from below. Likewise, as little regard to secure a stable foundation is shown.

At Horseshoe Lake in LaPorte County I found nests placed in the top of slender saplings fifteen feet above ground. The slightest shake of the thin trunk would spill out the eggs.

The bird is a graceful flyer and a stately wader. It would appear that the long legs and neck so indispensible most of the time would be most inconvenient for perching. However, it is quite remarkable how the bird can crawl among the branches and how gracefully it can settle down upon and leave the often precariously balanced nest without spilling out the precious, oval, pale blue jewels.

The largest green heron rookery I have found was located near Horseshoe Lake, Section 23, center township, LaPorte county. There were twenty-one nests here. They were all built in the tops of button bush not over six feet from the surface of the water in the slough. This slough was about one hundred feet in diameter, entirely surrounded by black and white oak timber and was isolated from any other marsh or swamp. Many of the birds nesting here fished along the margin of the lake one half mile away. Others flew in the opposite direction. The nearest fishing location of any size on this side was over three miles away.

I have frequently found the nests, three or four as a rule, in thorn apple thickets from six to ten feet from the ground. These nests were much more substantial in these cases and appeared to have been added to from year to year.

I tried to secure pictures of four half grown young from such a nest at one time. I climbed to the nest under very trying circumstances due to the wicked thorns and tangled branches. As I approached the youngsters betook themselves to keep out of reach. They climbed and crawled among the thicket very deliberately and with agility. I was surprised at the manner in which they accomplished this. Their ungainliness was not at all apparent. I did not succeed in getting within shooting distance of these youngsters so easily did they succeed in keeping at a safe distance from me.

I once found a green heron fishing from a little puddle about three feet in diameter. The puddle had been a good size pond for some time very likely for it was teeming with tadpoles, newts, and small fishes (top minnows and killfish). This was surely a "bonanza" for this heron. I should like to have known just how many days the supply lasted as the puddle dwindled in size and the prey became less difficult to secure.

It appears to me that the green heron follows the custom, to some extent, of some of its relatives, that of flying in the dusk. I have observed it silently wending its way homeward when there was barely enough light left for the human eye to distinguish objects. I have found them at their feeding grounds at daybreak. It is probably that they leave the heronry, in cases like this, before the first streak of dawn. I have heard their squawks in the air upon moonlight nights and it is possible that it prolongs the days fishing at these times. Perhaps fishing is best when the shadows are deepest. It is a fact that all fishes seek the shadows in the day time.

〰〰〰 DONALD F. BOYD (1882–1955) WAS AN ENTHUSIASTIC NATU-ralist who spent all his life in northwest Indiana. Never having attended college (as far as I know), he worked at the Standard Oil refinery in Whiting until illness caused him to leave. But natural history was what he cared about most anyway, and he found an outlet by becoming caretaker of the Boy Scout Camp in Portage. In his *Birds of the Indiana Dunes*, Ken Brock lauds Boyd for keeping voluminous and detailed notes on all facets of local natural history. He concludes with the statement that "Boyd's activities as a [bird] bander, in conjunction with numerous nesting observations[,] provide a solid picture of the Dunes Area birdlife during the first half of the twentieth century."

In addition to his notes, Boyd also composed essays on various birds. These are housed, along with Boyd's other writings, at the Calumet Archives located at Indiana University Northwest. The selection included here has never to my knowledge been previously published.

Source: Kenneth J. Brock, *Birds of the Indiana Dunes* (Bloomington: Indiana University Press, 1986).

Passenger Pigeon in Northeastern Illinois
Benjamin T. Gault

🕊🕊🕊 During late years, the Passenger Pigeon has become extremely rare in Northeastern Illinois, at least as far as the neighborhood of Chicago is concerned. My latest record was made at Glen Ellyn on Sunday, September 4, 1892. It was young of the year, very tame, and unsuspicious. It was discov-

Benjamin T. Gault, "Passenger Pigeon in Northeastern Illinois," *The Auk* 12 (January 1895): 80.

ered in the company of some jays and feeding about the piles of dirt recently made in excavating for the foundation of a house, well within the limits of the town, and was also observed to be picking the grain from some fresh horse-droppings, in which occupation it was harassed somewhat by the jays. But the day, location, and circumstances under which my observation was made precluded the possibility of the capture of the specimen.

Reminiscences of Early Experiences in the Chicago Area

Benjamin T. Gault

⌒⌒ ⌒⌒ ⌒⌒ Peace had come to a badly torn and disturbed nation when my parents moved from Decatur, Illinois, the writer's birthplace, to Manchester, New Hampshire, where we lived for about a year before coming to the then young city of Chicago. Had they gone to Logansport, Indiana, as they thought quite seriously of doing, the writer's career might have been materially altered. We lived in rented homes at first, but soon after the Chicago Fire of 1871 —witnessed on the two occasions by myself from a safe distance—moved to a permanent home on Washington Boulevard, then Washington Street. We came to Glen Ellyn in the fall of 1890. At that time Glen Ellyn bore the name of Prospect Park but prior to our coming the name was Danby. In the fall of 1867 we moved to Chicago from New Hampshire shortly following the assassination of President Lincoln, whom my father knew personally at Springfield in this state before he was elected to the presidency. I recall the wretched sanitary conditions existing at that time. Everything was wet and soggy, cholera was raging, and Chicago citizens were passing away at an alarming rate.

Recollections of my earliest years in Chicago, when we were living in a rented building on Wilson Street in the southwestern section of the city, are associated in my mind with the heavy flight of Wild Pigeons that took place at that time. It must have resembled in many respects similar scenes that took place during Audubon's day, for it reached from horizon to horizon in continuous flocks for practically one entire day. I sat on a moist packing box observing this great sight, barring meal times, during all the while it was going on.

Benjamin T. Gault, "Reminiscences of Early Experiences in the Chicago Area," *Audubon Bulletin*, no. 27 (1937): 19–21.

In the years following we had smaller flights off and on during the 70's and in the early 80's when they dropped off altogether, which was not surprising the way they were slaughtered for commercial purposes. The last specimen the writer saw was a bird-of-the-year in the early fall (See Chapman's Handbook for exact date) of 1893.*

The Ruffed Grouse is another bird that apparently has disappeared from these parts and it is questionable whether there are many of these noble game birds left in our state. Once they were fairly common here in DuPage County.

Riverview Park, then known as the North Woods, on the north branch of the Chicago River, in former days harbored many birds then migrating. At that time it was a heavy forested area, but scarcely any trees are now there and those that are I believe must have been planted. What is now known as Jackson Park, before being landscaped the year or two preceding the World's Fair of 1893, was another good locality in which to find birds and also the groves of woodland skirting the Des Plaines River. A few birds even nested at one time near Halfday and Wheeling and, I am told, north of Waukegan.

The Golden Plover is another bird I would like to mention. During the earlier years when we lived on Washington Boulevard it visited the closely grazed fields near what is now Humboldt Park both in spring and fall. Another favorite locality was in the vicinity of Lemont and Romeo on the Des Plaines River. But few of these birds were seen near Glen Ellyn when we moved there.

Although I learned considerable about the Prairie Chicken after coming to Glen Ellyn, my first experience with it was in Chicago. I accidentally stumbled on a nest in that portion of Garfield Park, then Central Park, south of Madison Street, a section left wild and unchanged. The nest contained fifteen eggs and I remember it as well as though it happened yesterday. A few of those eggs were taken and preserved but end-blown as was done in those days. The remainder were hatched and the chicks survived. "Chickens" then were fairly common on the virgin prairie lands all the way to the Des Plaines River and beyond toward Lombard and Prospect Park (now Glen Ellyn). At California Avenue (not then laid out) farm lands began. There was a corn field at California Avenue and for years afterwards cabbage fields were in evidence on Madison Street.

Those yet uncultivated prairie lands were rich with the original prairie flora and a fine sight to see in spring, summer and fall months. Upland Plovers

---

*More likely 1892. Editor.

or Bartramian Sandpipers bred there, delighting the ears with their whistling notes that should be heard to be appreciated. Even to this day scattered pairs are to be met with on the outskirts of Glen Ellyn, two pairs actually nesting and rearing young within the corporate limits of the village. Concerning the Calumet region, and especially Lake Calumet, I like to speak of the immense flight of Canvas-back Ducks once found there. The region was so famous that sportsmen from European countries came there to hunt them. At Sheffield, Indiana, just across the state line, a hotel was built especially for hunters. At one time I hunted on Calumet Lake myself. One trip is remembered well. My boyhood friend, John F. McNab (with whom I usually spend a part of the summer at a cottage we built at Bear Lake in Kalkaska County, Michigan), and I were at Calumet Lake one season when an immense flight of Canvasbacks took place. When the ducks arose from the center of the lake, keeping well out of gun range, the sound of their wings was like a heavy roll of thunder. Heavy gun fire could not have been louder. I have never heard anything like it before or since. It was a sight I never expect to see again.

On Calumet Lake there were also rafts of Coots which were never hunted or eaten in those days. They bunched together and could seldom be induced to fly and then only in short runs along the surface of the water. They certainly were fool birds and were a great nuisance to the hunters. It would have been a crime, I believe, to have shot them. It would have been the same in the case of the Mourning Dove. Now both of these birds in the eyes of so-called sportsmen are looked upon as legitimate game. Certainly some changes have come from what they were in those by-gone days.

We now hear a great deal about the Crow as being "Public Enemy No. 1" which I believe is a too hasty conclusion. Why not take into consideration his good points as well for he is not altogether bad. Of that I have some evidences and my opinion is that if we do not give him a fairer "break," crop diseases, especially those injuring corn, will be more prevalent in the future than they have been in the past. Cut worms will be more abundant and so will grasshopper scourges. The entomologists tell us that one of the latter is to be expected the present year, by the way. The underdeveloped hoppers or "nymphs" would make excellent food for the Crow and, in my opinion, would be much relished by him. No, the encouragement of "Crow Shoots," and dynamiting them at their winter roosts, is all wrong as it has been practiced in our state and by the Conservation Department. It is wrong, I think, for the harm it does psychologically to our growing youth by encouraging a murderous instinct.

➤➤➤➤BENJAMIN TRUE GAULT (1858–1942) WAS AN ONLY SURVIVING child out of six and obtained little in the way of formal education. At the age of eight, he received a small volume on natural history that may have sparked the passion that was to dominate his life. In 1902 he participated in an ornithological expedition to French Guiana that led the Field Museum to offer him support to continue his collecting activities in South America. He turned down the position because he refused to leave his mother alone (his father having died in 1905). Except for a two-year trip to Ireland immediately following her death in 1924, Gault spent the rest of his life in Illinois, birding, publishing notes on his observations, mounting the birds he collected, assembling his library, lecturing, and attending meetings of local nature groups. He was active in such scientific and conservation organizations as the Illinois Audubon Society, the Cooper Ornithological Society, the American Ornithologists' Union, the American Society of Mammalogists, and the Izaak Walton League.

Gault could well provide the model for some of today's most active birders and other naturalists: never married, he resided at his parents' home in Glen Ellyn for the last fifty-two years of his life and, according to biographer A. A. Chase, seems not "to have engaged in any routine income-producing activity." Unfortunately, this lifestyle apparently caught up with him in his later years when he became ill and found himself in financial hardship. One can only hope that his travails were lightened, at least in part, by the many friends that he had made through his enthusiasm, pleasant manner, and willingness to share information.

Source: Audrie Allspaugh Chase, "Benjamin True Gault," *The Auk* 65, no. 4 (October 1948): 647–48.

Geolycosa, the Wolf of the Dunes

Donald C. Lowrie

There ne'er were such thousands of leaves on a tree,
Nor of people in church or the park,
As the crowds of the stars that looked down upon me,
And that glittered and winked in the dark.

—R. L. Stevenson

Donald C. Lowrie, "*Geolycosa*, the Wolf of the Dunes," *Chicago Naturalist* 2, no. 1 (March 1939): 2–8. Courtesy of the Chicago Academy of Sciences–Notebaert Nature Museum.

⌇⌇⌇ As I walked along the beach in the dunes one spring evening, a sight which greeted my eyes brought this little childhood rhyme distinctly to my mind. There, winking at me in the dark—not from the sky but from the sand—as I flashed my head-lamp from spot to spot, were countless minute points of light. Uninitiated, one might well believe that he had imbibed too freely or for some other reason had jumbled his up and down directions. "What," you will ask, "would someone *initiated* recognize these topsy-turvy stars to be?" Nearly as startling as the imagined inverted firmament, the answer is that these myriad reflectors are spiders' eyes. Night is the time when most spiders do their hunting and *Geolycosa wrightii* is no exception. With eyes as good as any cat's, this sand-burrower waits at the door of its burrow for some incautious insect to crawl within reach of its cat-like leap. Perhaps I should say wolf-like leap for this spider belongs to the Lycosidae or wolf-spider family. Not the least startling thing about these hosts of spiders is their sudden, seemingly miraculous appearance in the evening. On a sunny summer afternoon any spot on the beach, fore-dune, or a blowout will show scarcely a trace of a burrow. The same spot in the evening will be well stippled with open burrows. What is the answer?

Let us sit down in the area in which we saw a pack of these "wolves" last night. As darkness descends, we watch the colorful glow of the sun on the clouds. On this particular evening we hope that the waves may be quite stilled, for we can look for more action from the spiders on a calm evening. As the sun slowly sets, in the flashing colors so well known in the dunes, we can speculate or dream—or catch a cat-nap—according to our inclinations. Our "wolves" will not show themselves until the quick-flashing, siren brilliance of the Lindbergh beacon begins to appear on the western horizon. Then, as the light becomes too faint for reading, we use a flashlight or headlamp. What do we see?

First of all little spider-stars suddenly begin to twinkle in the distance where only blackness was before. Is it magic? Do the spiders appear at nightfall like earthworms with the rainfall? Let us observe more closely. Near at hand we see a spot of sand suddenly dissolve. A small round opening appears and out comes our lustrous-eyed Arachne. This, then, is the answer to our question. As rain brings the earth worms from the sod, so does night bring the lycosas from their burrows. During the days these tubes usually are closed, especially if the wind is disturbing the sand. With her abdomen pointed upward, the spider weaves a curtain of silk over the entrance so that all signs of the opening are obliterated. If, at dawn, the sky is slightly overcast, she may leave the tube open, or if the night's catch has not been satisfying, she may not close her home, perhaps hoping for the approach of some unwary victim before she retires. Sometimes even when the burrow is closed, if no wind has disturbed the sand, the exact spot of the entrance may be determined, for the spider has left her foot marks radiating out in all directions from the nest. Each dash after a victim, successful or not, has left its mark in the sand.

If you wish to catch one of these Amazons (for the female is a warrior as well as a homebody, while the male is a carefree, burrowless roamer), you must become adept at treading lightly or at digging. When you spot one of these eyes, you must "walk softly and carry a big stick," as another nature lover has said in another connection. Of course, if you are out for the exercise, walk right up and start digging. Walking softly, however, you can approach to a position next to the hole where the spider is quietly waiting. Then with a quick slice of your "big stick" (in this case a trowel is really more helpful), you insert it beneath the spider, cutting off her means of escape to the bottom of the tube. The trowel full of sand can then be dumped out and the spider retrieved.

If your weight is not appropriate for the silent stalking method, you can creep up to the opening with a minimum of jarring and wait there for the impatient female to reappear. During the day the few spiders leaving their burrows may be caught in much the same way, though usually they will not come to the surface even when a bit of sand is slowly poured into the burrow. If during the day or night the spiders will not oblige by coming to the doorway, you must finally abandon strategy altogether and resort to digging in earnest for this buried treasure. You will find the technique quite an art in itself, and an interesting method as well, until the labor of digging overbalances the novelty of the reward. The excavation may be made in several ways—trial and error will teach you the best method. I have used the following to best advantage. A hole is dug close beside the burrow; then one wall of the tube is broken away with a twig, forceps, or similar rigid tool, exposing the burrow to the bottom. You will find that the burrow may not be straight. Typically, the first few inches slant a trifle. There is then an enlargement enabling the spider to turn around. From this point it drops vertically to the bottom. Often it will extend straight the whole way with only an enlargement at a depth of a few inches. You will lose many spiders at first, due to obscuring the burrow with falling sand, but if you finally reach the bottom, eighteen or twenty inches below, you will be rewarded with a large, scrambling, gray female. Her size will probably startle you, for this species, with a body nearly an inch long, is among the largest of our northern spiders.

If your interest extends beyond digging wolf-spiders out of the sand and dropping them into preserving fluid to be forgotten, you may find amusement and instruction by bringing them home for observation. If placed in a vessel as tall and narrow as a quart mason jar, one of these spiders, if not mature, may be induced to dig a fresh burrow, especially if a hole is begun in the sand. If full-grown she may have lost much of her tendency to burrow and may remain on top of the sand unless a hole is dug for her encouragement. Usually, however, with a little patient manipulating she can be stimulated to resume her natural habit.

The next problem is to provide food. She has a voracious appetite and, like all araneida, is entirely carnivorous. Insects are her staples and a few large cockroaches will suffice for several days. Flies, crickets, grasshoppers, beetles, bugs, or any other insects which can be easily procured, are very satisfactory, and raw beef can sometimes be substituted.

Some morning in late May or June a flattened ball of silk, a quarter of an inch in diameter may be found, hanging by a thread from the female's spinnerets. The construction of this cocoon is an interesting bit of the life of Madam Wrightii which must be left for another story. On examination the cocoon will be found to be filled with from one hundred to four hundred eggs. During any sunny day our lady may be seen at the entrance of her burrow with her abdomen pointed skyward turning the cocoon with her hind legs so that all surfaces are exposed to the full rays of the sun. If the eggs have been fertilized, you may again be surprised some morning late in June or July to see this spider appear much larger and fuzzier than usual. Closer examination will reveal this fuzziness to be a mass of young spiders. They have emerged from the cocoon and migrated to their mother's back where they crowd closely together, clinging to her fur and to each other. As she sits in her doorway, sunning herself and her young, you can observe these cannibalistic miniatures moving about and occasionally feeding upon their brothers and sisters. If the mother were still at large in the dunes, the young would begin to fare forth into the sandy wastes after a week or two of this life. Then we would find the pin-head-sized holes of the young in the vicinity of the larger burrow of the mother.

Occasionally the cocoons may yield a brood of small flies or wasps which have parasitized the eggs. Sometimes even the adult may be found shrivelled up with an innocent-looking fly pupa lying alongside. The study of this fascinating phase of life seems to have been sadly neglected for little is known of the details of this relationship.

After the young disperse they grow rapidly, enlarging their burrows with each molt, and become nearly adult by fall. The following spring, after one or two molts, they become mature, mate, and lay eggs by June. In the fall, and often in the spring, most spiders migrate by means of a "balloon"—a long, buoyant thread of silk, wafted by the breezes for great distances—an effective means of dispersal. The geolycosas probably do this although there has been no actual observation of such behavior.

At the approach of winter the female closes her burrow and hibernates at the bottom. Here we must ask more questions. Do the females of the second year have to be fecundated again? How many years do they live? We hope some day to find the answers. In the spring, as I have mentioned, when the male reaches adulthood, he will be found on the surface of the ground at night searching for the female. Again we ask, do they dig burrows when adult or do they pass the day under debris on the beach? Do they live only one year and the following summer, or do they hibernate the following year?

We have pried rather closely into the habits and home life of this lady. Therefore let us see, briefly, how this habit of living in a burrow originated. Burrowing is an old mode of life, being found in the two primitive suborders. In the higher families it is rare. The Salticidae or jumping spiders, as well as most non-web-building species, seek shelter beneath bark and in other protected places. Among the lycosids we find many individuals making a fairly permanent retreat in some protected place. Evidently there are all gradations of burrowing from a mere improved hollow on the surface of the ground to the deep sand burrows of *Geolycosa wrightii* and her relatives. Thus it would seem that this burrowing habit is a result of the change to a permanent abode. To insure the propagation of the species the male still retains a free-roving habit in adulthood.

Almost every fore-dune, cottonwood dune, or bare sandy blowout of the Lake Michigan shores of Indiana, Michigan, Illinois, or Wisconsin, or of the sandy areas of the Kankakee River Dunes is well populated with this particular species. The closely related *Geolycosa missouriensis* may also be found in the Chicago Area, but in the oak dunes and other only slightly sandy areas. Two other members of this genus are found on the east coast in comparable situations.

Ecologically this spider is one of great interest as it is clearly an index species for the fore-dune and cottonwood dune associations. The only other animal vieing for this honor is the sand-colpred grasshopper, *Trimerotropis maritima.* Another interesting consideration, which is still in the speculative stage, concerns the color of the spider. A first, superficial glance inclines us to say, "This is an excellent example of protective coloration." On closer analysis we can not be so sure for only infrequently does the spider show herself above ground during the day. However, we must also remember that usually she is abroad during moonlit nights when her color matches the sand even better than during the day. This question, as well as many others, must be left until experimentation and further observation give us the answers.

➤➤➤ DONALD C. LOWRIE (1910–2000) DEVOTED HIS PROFESsional life to the study of spiders. He attended the University of Chicago, earning a B.S. in 1932 and a doctorate ten years later. (His dissertation was entitled "Ecology of the Spiders of the Xeric Dunelands in the Chicago Area.") During at least some of that time, he worked at the Chicago Academy of Sciences. Lowrie served in the navy during World War II and the Korean War, rising to the rank of lieutenant commander. He then moved west, teaching briefly at the University of Idaho and New Mexico Highlands University before settling in at California State University, Los Angeles, from 1956 to 1972. After retiring from teaching in 1972, he relocated to Santa Fe, New Mexico, but kept very busy. At the age of sixty-nine, Lowrie joined the Peace Corps and served in Paraguay, and in the aftermath of the Yellowstone National Park fire of 1988, he was among the team of biologists recruited to assess the fire's impacts. Apparently, sometime around 1990 he was involved in a serious car accident, and he lost contact with fellow arachnologists.

Sources: Personal communication and e-mail from Donald Dewey, dean emeritus, California State University, Los Angeles; e-mail from David Richman, American Arachnological Society; e-mail from Lenny Vincent, American Arachnological Society.

The Opossum, Prophet without Honor
D. Dwight Davis

ᔕᔕᔕ If a European naturalist were asked to name the most interesting North American mammal, the chances are that he would reply "the opos-

D. Dwight Davis, "The Opossum, Prophet without Honor," *Chicago Naturalist* 2, no. 4 (1939): 99–104. Courtesy of the Chicago Academy of Sciences—Notebaert Nature Museum.

sum" without hesitating. His answer might well surprise those who have been taught to regard the opossum as a dull ratlike brute, relished as an article of diet by poverty-stricken southern negroes, and with a skin scarcely worth a quarter in a good fur market. But others, perhaps more familiar with the opossum's legitimate claims to respect, would probably agree that its uninteresting appearance and doltish behavior are only a sham. Is the opossum another proof of the truth of the adage that familiarity breeds contempt?

The appearance of this animal is familiar to everyone, although few may actually have seen it alive because of its secretive, nocturnal habits. Its appearance is not likely to inspire much admiration. The white, sharp-pointed face wears a vacant look in spite of the shiny black eyes, and this is likely to be heightened by the silly grin that is assumed when the mouth is opened in what is intended for a threat if the animal is cornered. A pair of naked patent-leather ears are perched rather absurdly on the head. The hair is coarse and dirty looking. The tail is naked, scaly, and ratlike, and is chiefly responsible for the popular misconception that the opossum is a first cousin of the rat. With these esthetic handicaps the poor animal generally has two strikes on it the minute anyone sees it for the first time.

Naturalists say of the opossum that it is "primitive" and that it is a "marsupial." We may well inquire briefly into the meaning of these two words, for they are the key to the opossum's claim to more than passing interest.

The ancestry of most back-boned animals is a history of slow improvement and gradual perfection. Brains grow larger and more complex. Limbs become graceful and mechanically efficient. Teeth develop into amazingly effective tools for grinding or chopping or gnawing. The whole organization of the animal is stepped up into a smoothly functioning unit that "clicks." Comparing one of these modern mammals with one of its remote ancestors would be much like comparing a 1939 automobile with a horseless carriage from 1900. But occasionally an animal may fail to "improve," and may remain in almost exactly the same condition for extraordinary lengths of time. The reasons for this strange lack of evolutionary ambition are mostly unknown. Fossils show that the opossum is one of these—fossil opossums from the Eocene (almost 50 million years ago) are nearly identical with opossums living today. Hence such animals are often referred to as "living fossils." They are of extraordinary interest, since they provide a clue as to how animals must have looked and acted millions of years ago. Such, then, is a primitive animal—it is like a 1900 model automobile that is brought out of storage and paraded with this year's streamlined models. For some reason people associate strange creatures with remote places, and it is something of a shock to find that one of the strangest lives at our very doorstep.

Everyone associates kangaroos with Australia, and most people know that the female kangaroo carries her young around in a pouch. Few know that all the mammals of Australia are provided with pouches similar to that of the kangaroo, and probably very few indeed realize that our native opossum shares this strange character with the mammals of Australia. This pouch, or *marsupium*, is only one of the many features that these animals have in common, but it is the most conspicuous and it is important because it is the one from which the group derives its name. The marsupium is associated with a pair of "marsupial bones" on the pelvis. Careful study of any part of a marsupial's body generally reveals fundamental differences between them and all other mammals. Some have been so impressed by these differences that they deny that marsupials are mammals at all, in the very strictest sense of the word. Under any circumstances, a rat (or a lion or an elephant, for that matter) is uncomfortably more closely related to a man than it is to an opossum. And this is what a naturalist means when he says that the opossum is a marsupial—a curious mammal so distantly related to the other mammals with which it lives that it can be regarded as a stranger in a foreign land.

Under these circumstances I trust that I may be forgiven for keeping an opossum alive in my basement for several months—much to the disgust of my dog. The animal had been treed in one of the suburbs of Chicago, and it seemed that if opossums had survived this extraordinary history they should be worth a better acquaintance. Dozens of questions arose that seemed worth looking into. I had no desire to make a pet of the creature in the usual sense of the word—indeed few animals are more ill-suited to taming. I had a hearty respect for its teeth as a result of a disagreeable experience several years ago when an opossum "clamped down" on my hand and held fast with the proverbial tenacity of a bulldog; its jaws finally had to be pried loose. But an occasional hour spent watching this sulky captive, with a notebook and pencil to record bits of behavior, sometimes supplemented with a camera, soon produced a surprising record of its habits.

It is common knowledge that opossums are arboreal, but I was scarcely prepared for the almost monkey-like agility with which mine hoisted itself up a vertical one-inch water pipe, using its awkward-looking forefeet as very efficient grasping hands. This at once raised the question of how it climbed trees that are too large to be grasped, and the matter was promptly settled by placing it on a good-sized tree trunk and awaiting events. The resulting spread-eagle posture, with the claws digging into the bark, was vastly different from the grasping used on smaller branches, but it was effective. Like many other arboreal animals, opossums have prehensile tails that can be wrapped tightly around objects and thus function like an extra hand when the animal is climbing. Some have denied that an opossum can support its weight from its tail unless more than half of its tail is wrapped around an object, but mine could hang from a broomstick with only a couple of inches of the tail tip wrapped around it. It could not lift its heavy body up to the broomstick by means of its tail, however, but it solved this engineering problem very simply by turning and climbing up its own tail! The ludicrous sight thus presented may be imagined.

If the opossum looks at ease and not too unskillful when it is climbing, it is quite the reverse on the ground. There it sits hunched up, the toes of its forefeet splayed out absurdly so that each foot looks like an animated asterisk. It walks with an awkward shuffle, the hind feet pointing out in Charlie Chaplin style. No matter how hard pressed it refuses to break into a gallop or even a trot, but merely moves its feet faster and faster until often it falls over from sheer clumsiness.

Opossums are far from choosey about their food. Mine would eat almost anything within reason except salad greens like lettuce or celery, but it was particularly fond of canned dog food. Feeding was accompanied by loud and unbeautiful piglike smackings of the lips. Opossums have 50 teeth, compared with our 32, and the many cusps of these teeth are used in masticating very thoroughly and deliberately. The clumsy-looking forefeet were again used expertly during feeding. Whenever it picked up a piece of food too large for a single mouthful, the protruding end was invariably grasped deftly in one hand and held until the rest had been eaten. On the other hand, it never picked up food with its forefeet to convey it to its mouth.

Simple experiments showed that the senses are very unequally developed. The beady black eyes must be much less sharp than they appear. If I waved my arm vigorously within three feet of its nose, being careful meantime not to make any noise, there was no response whatever even though the animal might be looking straight at me. Smell, on the other hand, must be very acute, for it continually pointed its nose in the air and sniffed vigorously; it always located its food by smell. The most interesting results, however, came when I tried a crude test of its sense of hearing. Any unusual sound caused it to flinch in a ridiculously exaggerated way, as if it had been struck or a cannon cracker had gone off under its nose. Taking advantage of this fact, I stepped back about four feet and gently scratched my trouser leg with my finger nail. Although I strained my ears—which were less than two feet from the source of the sound—I could hear nothing. But the opossum jumped as if a bomb had fallen beside it!

This creature is one of the most silent of mammals. The only evidence that mine had a voice at all was when my dog happened to corner it on a couple of occasions and once when it started to disappear into a hole and I pulled it back by the tail. Then it emitted a very low moaning growl with an almost eerie quality that startled me the first time I heard it.

My opossum lived up to the reputation of its kind for being nocturnal and always spent the day asleep. The sleeping postures adopted were often extraordinary. Several times I was positive at first glance that the animal was dead—it lay on its side, with its head twisted around so that it rested on its top surface and the legs stuck out at stiff and awkward angles. Just as often one of its feet had a firm grip on the edge of its water dish, and I wondered if this had anything to do with the necessity for holding fast when it normally slept in trees. Then one day when I had stolen up on it while it slept I noticed that the ears, which stand erect when an opossum is awake, were folded forward so that they fitted snugly over the ear openings. No particular note was taken of the fact, as it seemed likely that their strange position was only an accident. But each time the animal was discovered asleep the naked black ears were in the same position. It was apparent that this curious bit of behavior was one of its normal habits, although it is hard to imagine why it should be necessary. This "folding up" of the ears does not seem to interfere seriously with the opossum's hearing, and the least sound causes them to fly back to their usual position. Making [a] photograph...consequently posed a pretty problem, which was finally solved by setting up lights and camera, then waiting for the animal to go back to sleep and stealthily clicking the shutter.

By the time I had completed my leisurely series of observations on its behavior my opossum had grown fat as a puppy, although it was still as unfriendly as the day it was captured. Holding it captive any longer was pointless, and certainly not to the opossum's liking. One night it was bundled into a sack, taken to a forest preserve that provided ample cover, and unceremoniously dumped out on the ground. It stared dazedly into the beam of a flashlight for a moment, then turned and scampered off into the gloom.

Out of curiosity I decided to check my notes against observations that others had made on opossums. A search through the literature produced rather surprising results. Dozens had killed opossums, laid them on the dissecting table, and studied their structure. Other dozens had studied the strange embryology of this animal, which introduces tiny grub-like young into the world after the unbelievably short gestation period of only twelve to thirteen *days!* But no one had recorded more than bits of the behavior that go to make up the daily life of this remarkable animal, and which are just as important as its structure in making the opossum what it is. Surely the possum is a prophet without honor in his own country.

◆◆◆◆ D. DWIGHT DAVIS (1908–1965) BEGAN AND ENDED HIS CA-reer at the Field Museum, twenty-four years of which was as curator of anatomy. During that time he did hold visiting professorships at the California Institute of Technology and the University of Malaya. Born in Rockford and educated at North Central College in Naperville, he published not only in his field of comparative anatomy but in mammalogy and herpetology as well. He produced over fifty scientific papers, as well as many book reviews and general interest articles. His magnum opus, "A Morphological Study in Evolutionary Mechanisms" (*Fieldiana: Zoological Memoirs*), was a treatise on giant pandas that was published just months before he died. It was a project that he started in 1938, when the Brookfield Zoo conveyed to the museum the recently deceased Su Lin. At that time it was unclear where the giant panda belonged in the order of carnivores. Over the years Davis assembled the evidence that definitively established that the animal is a bear, although not an ordinary one but rather "an exaggerated bear" that had evolved from being omnivorous to strictly vegetarian. Adding to the study's significance, he used the panda as a model to tackle other issues of comparative anatomy.

Biographies written on the occasion of the subject's death, and especially when authored by a close friend of the deceased and published by the institution with which the subject was associated, tend to be as complimentary as possible. It was therefore surprising to read in Rainer Zangerl's article that Davis, although "meticulous" and a "perfectionist," could be quite the unpleasant fellow—"as a colleague, he was often difficult, unapproachable, sometimes caustic." Only his closest friends were treated to his warm, generous side.

Source: Rainer Zangerl, "D. Dwight Davis," *Chicago Natural History Museum Bulletin* 36, no. 3 (1965).

Pyramids of Palos

A. S. Windsor

"The Ant, as the Prince of Wisdom is pleased to inform us, is exceedingly wise. In this Light it may, without Vanity, boast of its being related to you, and therefore by right of Kindred merits your Protection."
—William Gould, An Account of English Ants, 1747

For almost ten years I have watched the spectacular mounds of *Formica ulkei*, a sparsely distributed but exceedingly interesting ant, endure the wear and weathering of the seasons near Palos Park. Even when judged by human standards a decade is not brief, and for the ant it must be infinitely longer. Who knows but that to the worker which lives but three or four years, or even to the queen whose mature life span may reach ten or fifteen, these mounds—majestic monuments to their tireless endeavor—may be as age-resisting and enduring as the pyramids of Egypt! As the great monuments along the Nile took shape, stone upon stone and life upon life, the importance of the individual was dwarfed to insignificance, yet the combined effort of numberless workers produced structures which may resist the passing of a million years. So with the ants: a continuous succession of individual lives, each carrying to the surface her tiny burden—a pellet of earth—each worker replaced, when she dies, by another and then another, similarly imbued with the same constructive instincts.

Standing before one of these insect sanctuaries, I recall, invariably, the pleasure which accompanied my first visit to these large mounds. My companion on this occasion was Dr. T. C. ("Ted") Schneirla, now of New York University. Having been told of the aggregation of mounds we wandered erratically over a considerable area in our efforts to follow rather vague directions. At last we came upon one, rising sharply out of the side of a ditch bank. The size and appearance of the mound alone were enough to convince us that we had found an ant worthy of our diligent search. The trowel pierced the sun-baked periphery of the cap and only a glance at one of the forceps-held workers was necessary for Ted to assure me that it was *ulkei*. A small complement of workers was placed in vials of alcohol, a superficial examination of the tunnels was made, then the earth was returned to the wound in

A. S. Windsor, "Pyramids of Palos," *Chicago Naturalist* 2, no. 3 (October 1939): 67–72, 91. Courtesy of the Chicago Academy of Sciences—Notebaert Nature Museum.

the mound and we left it very little the worse for our inquiry. A short time later that year I returned to this mound and, taking the only direction in which we had not searched before, came upon a semi-circular clearing at the edge of a dense oak woods. Dotted here and there among the scrub bushes between the towering oaks and the grassy clearing were eight or ten huge conical mounds. Upon looking more carefully about, I saw dozens of mounds, each one placed rather equidistant from the others. There was no doubt that I had come upon "ulkei-land"—the "promised land" to him who inquires into the ways of the interesting ants. On August 13, 1939, Ted and I again went to ulkei-land, and even to the site of the first mound we had found together in 1931, which, since ditch-widening operations had dealt less tenderly with it than had we, had long been abandoned.

Be assured that in talking of ant mounds the use of such adjectives as "huge" is strictly relative. We who roam the meadows and the waste places to become familiar with the habits of ants consider a mound eighteen inches in diameter and ten or twelve inches in height as "large." Thus, in comparison, a mound seven feet in diameter at the base and towering up to a vertex three feet above the ground level would have to be referred to as "gigantic." Several of the ulkei-land mounds are of such dimensions—all are distinctly "large" except the young ones, for the different nests tend to the same architectural style.

Several years ago a census was taken of all mounds in this area of not more than a few acres, and the total number was over four hundred. Over half of the mounds bordered the oak forest, about a third were found in small clearings in the forest where the sun could penetrate effectively even in the summer, and the rest were out in the open lowlands with no trees about them. Truly a wondrous aggregation of ant communities! There are but few other sites where this ant is found in such great concentrations in the Chicago Area. It is, in fact, to be considered a rare ant. The few localities in the region where it has been taken present the same general type of communities although on a more modest scale. A somewhat similar group of mounds has been found near Palatine, but recent visitors have said that they are rapidly declining. Near Hammond, Indiana, I came upon a group of these mounds in the early spring of 1938. These were by no means as numerous and the size of the individual mound was not particularly spectacular.

The superstructures of the nests of *Formica ulkei*, the mounds in the Palos region, are of an interesting composition. Around the edge of the forest they are made up principally of clay, with small twigs, leaf petioles, and the like, mixed in. The surface is fused by the rain and becomes quite hard. This outer cap, usually about an inch in thickness, effectively protects the honey-combed maze of earth tunnels beneath it. The much-used entrances are more numerous around the base, and here since the galleries are being constantly extended or repaired is inevitably an accumulation of fresh earth pellets brought out from the interior. During the most active summer period the accumulation may even cover the entire mound. The more vigorous mounds are usually entirely bare, although on some a few tufts of grass have escaped the "home improvement brigade." Grass and small shrubs yield to the onslaughts of these industrious insects and it seems probable that the ants exert some restraining influence on young trees that are wont to compete. If vegetation were to become dense about a mound, the ant community would soon be choked out, for ants must be hot to be most active and the temperature of the earth inside the mound is a most important determiner of how rapidly the broods are to be matured. In short, a low temperature brought about by a damp shady location would restrict the population of the mound and thus cause the community to decline gradually, if indeed not bring it to total extinction.

Preëminent students of ants have given us a most pleasing concept of the ant colony: that it should not be thought of in terms of its individual units, the ants, but that the entire community be considered a single organism. This is not entirely new to us, for one seldom thinks of the individual cells that make up the human body—but of the body as a whole. In the ant colony the reproductive functions are cared for by the fertilized queen and her ephemeral consort, nutriment is brought into the "body" by some of the workers, and the excretory functions are performed by certain others. The analogy is obvious. Therefore, in looking over some four hundred of these colonial organisms, we can see interesting individual differences. There is the young mound, not more than a foot in diameter and of similar height. We know that its population is strong and most active because over the mound is a virtual covering of fresh earth crumbs, sticks, and other chippage. Then there is the colony at the very climax of its life, the large conical mound, sturdy, stalwart, and actually of sufficient solidarity to deflect as large an animal as a cow. In other mounds we find that senescence has begun. Instead of the piercing crown the apex has melted away until only a plateau six or eight inches high is left. All activity may have ceased except in one small area where a few individuals struggle valiantly but hopelessly to restore the former rugged strength of the colony. In another case, we see tell-tale moss as it enshrouds a victim of adverse circumstances. The final stage is seen when grass regains complete possession and scarcely any elevation of the earth can be detected at the spot where once trod tens of thousands of hurrying, busy workers in earthen temples.

The individual *ulkei* workers are of sturdy stature, not the largest in the region, but definitely of a large size, about one fourth of an inch long. They have moderately long legs so that they get about with great agility, and their trim thoraces and gasters suggest great endurance. Their heads and gasters are deep brown or black in color and the thoraces a pale reddish brown. The bite of these ants is rather forceful—especially if it be in a tender spot—and they cling tenaciously! To stand beside the mound in a seething mass of these ants is far from comfortable, and browsing animals have undoubtedly found it best to forage at a safe distance from the main trails lest their nostrils become plagued with tenacious defenders.

Valiant though these ants be and proud their pueblos, there is to be recounted one surprising but undoubted weakness. Because of the close affinities of this species to others which are better known, it is most commonly thought that F. *ulkei* is a temporary parasite on another species. That is, the young *ulkei* queen, after her marriage flight, in all probability enters the nest of a very young colony of another species. She may not be received amicably at first, but may succeed in becoming adopted. (Think of the thousands of queens that will not survive such a rigorous vicissitude—only an occasional one being successful!) The host queen is killed and the host workers care for their new center of interest as zealously as they did for their own mother. Her eggs are tended, they mature, and gradually a goodly number of *ulkei* workers become active in the mound. Old age gradually eliminates the workers of the host species so that eventually a pure *ulkei* colony results. Whether or not this actually takes place we do not know; more observations on *ulkei* will have to be made. Once a colony becomes well established it may give rise to other colonies by budding, the precise method of which, in this species, is not exactly known. This phenomenon, however, probably explains why the mounds are so numerous in a small area.

These vast hordes of hurrying communal organisms require a never failing food supply. Hordes? Yes, and then some! My guess as to the total number of ants of this species in the Palos group is sixty millions! This is only a rough figure, of course, and one which might vary either way by fifty per cent. You would probably arch your eyebrows if told that these ants eat oak trees, but, indirectly, such trees do furnish a considerable proportion of their nutriment. The intermediate agent, an all-important ally, is the ubiquitous aphid or plant louse which feeds by piercing the epidermis of plants and sucking the sap. The droplets of honeydew (undigested sap) given off by aphids, are sweet and delectable to the ants, and thus the plant lice of the oaks provide much food for our mound builders.

Upon a recent visit to the Palos mounds, I saw a stream of ants communicating with an oak along a trail some three inches wide. The trail began at a large mound and connected with the tree. Great numbers of ants ascended the trunk and went out to the leaves which were curled over groups of plant lice. When heavily laden, they returned their sweet cargo over the same trail to workers in the nest and to the masses of brood. This was not the only evidence of feeding, of course; many workers brought in grasshoppers, caterpillars, and other small insects.

For the *ulkei* the summer months are characterized by great activity, only a part of which is food-getting. The galleries have to be repaired and extended; the cap may need reinforcement where it may have been damaged by rains or animals; young have to be reared; and every individual must work toward the grand climax, the wedding day, when the young queens seek to establish new colonies. There are endless duties, in short, and no ant shirks that which is her responsibility by virtue of having been born into such a complex social system. But there is a time when all workers can rest:—winter! The first cool days of fall keep all but the most ambitious inside their sheltering mounds. Later the penetrating cold drives them deeper and immobilizes them for their long, practically motionless, rest. They cluster together in their burrows, from frost line down two feet or more below. This hibernation can be interrupted momentarily, for blowing warm breath over a few workers held in the hand causes them to become as active as they would normally be in summer when such temperatures prevail.

More than eight years have passed since I first looked upon the *ulkei* mounds of Palos and their sanctuaries are in as efficient a state now as when I first saw them. As long as man does not interfere, all will probably remain well along the *ulkei* sector. Because of their superior ability to cope with their environment, these ant communities have doubtless witnessed the development of other societies in the same region, the human members of which could have consulted them with profit for a better understanding of how to live together peaceably. May the *ulkei* long continue to dominate the area encircled by the "Pyramids of Palos"!

〰〰〰〰 I KNOW ALMOST NOTHING ABOUT A. S. WINDSOR. AT THE time of his article, he worked for the General Biological Supply House in Chicago. In a later piece appearing in the *Chicago Tribune*, he is identified as secretary of the Chicago Academy of Sciences, but neither of the academy's two most recent archivists nor a past president had ever heard of him. Naturalist Laura Rericha, an expert on ants (and many other facets of natural history) with the Forest Preserve District of Cook County, knows his name only from the specimens he collected and the assistance he lent to more prominent myrmecologists. But he was instrumental in a fascinating sequel to his *Chicago Naturalist* article that was documented in a *Chicago Tribune* story of October 19, 1941.

The eighty-acre parcel holding the Palos *Formica ulkei* colony was acquired by the YMCA in 1940 for the purpose of establishing an overnight camping facility replete with cabins. But YMCA personnel soon discovered the presence of the ants, which they deemed incompatible with their plans. To determine the best way to rid themselves of the tiny insects, YMCA official J. C. Vynalek wrote a letter to the Illinois Natural History Survey. Fortunately, the query reached David Thomas, who recognized the location as being the home of the *Formica*. He then contacted Windsor, who launched a campaign to protect the ants. Experts from around the country urged the YMCA to accommodate their diminutive neighbors. To their great credit, the Palos Hills Camp Development Committee of the YMCA agreed to declare the area an ant refuge, probably the world's first if not only. YMCA personnel placed signs and incorporated the ants into their nature-study program. The *Tribune* article reported that campers found the visits to the ant mounds a highlight of their stay and that "their sudden interest in watching the busy *formica* at work overcame the natural boyish inclination to poke the mounds with sticks or to throw rocks at them."

Sixty-five years later there is very little record of any of this. The YMCA has no information on their former ant refuge. They conveyed the property to the Girl Scouts in 1976, and it is known as Camp Palos. A Girl Scout staffer with the South Cook County Council has no recollection of the ants ever being mentioned at the time her group acquired the land. Although familiar with the property, she is unaware of any pyramids.

Laura Rericha tells me that the Palatine ants have also disappeared; the only two places in the region that still harbor the species are in Berrien County and Kenosha County. And at neither site are there the large telltale mounds. The best way to see them, Ms. Rericha said, is to scatter bits of Pecan Sandy cookies, of which they seem particularly fond.

Source: Arthur Middleton, "Science Beats a Pathway to Ant Sanctuary," *Chicago Tribune*, October 19, 1941.

On the Trail of the Chat
Amy Baldwin

᭓᭓ ᭓᭓ ᭓᭓ There is a real treat in store for anyone who has an opportunity to study the yellow-breasted chat, either in migration or on his breeding grounds....

For some years it was possible for me to see the chat only occasionally in migration, several times in Jackson Park, once a fleeting glimpse in Waukegan, and at other places around Chicago. I have never forgotten my first sight of him in Jackson Park. That was truly a red letter day. Just by chance as I was on my way elsewhere with a little time to spare, being near the park I thought I would just hurry in to the Wooded Island for a short time. While standing quietly at the edge of the lagoon, looking across at a small island, the chat came out into full view, not for long but long enough for me to get a splendid view of him. He gave no call, which seems to be typical in migration so far as I have noticed.

Whether it is that he is very shy or that he is just fond of playing tricks on one, I just can't make up my mind. You can try ever so hard to see him on his breeding territory where he gives his calls or song and he will evade observation by creeping slyly from one bush to another. He seems to see you, but the leaves hide him from view. On better acquaintance with him and by sitting down and being quiet, he will eventually come out and be seen, probably at an entirely different place, oftentimes flying to a high place on a small tree near the nesting site or to a taller tree a little farther away. It is fortunate if you see him fly up there for otherwise you would not know where to look for him. A friend "put me wise" to this trick of his. As though this were not enough to make him intriguing he is also a very good ventriloquist. Be sure you have both time and patience, for you cannot hurry Mr. Chat; then you will be fully rewarded as he is really an interesting and handsome bird.

Amy Baldwin, "On the Trail of the Chat," *Audubon Bulletin*, no. 47 (September 1943): 5–8.

Once this spring in the latter part of May, three of us were looking for warblers. I had a feeling that the chat was just ahead in a low bush, but look as we would we could not see one, so decided that I was mistaken. We changed our location, two being outside the swampy area and the other going in another direction. The latter flushed the chat and got a splendid view, giving him a new life-list bird. We circled around and the chat flew to a low branch, then down to the grass and we all had a most satisfactory study of him. From there he flew to a large stone and gave us a fine exhibition of himself. How did that chat get away from three watching for him when he had to cross an open place to get over to where he was found? This makes me feel that he is tricky instead of shy, for why will he be so evasive one time and come out and be so bold within such a short time? Incidentally, this day the blue grosbeak was added to my life list.

I thought I had his breeding habitat all settled in my mind, for in three different localities the area was substantially the same; then I found out I was wrong. This is one reason that the study of birds is so fascinating: the longer you study them the more there seems to be to learn. One day at Orland Wildlife Sanctuary I was hiking alone when I heard a peculiar call, very ventriloqual, for it sounded as though it was right under my feet. I had been told that the chat had a number of different calls, but up to this time had not heard them so did not realize what it could be. I thought I was stepping on a mouse, snake, or rodent of some kind. It was a queer sort of *meow* repeated several times. Not finding anything under my feet I looked up to see the chat looking at me, and all I could say was, "You rascal!" Here I learned the chat's calls and from then on was to know them whenever heard and not be deceived by him again.

I was to see him next at Mrs. Smith's, on the Desplaines River, and become still better acquainted. Standing semi-hidden, I could see him and watch him give the calls over and over again, among them the whistle for the dog, which some of the books tell about. A trip to Depue, on the Illinois River, was rewarded by seeing and hearing several birds calling back and forth. On a trip to Mr. Ridgeway's "Bird Haven" at Olney, Illinois, I heard the chat calling but was unable to venture into the swamp because of briars and thick undergrowth. It was best not to go into a strange location without boots.

Up to this time each habitat had been the same: swampy, with lots of brush, trees, and swamp weeds and grasses, a good place for snakes. This environment must provide food to his liking. Then on a trip to Waukegan a friend and I found him nesting on the edge of a swamp, adjoining a ravine and a bluff.

This was getting away from the center of a swamp a little. This bird evaded us to distraction, and as this friend had not seen the chat for years it was disappointing to know that he was only a few yards from us and still not be able to get a glimpse of him. No means we used was of any avail, so we practically gave up trying and started away, when all at once he came out in full view on the path ahead of us, and then was in no hurry to leave us. Here we were also to see the hummingbird swinging in the arc of his nuptial flight.

At the Arboretum, while riding along the extreme eastern side on Prairie Road, friends and I were surprised by a peculiar call unlike any we had heard and could not figure out what bird it could be until up flew a chat to the top of a tree. He sat there looking at us, giving the same call over and over, and though we watched for some time he did not vary it with his other calls. The next day three other friends and I were there to see him and he was very accommodating, giving many, if not all, of his calls. There was a dense green hedge of low bushes along the fence which divides the Arboretum from pasture land beyond. Two weeks later other friends were with me and the chat was still calling, but now the green hedge was a bower of wild roses and it was a beautiful sight. No wonder the chat chose such a spot for his nest with his mate and three or four babies. This breeding ground was different from any of the others and was the driest of all, though a small sanctuary lagoon was not far away and a small drainage stream flowed close by.

Here we saw bluebirds, robins, doves, towhees, indigo buntings, Carolina wren, chickadees, and heard the short-billed marsh wren just over the fence. Over the pasture we saw an American bittern and crows flying. Many of these birds were still singing though it was now the middle of July. The chat had lots of company, birds as charming as he. What is more lovely than a bluebird or indigo bunting in full color?

This June while at Edinburgh, Indiana, a friend took me exploring to many interesting places around her home. Here the chats were common for they were heard in a number of different places. While I was there they seemed to be quite secretive, but she wrote me later that they had become much more friendly.

Without doubt there is much more to learn about the chat, but these have been my observations up to this time.

➤➤➤ AS I HAVE WORKED ON THIS ANTHOLOGY, I HAVE BECOME acutely aware of how little information exists for most of the stalwarts who made up the local birding community from the 1940s through the 1960s.

They planted the seeds that helped produce today's multitudinous crop. One of these people was Amy Baldwin (1885?–1980?), whose approximate dates are the best that recollection can provide. A nurse by profession, she and her husband bought a home in the Woodlawn area of Chicago's South Side early in the twentieth century. He died many decades before she did, but she stayed in the house until ill health forced her to retreat to a nursing facility.

I recall her saying that she started birding in 1911. Doug Anderson told me that she became the first woman member of the Chicago Ornithological Society when they allowed her to join in 1920. About the same time she ran into Nathan Leopold as they birded Jackson Park, and she attended some of the walks he led. She had an indomitable spirit that enabled her to brush aside adversity in pursuit of her great love of birding and, later when her eyesight began to falter, wildflowers. (Mr. Anderson refers to her as "a spunky gal who was determined not to let things bother her.") She tried to bird most every day, at least when weather permitted, but was a conspicuous and easy target for neighborhood youths, who robbed her and her home repeatedly over the years. My clearest memory of her was when she unhesitatingly headed into knee-deep water, her hand holding my arm, to see a family of piping plover.

Cliff Swallows

Violet F. Hammond

⌘ ⌘ ⌘ To us who live in Beverly Hills the southwest section of so-called "Chicagoland" is truly lovely—forest preserve, bridle path, thicket of crab-apple overshadowing woodland stream, and meadow studded with bluets and violets, purple and white.

Across one of these meadows, on a May evening in 1930, we strolled, listening to the plaintive whistle of an Upland Plover. Here for the first time we noticed numbers of cliff swallows flying low over the field, then vanishing in a southerly direction. This could mean but one thing—a nesting colony. But where?

Investigation of one or two farms answered the question. A fairly well-established colony was found on the wall of an unpainted farm standing well back from the road. Colonies of cliff swallows are extremely rare in Cook

Violet F. Hammond, "Cliff Swallows," *Illinois Audubon Bulletin* 31 (September 1939): 4–6.

County, hence, when we discovered this one the event was marked in "red" upon our calendar. Neither was the setting lovely, not the building itself at all attractive, but something in the general set-up answered the purpose of these very clever little masons. A pond close at hand undoubtedly offered mud of precisely the correct consistency. Here they worked and we visited them from time to time until a series of untoward circumstances inspired this little story.

To begin with, the eaves 'neath which they took shelter were extremely shallow, and the cross strip to which the nests were attached afforded but poor anchorage. Nest after nest had to be rebuilt after being washed down by heavy rain.

In 1931 misfortune of another sort was theirs. The farmer knew well the meaning of the word "Depression" and was forced to eke out a meager existence by promoting a series of barn dances, which continued the entire summer. In spite of the confusion which invaded the once peaceful barn yard, our swallows persevered beneath the eaves, to the weird accompaniment of accordion and fiddle. Incubation was well under way and in spite of din and disorder the parent birds refused to leave until the young were on the wing. This accomplished, old and young flew southward after a season of unhappy experiences. The next two springs, 1932 and 1933, no birds appeared.

However, the instinct to return year after year to the same nesting site still prevailed. In 1934, back to the inadequate eaves and poor anchorage, back to the scene of the barn dance with its hideous dither came a few straggling pairs of swallows. Misfortune was again theirs for this was a spring when the pond was nearly dry. Mud was almost impossible to find. This was a real hardship and our colony dwindled to perhaps not more than a dozen pairs of nesting birds. Their effort to obtain building material was really pathetic. Nest after nest fell from the wall for lack of proper moisture. At last a very few imperfect nests were completed, most of them little open saucer-like affairs in which the young were finally hatched. By this time the colony presented a very sad appearance. Failure seemed inevitable. May 1935 came, but no swallows.

Two years later, again on a May evening, in the same meadow before mentioned my brother and I noticed cliff swallows in numbers much greater than ever before. To our surprise and delight not half a mile away we discovered the new home-site, much more attractive than the last.

A barn belonging to an old farmhouse now stands at the very edge of a small cemetery. The building itself is picturesque, having a new roof and deeply overhanging eaves which drop low at the rear. There we counted upwards of seventy nests, each a perfect flask-like structure, built of the usual pellets of clay. The whole arrangement affords splendid opportunity for observation, the nests being scarcely ten feet from the ground, tucked in symmetrically against each rafter. The set-up seems quite ideal—the barn is scarcely visible from the road, a heavy gate discourages intrusion from the cemetery side, and the people at the farm are not the "barn dance type." The swallows undoubtedly came under the leadership of the survivors of the original colony and are prospering under better living conditions. Long may they thrive.

An Old Colony of Cliff Swallows

Alfred Lewy

ᥕᥕᥕ Old colonies of cliff swallows like the one at San Juan Capistrano, which allegedly dates back to the arrival of the Franciscan Monks in California in the sixteenth century, are well known. The one I wish to describe is in the metropolitan district of Chicago, located on the Mulderink farm, between Mt. Hope Cemetery and the town of Blue Island, which borders it on two sides, the third side being more or less open country with relatively few houses nearby.

According to John Mulderink—six feet tall, barrel chested and stoop shouldered, 75 years old November, 1942, who says that "work never hurt anybody" and looks as though he spoke from experience—his family moved to this farm in 1868 when he was about six months old. They built a combination house and barn which was immediately used by the cliff swallows. Of course this was told him by his parents, but he says that he remembers the swallows as early as he remembers anything.

The family subsequently built a separate better house, using the original building entirely as a barn. There are now four barn and shed buildings on the premises (not counting the corncrib), unpainted, and all used by the swallows, for whose convenience Mr. Mulderink has put up many little supporting brackets. When I suggested that nailing a little strip along under

---

Alfred Lewy, "An Old Colony of Cliff Swallows," *Chicago Naturalist* 6, no. 3 (1943): 67, 72. Courtesy of the Chicago Academy of Sciences—Notebaert Nature Museum.

the eaves would be easier, he answered that the swallows seem to like the brackets well enough.

Next door to the farm is a clay hole owned and used by the Illinois Brick Company, which does not collect anything from the swallows for the use of their material. This company also now owns the farm property, and while adjoining property is being prepared for possible industrial use, the farm is still being operated as such.

This spring all the old nests were knocked down by Mr. Mulderink to keep the swallows from preempting them. He has done this before and it seems to work out all right. Without counting about a dozen nests that were damaged by a recent heavy rain, there were on the four buildings altogether 125 new nests apparently in active service. Thirty-four of these faced south and sixty-eight east, inward on the house yard; twenty-three faced west, which is outward toward the cattle yard and open country. One young bird at the time of our visit was wedged between two of the nests and had to be rescued. He was about ready to fly and had wedged himself in so tightly that it was quite a job to get him out without injury, and naturally any cooperation on his part was in reverse.

The parent birds were frequently both at the nest at the same time, made frequent trips, and usually remained only a few seconds, apparently carrying food to only one offspring at a time. Most of the fledglings as seen at the nest entrance had white throats, streaked with black, and the same markings on the forehead. The rescued juvenile had a dark throat, as have adults, with a blurred white on the forehead.

This is the only colony I know of in the Chicago area at present. Another about three and a half miles further west was given up several years ago, for reasons I do not know.

◄◄◄► THESE TWO SELECTIONS PROVIDE GLIMPSES OF THE TRIALS and tribulations that befell a colony of cliff swallows. It seems that the first colony Violet Hammond describes was an offshoot of the colony to which they returned and which was subsequently studied by Alfred Lewy. In his *Birds of the Chicago Region*, Edward Ford advances the story to its conclusion: "There was until recently a colony in Blue Island, which according to Karl E. Bartel, numbered 76 occupied nests in 1944. Mr. Bartel writes that the last nesting was in 1948 (22 nests) when the barns were destroyed. [Philip] Du-Mont gives the figures as 100 nests for 1946 and 39 for 1947." As a local breeding species, this swallow was much rarer in the 1930s and '40s than it is now.

Unfortunately, I can find no information about Violet Hammond, but Dr. Alfred Lewy (1873–1958) did leave something of a record. At the time of his death, he was Chicago's oldest otolaryngologist still practicing. He had graduated medical school (Rush) in 1898 and was affiliated with a number of Chicago hospitals, including Mount Sinai, Cook County, and Chicago Memorial. Lewy's most noteworthy achievement as a physician was developing a technique to treat spinal meningitis before antibiotics had become available.

Lewy's interest in nature dated back to his childhood, having been influenced by a botanically oriented mother. As an adult, he was an active birder and conservationist who served on the boards of such organizations as the Chicago Ornithological Society (indeed, he was a charter member), the Illinois Audubon Society, and the Wilson Ornithological Club. Up until shortly before his death, despite having a leg problem and impaired hearing and vision, he remained an enthusiastic field birder. Dr. R. M. Strong remarked upon his stamina: "When, in the late afternoon, some of us would be ready to go home, he would still wish to visit other favorite places. He was indefatigable in tramping about these areas, even when in recent years, a leg difficulty made walking painful."

Sources: "Services Tuesday for Dr. Lewy, 85, Ear, Nose Specialist," *Chicago Tribune*, December 16, 1958; Edward Ford, *Birds of the Chicago Region*, Special Publication No. 12 (Chicago: Chicago Academy of Sciences, 1956); Reuven M. Strong, "Dr. Alfred Lewy, 1873–1958," *Audubon Bulletin*, no. 109 (March 1959): 2–3.

Wildlife of Elk Grove Preserve

Gordon Sawyer Pearsall

᭞᭞᭞ The Elk Grove Preserve of approximately 1400 acres, located in the northwest portion of Cook County, 22 miles from the "Loop" of the City of Chicago, is perhaps the wildest and least disturbed of the 36,000 acres of holdings in the Cook County Forest Preserve District. It is one of the original prairie "oak groves." ...

Higgins Road, State Route 72, divides the north half of Elk Grove from the south half. The westerly portion of the south half contains a winding drive,

Gordon Sawyer Pearsall, "Wildlife of Elk Grove Preserve," *Illinois Audubon Bulletin*, no. 49 (1944): 11–14.

two small shelters, and a few picnic areas. But most of the east half is comprised of woodlands so dense and so wet that they are rarely penetrated by persons other than the nature lover or student.

Fundamentally, Elk Grove is unique, not only for the diversity and abundance of all forms of wildlife, but, and more astounding, for the very presence of this wildlife within 22 miles of the heart of Chicago and within a metropolitan area of five million people....

During the fall and winter the whole area was cover-typed for tree types, and the drainage, streams and ponds mapped. This was accomplished by dividing the area into some 23 smaller areas. When these 23 smaller maps are put together they form a complete map of Elk Grove. During the spring and summer these smaller areas were visited often enough so that each was covered once a week. During the winter months the sharp thorns of the prickly ash shrubs were the biggest drawback, tearing clothing and skin. In the spring the water in the low grounds made getting into them almost impossible. From June to August, the immense swarms of mosquitoes, especially in the swampy areas, made tarrying in those areas impossible. During the latter part of June and early July, I had to wear a mosquito-netting veil dropping from my hat down to my shoulders, a "puncture-proof" light jacket and gloves, to be able to do any work in the low areas at all. The mosquitoes were the largest and most numerous I have ever seen.

MAMMALS

Different parts of Elk Grove seem to be particularly adapted to different kinds of wildlife. The larger mammals, particularly the predators, seem to prefer the swampy ground—wet woods with heavy underbrush of prickly ash, hawthorn, viburnum, cornus and hazel, interspersed with ponds partially choked with buttonbush. This is the favorite hunting ground of both red and gray fox. They seem to skirt the edge of the ponds, seldom leaving the cover of the heavy underbrush. Elk Grove is the only area where I have found the gray fox at all common. In most parts of our area it is almost extinct. During the winter months when snow was on the ground I saw gray fox tracks almost every day, but always in the swampy ground. They seemed to have more or less regular trails which they followed. I saw a gray fox on at least a dozen occasions during the winter, either on snowy or cloudy days. A gray fox was seen on several occasions by August Busse crossing Township Road. In June a gray fox kit about a month old was picked up by Ernest Tagge near the chicken coop on his farm. It was placed in an outside wire cage near the edge of the woods. Although it would not come out when anyone approached the cage, but would hide in the back of its house and growl, it ate readily and grew rapidly. When first put in the cage it was about the size of a young kitten and its fur a reddish-gray. By mid-July it was about the size of a large fox terrier dog and had a fine coat of fur of typical gray fox color. Only Mr. Tagge could get it to come out of its house when he was around, and finally to take food from his hand. In July it either escaped or was let loose, and Mr. Tagge says he still sees it occasionally in the edge of the woods.

Red foxes are quite common in Elk Grove. I often saw their tracks on the higher ground, generally hunting in pairs when they were hunting rabbits, but they always returned to the lower ground. When I was mapping the swamp ground northwest of the caretaker's house, a red fox became curious as to what I was doing and followed me for about an hour, at about eleven o'clock, for three days. His tracks told the story. The third day I decided to follow him. He evidently thought I was playing a game. He would keep some 25 or 30 feet ahead of me. When I stopped, he would stop, get up on a stump or log and watch me. When I moved forward again, he would move on ahead of me. I followed him clear out to the northwest corner of the Grove. When he came out to the open fields, he raced across them at a rapid rate, cutting back south to a cornfield where he slowed down to a trot again. He worked back southeast along the edge of the woods where I finally lost him when he crossed the elk pasture. The next day he came back for another game of hide and seek. The third day he had become more adventuresome and would allow me to get within a dozen feet of him before moving ahead. The fourth day we both got a surprise. I was following him along the edge of a back road, moving slowly because I could see him standing on a low stump behind a bush six feet in front of me. Suddenly a rabbit made a mighty leap from in front of me, landing within a foot of the fox. This surprised the fox so that he made twelve feet at that first leap. Evidently he had been so interested in me that he had not noticed the rabbit at all. He raced away for about a hundred yards, then slowed down to his trot again, finally waiting for me to catch up. For two more days we played our game of hide and seek, then he disappeared.

One April morning I surprised a red fox sleeping in the sun on the east bank of Salt Creek north of the elk pasture. I was moving west toward the creek, moving upwind. I did not see him as I was watching a cock pheasant across the creek. We both saw each other when I was within about a foot. He made a mighty leap toward the opposite bank fourteen feet away. He landed with a splash some three feet from shore, swam out and dashed about one hundred yards across the meadow, then stopped, shook himself, looked back to see what kind of creature had disturbed him, then trotted off into the underbrush. During the summer I found two fox dens and had the opportunity of watching their families from the vantage point of neighboring trees. One den was north of the elk pasture on a little bare knoll in the heavy woods; the other was south of Higgins Road near the eastern boundary of the preserve.

Raccoon, opoossum, mink and weasels are also fairly common in this same area. I found raccoon dens on five different occasions in the spring, and once saw a parent raccoon leading five youngsters about the size of a big kitten down to hunt along the edge of the ponds where frogs were particularly common. This was in the late afternoon on a cloudy day. While mapping the area northeast of Higgins Road, I saw a weasel catch a meadow mouse within a dozen feet of me. A quick spring, a single bite at the base of the skull, and the mouse was food. The weasel looked at me, picked up the mouse and bounded away, not at all frightened by my presence. On another day in the same area, while eating my lunch on a fallen log a very large mink suddenly appeared from nowhere, under a stump. I squeaked like a mouse. The mink stopped and sniffed the wind and looked all around. I squeaked again. The mink loped toward me, stopping when about six feet away to listen. I squeaked again. Now it saw me and moved toward me boldly. It smelled my boots, climbed up and smelled my lunch bucket, looked me over with its gleaming eyes, ran the length of the log and disappeared into the brush, showing no sign of fear.

During July the ponds near the northern boundary became choked with frogs and crayfish, literally hundreds of them. In a few days I noticed paths worn in the tall grass along the shore and leading back to the woods. At first I thought that humans must be coming in to catch these frogs and crayfish. Then I noticed piles of scat of raccoon, mink and opossum all along these paths. They also had discovered the frogs and crayfish and were coming down to these ponds to gorge themselves, probably squabbling among themselves just like humans. I think that most of the mammals in this area are permanent residents with the exception of the bats.

➤➤➤➤ GORDON PEARSALL (1905–?) WAS A WELL-ROUNDED NATU-ralist who worked at various times for the Field Museum, the Forest Preserve District of Cook County, and the Illinois Department of Conservation. Among the places he resided were Batavia and on Harvey Avenue in Oak Park. As to his education or what he did later in life, I know nothing.

Back in the 1930s and '40s, there were relatively few working naturalists around, so Pearsall was frequently tapped by the press to comment on nature subjects, particularly those related to phenological events. (The *Chicago Tribune* mentions him twenty-four times between August 1932 and May 1945.) An article on a population of several thousand cardinal flowers growing along a half-mile stretch of stream near Wheaton, DuPage County, quoted him as opining that it was the largest stand in the state.

During his tenure as director of the Forest Preserve District's Trailside Museum, the *Tribune* gave Pearsall $5 for winning their politeness award. He earned the accolade for his deft handling of an irate "little old lady" who stormed into the museum threatening to report him to the Humane Society for keeping various live animals. He explained with great patience that the wards of the menagerie had all arrived either injured or abandoned, and thus were ill-suited to survival outdoors. In the face of facts kindly offered, the would-be crusader meekly abandoned her cause, admitting that she "talked too much."

Sources: Doris Lockerman, "His Politeness Turns Away a Lady's Wrath," *Chicago Tribune*, October 11, 1937, 1; Ben Markland, "Day by Day on the Farm," *Chicago Tribune*, July 31, 1944.

My Secret Places: One Man's Love Affair with Nature in the City

Leonard Dubkin

THE GROTTO

᭤᭤᭤ I discovered my second secret place when I was fifteen years old and a student in high school. I still had my collection, and I still wanted to be a naturalist when I grew up. But I was not keeping up my notebook, for I was engaged in a higher form of literary expression.

I wrote little articles about my collection and the nature I observed for the children's page of the *Chicago Daily News*, and most of them were published with my by-line. The children's page was called "The Wide Awake Club," though it was not really a club, just a page every Saturday with articles

---

Leonard Dubkin, *My Secret Places: One Man's Love Affair with Nature in the City* (New York: David McKay Co., 1972), 15–18, 27–40. Courtesy of Random House.

contributed by children. The editor of the page was Mrs. Frances Dodge, an old lady in her seventies, who had written a number of children's books.

Of course I was not paid for the articles I wrote, but the sense of satisfaction I felt in seeing my name in print, and the knowledge that I was imparting to the children who read the "Wide Awake" page some of my enthusiasm for nature was reward enough. Every month Mrs. Dodge gave a prize for the best article published during that month, and I won a number of the prizes. One month I won a camera, another time a bird guide, and then a telescope.

I wrote my articles at home, in pencil, but of course I could not submit them to the newspaper in such an amateurish form. They had to be typewritten or Mrs. Dodge would not even consider them for publication. So, as soon as I finished an article I took it to Hull House, a settlement house just a few blocks from home, where they let me use one of their typewriters. First, I had to ask the head lady for permission. She was very kind; she would always take me to an empty desk and I would sit there and type my article. But she did not know what I was typing, nor what I did with my articles, until one day she stood watching me as I put a piece of paper in the machine.

"Son," she said, "what are these things you come in here to type?"

"They're my articles for the newspaper," I said very proudly.

She seemed incredulous. "You mean your articles appear in a newspaper?"

"Sure they do. Here's one that was in last Saturday's *Daily News*." I took out of my pocket a folded-up page of the "Wide Awake Club" with my article in it. I always kept my latest article in my pocket for a week or so, so I could read over it in my spare time. It would also be helpful, I thought, in convincing someone that I was practically a full-fledged naturalist—though of course it had never come to that.

She read my article, which was about the migratory instinct in birds. "Do you always write about nature?" she asked.

"Yes," I said, "I'm going to be a naturalist when I grow up."

"Don't you think you need a typewriter to be a naturalist?"

"Sure I do. And some day I'm going to be able to afford to buy one."

She asked me where I lived, and after I told her she walked away. A few days later a man delivered a package to our house, addressed to me. Inside was a brand new typewriter from the kind lady at Hull House. Her name was Jane Addams.

I discovered my second secret place quite by accident. While riding on a streetcar to take a dress my mother had made to one of her customers, I saw through the window a wide expanse of open land, with a few trees on it. This, I said to myself, would be a fine place to go hunting for specimens, and it was only about a half-hour's ride from my home. Where as my first secret place had been on the outskirts of the city, this new place was right in the heart of it, at California Avenue and Roscoe Street.

The very next Sunday I gathered my butterfly net, cigar box, a sandwich for lunch, and took a streetcar to the place I had seen through the window. It was beside the north branch of the Chicago River, and on the opposite bank was an amusement park. I could see people in the park across the river going on the various rides, the bobs and the chute-the-chutes and the ferris wheel, and jumping off a high tower with parachutes; and I could hear their shouts and laughter, and their screams whenever one of the rides went around a sharp curve....

One day about the middle of September I was searching for cocoons along the river bank when I saw a strange looking tree a short distance away. I went toward it and stood staring at it for a long time. It was a large maple tree, but the bottom branches hung down close to the ground, and all sorts of vines and creepers had grown up around these bottom branches to form a sort of dome-shaped igloo. What could there be inside? I wondered. Was it the home of a hermit, or were there animals or birds inside?

I walked around the tree, looking for some way I could get inside, but the vines and creepers were so thick that I would have had to cut my way through, and I did not have a knife with me. I decided to climb up to the top of the dome and see if I could get in that way, perhaps near the trunk. I felt about until I found a branch of the tree, then pulled myself up as if I were climbing a mountain. When I got up near the trunk I spotted a large hole in the dome, and put my face down in it to look inside. It was dark in there, and I could see nothing at all. Suddenly something soft and furry brushed against my cheek. At that moment I lost my balance and fell inside.

There was a thick covering of leaves on the ground under the tree, and these cushioned my fall so I was not hurt. I sat up and looked about, but still could see nothing in the dim light. Then, as my eyes gradually became accustomed to the darkness, I made out a sort of ghostly movement around the trunk of the tree—a circular movement as though hundreds of leaves were being whirled at great speed around the tree trunk. As I peered through the dim light at the movement, slowly I began to see that it was not leaves whirling around the trunk, but little furry bodies borne on wings that beat up and down. This was a grotto filled with bats!

I got up and walked over to where the bats were circling the tree trunk. I had left my butterfly net and my cigar box outside, so I put my hand into the stream of flying bats and tried to catch one—unsuccessfully. Every one of them broke his flight pattern to avoid my hand. Finally I sat down at a little distance from where they were circling the trunk and watched them.

They were small, about the size of sparrows, they were brown, and they were uttering quiet little squeaks as they flew around the trunk. They had probably been frightened when I fell into their roosting place, but if I sat here long enough perhaps they would calm down and go back to sleep, and then I would be able to catch one.

I sat quietly for about a half hour, and I could see that the stream of bats circling the tree trunk was thinning out. There must have been at least 200 of them when I first fell under the tree, but now there did not seem to be more than about fifty or sixty. I continued to sit quietly, and fewer and fewer bats circled the tree, until finally there were none at all. I could still hear the little squeaking sounds coming from all over the grotto, so I knew the bats were still there, but where?

I got up and walked slowly around the tree trunk, looking all about me. Then I looked up and I saw them. They were hanging upside down from the branches and the vines and creepers, crying their quiet little squeaks. But every time I put my hand up to capture one it flew off and hung itself up somewhere else.

I decided to leave then, but I had no intention of trying to climb up to the hole in the top of the grotto where I had fallen in. It would be too difficult, and it would disturb the bats. So, instead, I searched among the vines and creepers near the ground until I found a spot that was not too thick. I pulled the branches aside and slipped through, noting the spot very carefully, for I intended to come here again next Sunday, and every Sunday that I could.

I had read a few articles in magazines about bats, but scientists knew very little about them. People had all sorts of superstitious notions: that they were in league with the devil, that they got entangled in people's hair, and that they possessed some sort of sixth sense by means of which they could avoid objects in their path. Now, I thought as I rode home on the streetcar, I would have an opportunity to observe and study a species of animal about which very little is known, and perhaps I could make some contribution to science. I would study their behavior, perform experiments, and learn their most secret habits.

The following Sunday I returned to the grotto. This time I got in at the side, at the same place I had let myself out the week before. And I remembered to take my butterfly net and my cigar box in with me. Only a few of the bats were disturbed by my entrance this time, and they flew around the tree trunk again. I tried to catch one with my butterfly net, but even when I put the net right in their path, they flew over it or under it, evading capture. I decided that maybe they did have a sixth sense after all, for otherwise how could they tell when a thin butterfly net was placed in their path?

I adopted a new tactic. I looked about under the branches until I saw a bat hanging upside down. Slowly I raised the net until it was directly under him, and, with a quick upward sweep netted him. I knew he was caught, for he began squeaking loudly and thrashing about in an effort to get free. His commotion aroused the other bats, and soon they were all circling the tree again, about two hundred of them, whirling around like leaves caught in a tornado.

When I tried to take the bat out of the net he sank his sharp little teeth into my finger, and I pulled my hand away. Then I sat down on the ground to think over what I would do. Why did I want a live bat anyway? I would not learn anything by taking him with me, and there was a chance he would die in the cigar box before I got him home. What I really wanted to do was to study the bats in their natural surroundings, here in the grotto. This decided, I turned over the net and dumped the captive bat out on the ground, at which point he immediately flew up and joined the throng circling the tree trunk.

I went off to one side and sat quietly on the ground. Slowly the bats circling the trunk began to thin out. They were hanging themselves up from the branches and going back to sleep. How long would they stay in the grotto, and when would they go out, and when they went out would they stay together, or would each bat go off by himself? Were there both males and females in the grotto, and when did they mate, and where were all the young bats? What did they eat, and how did they find their food, and how did their sixth sense operate? My mind was filled with questions to which I could only find the answers by observing and studying the bats here, in the grotto.

I stayed in the grotto all day, even though I had not brought my lunch. For the rest of the morning and all afternoon I sat there waiting for something to happen, and nothing did. Besides the muted squeakings of the sleepy bats, there was not a sound to be heard. The only light that entered the grotto was sunlight that filtered between the leaves, and a shaft of sunlight that came down through the hole in the foliage where I had fallen in last Sunday, and danced about on the grotto floor.

It was getting late, and I was hungry and tired of sitting motionless in the darkening grotto. I decided I would go home now; next Sunday I would bring my lunch, and then I could stay all day and not get so hungry. I got up, picked up my net and cigar box, and pushed my way out. I was walking away from the tree when I happened to look back and saw black smoke pouring out of the top of the grotto. Had it caught on fire? I hurried back to the tree and looked closely at the smoke. It was coming out of only one spot, the hole near the tree trunk, so I did not think there could be a fire inside. And then I realized that it was not smoke at all: it was the bats coming out of the grotto. They swarmed out of the hole, and as soon as they were clear of the tree they all went in one direction, toward the river.

I hurried after them, and noticed the bats flying back and forth across the river just above the surface of the water. I watched them closely, and when they came toward me, I saw that as they skimmed over the water they stuck their tongues down to get a drink. Every time a bat touched the water he caused a little ripple that spread out in all directions in the shape of a circle. There were so many circles in the water from so many bats that they collided with each other and broke up. When the bats had had enough water, they flew up into the air and disappeared.

I went back into the grotto and looked about under the branches. There was not a single bat inside, they had all flown out together. I left and started back toward the streetcar line, for it was getting quite dark. It seemed to me that I had just witnessed an event that was very rare and very beautiful. How many people in the whole world have ever seen a swarm of bats drinking from a river as they flew over it?

All that week I pleaded with my mother to let me spend the following Saturday night away from home; it was a scientific experiment, I told her. Actually, I wanted to spend a night in the grotto, so I could see the bats return, and observe how they behaved, whether they came in and immediately hung themselves up on the branches, or if they circled the tree trunk for a time before going to sleep. Did they all come home together, as they went out, or did they straggle in one at a time?

Finally my mother agreed to let me sleep out if I took a blanket to lie on, a thermos bottle of milk, and some sandwiches. When I arrived at the grotto

I found it deserted. It was late afternoon, but much too early for them to have gone out for the night. I slept on a blanket in the grotto that night, and remained there until noon the next day, but I saw no sign of the bats. Had they deserted the grotto and gone somewhere else to spend their days? Was it my presence in the grotto that had disturbed them? Or did bats migrate in the fall like birds? It was October now, and the migratory birds, the robins and the bluebirds and the cardinals, had all gone south. Maybe the bats had gone south too.

That winter I spent as much time as I could in the public library searching for information about bats. I found only one book, and that was of no help to me, for it was mostly about South American species. Here and there in magazines, however, I found a few articles about bats of North America, and I read them avidly. I learned that the bats in my grotto had been little brown bats, *Myotis lucifugis*, and that they do indeed migrate in October, when they fly south to caves in southern Illinois and Indiana.

They do not have a sixth sense; it has been proven scientifically that they use what is called "echo location" to detect objects in their path and even to catch the insects on which they prey. It is a sort of natural radar; while they are in the air they emit supersonic beeps that bounce back from objects in front of them, and from insects in the air, and enter the cochlea of their ears. This was how they had managed to avoid my butterfly net when I held it up in front of them in pitch darkness.

In May, when the snow had melted from the ground, I went back to the grotto and pushed my way in. It was brighter now than it had been last summer, for instead of the thick foliage obscuring the light, there were now only little spring leaves. As soon as I was inside I knew that the bats were back, for there was a continual high squeaking coming from all parts of the grotto. As I walked about looking at the bats hanging from the branches that formed the roof of the grotto, I noticed a strange thing. Every one of the bats had a baby bat clinging to the fur on the underside of its body.

Was every bat in the grotto a female? It seemed highly unlikely, and yet I did not see a single bat without a young one clinging to its fur. Did the males nurture the young too? No, that was impossible. The only possible conclusion was that they were all females.

I had brought my butterfly net with me, and I caught one of the bats as she hung upside down from the roof. Very carefully, while she was still in the net, I pinned her wings behind her with one hand, and took the baby from her fur with the other. What a cute little rascal he was—about two inches long and pitch black, flashing his wings and squeaking piteously. And, I learned, he could bite too, and his little teeth were like needles.

I came back to the grotto every Sunday, and a few times I came Saturday afternoon and slept there so that I could observe them leave the grotto in the evening. For about a half hour they would circle the tree trunk, then, as though one of them had given a signal, they would zoom up to the opening at the top and stream out *en masse*. They carried their babies, who clung to the fur on their stomachs, with them.

In June, either because their babies were becoming too heavy to carry around all night, or because now they were old enough to fend for themselves, the bats began leaving their babies in the grotto when they went out at night. The young were not yet able to fly, but they would crawl about among the foliage, squeaking loudly and, I suppose, playing games with each other. Sometimes a young bat fell to the ground from the roof of the grotto. Then he would crawl slowly, using his hind legs and the joints of his wings, to the edge of the grotto and climb back up into the foliage.

In a few weeks the young were able to fly. How they managed to learn I do not know; they must have practiced at night when they were left alone in the grotto, and when, even if I slept there, I could not have seen them because of the darkness. But now in the evening, after all the adults had left the grotto, the young bats, not yet proficient enough to stay out all night with their parents, would circle the tree trunk. They flew as swiftly and with as much confidence as the older bats.

The next time I came to the grotto the young bats went out with their parents; they were old enough to stay on the wing all night. The grotto was a ghostly place at night, there was not a sound, not a squeak to break the silence. In the morning, just as it was beginning to get light, the bats returned, seemingly all at once, as though they carried watches with them and knew what time it was. They flew around the tree trunk a few times, then hung themselves up on the roof. They no longer seemed to mind my presence in the grotto; they had become accustomed to me. In the first part of October they all disappeared. And I knew they had migrated again.

I wrote an article for the "Wide Awake" page about the bats, telling of the grotto without mentioning where it was. After my article appeared I got a note from Mrs. Dodge telling me to come to see her that Saturday, that it was very important. When I called at her office she had me sit down while she delivered a lecture.

I had been greatly honored, she told me; nothing like this had ever hap-pened to a contributor to the "Wide Awake" page. The editor of the paper, James Baird Warner, had read my article about the bats, and he had been very much impressed. Now he wanted to meet me, to have a talk with me. It was a great honor, she repeated, and if I conducted myself properly during my talk with Mr. Warner my future would be assured.

She escorted me to the elevator and up to the next floor, where Mr. Warner's office was located. I was a little dubious about this interview, and what Mr. Warner could do for me. I wanted to be a naturalist, and I didn't understand how the editor of a newspaper could help me to achieve that goal.

Mr. Warner was a tall, thin, stern-looking man with a little gray mustache. He asked me to sit down beside his desk, and Mrs. Dodge left us alone.

"I read your article about the bats," he began, "and I liked it very much. You have a great deal of curiosity, and the ability to set down what you see and hear in simple English. I think you'd make a good newspaperman."

"I want to be a naturalist when I grow up," I told him.

"How old are you?"

"I'm sixteen."

"Sixteen years old, and you want to be a naturalist. Do you know how long that will take you, and how much money it will cost? You'll have to go to college, and you'll be an old man before you achieve any recognition as a naturalist. As a newspaperman—now that's different. You could become a reporter right now, and people would be reading your stories, and you might become nationally known before you're twenty. Besides, any fool can be-come a naturalist and study the lower forms of life. It takes a man with your curiosity and your ability to write to become a good newspaperman. You know, the Greeks had a saying, 'The proper study of mankind is man.' Here on this newspaper we study man, we study his foibles and his fallacies, his strengths and his weaknesses, and we report it to the people. That's what you should be doing."

He was right, I thought, he was absolutely right. I had thrown away my boyhood years studying nature, making collections of things that would never be of any use to me, when I should have been studying people, learning how they thought and acted. Suddenly it all seemed very clear to me, I must give up nature and put all my energies into observing people.

To Mr. Warner I said, "Sir, I believe you're right. I would like to become a newspaperman."

He gave me a job as copy boy on the paper after school and on Saturdays, and he promised that when I graduated from high school he would make me a full-time reporter. And he did.

〰〰〰 LEONARD DUBKIN (1905–1972), WHO IMMIGRATED WITH HIS parents to Chicago from Russia at the age of two, was one of the region's most prolific nature writers. (He also was responsible for the trade publications *Chicago Talent Directory* and *Radio Guide*, the latter appearing in thirteen editions in cities throughout the country.) He wrote five books, as well as a nature column at various times for both the *Chicago Tribune* and the Lerner suburban newspapers. The territory he claimed—at least in *My Secret Places: One Man's Love Affair with Nature in the City*, *The Natural History of a Yard*, *The White Lady* (this book about an albino bat is said to have been his most popular), *Enchanted Streets*, and *The Murmur of Wings*—was all within Chicago's municipal limits, with one inadvertent exception: one of his secret places turned out to be just beyond the border in Niles.

Many of his adventures involve the conflicts arising from his need to interact with a human world that shuns the arthropods, rodents, pigeons, and other urban denizens that are his obsession. (One chapter in *Enchanted Streets* is entitled "I Learn That I Am of No Value to Humanity.") A key event in his life was being fired as a reporter because he missed a big story. This occurred when his editor sent him to interview a murder suspect in Oak Park. The conversation quickly turned to the squirrels that resided in the man's attic, and eventually Dubkin was left observing the rodents. Meanwhile, the suspect ran an errand: to retrieve his glasses from the scene of the crime, an act that resulted in his arrest and confession.

There is an odd sensibility running through these books that is difficult to characterize, yet rings discordant to a modern naturalist. No one can doubt Dubkin's patience, passion, and great skill as a writer, but his knowledge seems to be pridefully superficial, as if he resents experts and technical accuracy. A thrush singing in Niles, Cook County, in June is not a hermit, although it might be a wood or veery. Weasels do not form pair bonds—adults rendezvous for one purpose only and when that act is completed they go their separate ways. Chicago's winter avifauna is, as any birder will tell you, far more extensive than the house sparrow, starling, rock pigeon, and herring gull that Dubkin lists. He knows but a few warblers and has no interest in learning others because they are country birds and not city birds, a dichotomy both intrinsically silly and patently untrue, given the concentrations of migrant warblers on the lakefront and numerous other places across the city.

He does meet bird-watchers twice but finds them "pathetic" because in his view they should be content with the "city full of pigeons and sparrows and starlings and robins and grackles and nighthawks," rather than having the need to visit the wilds of Northerly Island on the downtown lakefront to seek less common species. According to his daughter, Dubkin was fired from the *Tribune* by publisher Robert McCormick for "denigrating" bird-watchers, because Mrs. McCormick actively engaged in the pursuit herself.

Sources: "Leonard Dubkin," *Chicago Tribune,* November 20, 1972; Pauline Dubkin Yearwood, "The Urban Nature Lover," *Chicago Jewish History* 29, no. 24 (Fall 2005).

<hr />

Life Plus 99 Years

Nathan F. Leopold Jr.

ᢙᢙᢙ… I acquired my first pet, too; I've had a number since. One of the boys in the Fiber Shop, as the old Rattan Shop was now called, a Mexican, had a pet horned lark, which he had picked up in the yard as a fledgling. It was now quite grown, looked like a regular bird instead of a powder puff on legs, the way the youngsters do. It was quite tame. I bought the bird from its owner for ten sacks of tobacco, our medium of exchange. I used to transport the bird between my cell and the shop in a little cardboard box with holes or sometimes just in my hand. In the evening he had the freedom of my cell; during the day he had the whole stockroom to play in. Sometimes he'd fly to you when you whistled; more often you'd have to go and pick him up, but he never flew away when you approached him. Usually I took him back to the cell for the hour rest period at noon. But one day I left him alone in the stockroom when we went to dinner. I had forgotten that we had set out a number of mousetraps to catch some of the myriad mice in the place. The little bird's curiosity got the better of him and he pecked at a piece of cheese with which the trap was baited.

Several times I have had baby horned larks as pets. The little fellows can run soon after they are hatched. They are hardy little creatures and make

<hr />

Nathan F. Leopold Jr., *Life Plus 99 Years* (New York: Doubleday, 1958), 166–68. Courtesy of Random House.

excellent pets. They have to be force-fed the first time or two; one inserts his thumbnail between the mandibles and forces the beak open. Then, with a toothpick held in the other hand, one shoves a piece of worm or other succulent tidbit well back in the throat, behind the tongue. This takes some doing if one is alone, for of course the struggling bird has to be held in the hand which is prying the mouth open. Water is given with an eye dropper or, in the absence of that, by merely wetting the index finger and letting a drop roll off into the open beak. But after two or three such feedings the little fellow gets the idea, and then your problem becomes that of keeping up with his almost insatiable demands. The little bird comes running as soon as he catches sight of you, screeching his lungs out to be fed. When full of worms, he'll fluff out his feathers and cuddle up hard against your shoe. It's pretty easy to lose your heart to one of the soft little things.

But perhaps the most interesting bird pet I had was a robin, whom I named Bum. I was working in a division in E House then, just after the riot of 1931, and Bum was given to me when he was about three weeks old. It didn't take long for him to lose any fear he had of humans and to open his mouth when food was presented. In fact, young birds, or young animals of any kind, don't have any instinctive fear of human beings; fear is a reaction they learn later. Bum was quite a pet.

I was celling with two other fellows, Dan and Charlie, and I asked whether they wanted a bird in the cell. They were both enthusiastic. Neither of them had ever had any experience with a wild bird and they were devoted to Bum—this in spite of the fact that a bird the size of a robin presents some problems in sanitation. Young birds are prodigious eaters and their droppings are both much more frequent and much greater in volume than those of adult birds. That didn't matter; we'd simply spread newspapers on the floor and on the bunks. Bum had such winning ways that they were both glad to take the necessary trouble.

But after a bit that wore thin. If Bum would only stay off the bunks! He had been permitted on the bunks for a week and definitely wasn't house-broken. But I thought I could condition him to stay off the bunks. That evening, when Bum flew up on the bunk as usual, I batted him off with my hand. Not hard, but hard enough to knock him off the bed. He didn't get the point. Soon he was back on the bed. This time I snapped my finger and hit him on the beak with my fingernail. Bum didn't like it one little bit; he eyed me reproachfully and flew up on the shelf over the door. Here he sat all evening, head toward the wall and tail out. He wouldn't rap to any of us. I've never seen a bird pout before, but certainly Bum pouted that evening. Dan and Charlie couldn't stand it. Let Bum fly onto the beds if he wanted to; so I stopped his lessons in deportment.

After a week or so I let Bum have the run of the whole cell house. He knew where he lived: when he was hungry or thirsty or wanted a bath, he'd fly right to cell 126. If the door wasn't open, he'd come in through the chuckhole. At night he roosted on the rung of one of our stools. But all day he went visiting. If I wanted him, all I had to do was step out on the circular flag and whistle. If that didn't bring him at once, rattling a red Prince Albert tobacco can in my hand would. For his raisins were kept in a Prince Albert can and Bum was inordinately fond of raisins. Down from the rafter fifty feet above my head he'd dive-bomb and perch on my shoulder or hand.

Bum was everyone's friend; he went visiting in all the cells. And if he came hopping down one of the galleries and some con didn't step aside for him, he'd snap his bill at him, ready for a fight. He had absolutely no fear; he was quite willing to spot you a hundred and fifty pounds.

But it was his friendliness and his total lack of fear that led to his undoing. One day Bum was missing, and when the trash was swept up from the flag that evening, one of the boys found his body in a brown paper bag. Someone had wrung his neck. I suppose he had flown into someone's cell and started worrying apart some rolled cigarettes, as he was in the habit of doing; or maybe he was messy on someone's pillow. If the fellows in E House could have got hold of the man who killed him, I think they might have wrung his neck. For everyone liked Bum. As for me, I decided then and there that I'd think twice before I prevented another wild bird from acquiring his natural fear of humans. I could guarantee the manners of the bird, within limits, but I couldn't answer for my fellow humans.

〰〰〰 OF ALL THE AUTHORS IN THIS ANTHOLOGY, THE ONE WHOSE existence probably registered on the consciousness of more people than any other is Nathan F. Leopold Jr. (1904–1971). Unfortunately, the act that gained him international fame was his participation in the murder of a fourteen-year-old boy. On May 21, 1924, Leopold and his confidant Richard Loeb abducted Robert Franks and then bludgeoned and suffocated him, before placing the corpse in a culvert connecting Hyde Lake with Wolf Lake on Chicago's southeast side.

There were several aspects of the case notorious enough for some at the time to call it "the Crime of the Century." Foremost among these was the nature of the perpetrators. They were products of wealthy and respected families, and they distinguished themselves as excellent students. Leopold had just finished his first year of law school at University of Chicago and would have transferred to Harvard in the fall if he hadn't been sidetracked by the extracurriculars. The only motive ever offered for the killing was that Loeb and Leopold wanted to demonstrate their intellectual superiority by committing the perfect crime, which meant of course eluding capture and punishment. But Leopold inadvertently dropped a pair of glasses at the crime scene that was easily traced to him. Soon after their arrests, both young men confessed. And finally their families engaged the services of Clarence Darrow, one of the most celebrated American criminal lawyers of the twentieth century. This would be his last case, and his sound tactics and eloquent oratory probably prevented the execution of his unrepentant clients. Instead, Loeb and Leopold were sentenced to "life plus ninety-nine years."

For the purposes of this anthology, however, the important fact is that Leopold was an avid birder and had been since the age of five. The spot where the murderers deposited Franks's body was in an area Leopold had visited just a few days earlier in search of Wilson's phalaropes. Leopold had published a paper in the *Auk* regarding his collecting the third specimen of the long-tailed jaeger from the southern end of Lake Michigan. His greatest ornithological achievement, though, was the discovery and collection of one of the first Kirtland's warbler nests ever found. He presented his paper on that experience to the annual American Ornithologists' Union meeting in Boston; it was later published in the *Auk*.

Leopold was eventually released from prison in 1958, having served thirty-four years. He moved to Puerto Rico, where he worked at a religious mission. (He also was to publish *Birds of Puerto Rico*.) Loeb, on the other hand, never tasted freedom again. He bled to death in 1936 from fifty-six slashes inflicted by a razor-wielding inmate. Although there is no question that Loeb's killing was the result of a premeditated attack and not a sexual overture on his part, an early account by Edwin Lahey in the *Chicago Daily News* contained one of the best lines ever to begin a news story, even if it was incorrect: "Richard Loeb, a brilliant college student and master of the English language, today ended a sentence with a proposition."

Sources: Hal Higdon, *The Crime of the Century* (New York: G. P. Putnam's Sons, 1975); Nathan F. Leopold Jr. Papers at Chicago Historical Society, box 34, file 5; Nathan F. Leopold Jr., *Life Plus 99 Years* (New York: Doubleday, 1958).

Water World

Letter to Spencer F. Baird [Regarding Crayfish]

R. P. Stearns

〰〰〰〰〰〰〰〰〰〰〰〰〰〰〰〰〰〰〰〰〰〰〰〰〰〰

დ დ დ To Spencer F. Baird. Some observations confirming those made the last year, upon the Amphibious habits of the Acastus.

A new, and to me very interesting fact, in relation to the habits of this shell fish, I accidentally discovered on the prairie near Chicago. On the ridge dividing the waters of the Chicago River and the Aux Planes [Des Plaines], is a wet marsh. It lies near the bed of bituminous limestone. The marsh at the time of my crossing it was quite dry. Exam[in]ing for lacustrine shells, I soon discovered a large number of the paths of the Acastus, some of them lately travelled—fresh trails. I had not seen any water since crossing the Chicago River, two miles distant. I was stimulated to follow up one of the freshest trails, until I discovered the local habitation of my friend the crawfish. Never was a Potawatomy [*sic*] more intent on his pu[r]suit of his game, than I. Pursuing my examination along one of the freshest trails, having first crossed + recrossed the marsh in search of some little pool of stagnant water, but finding none, I was soon delighted to find an Acastus, leisurely travelling along his own highway, his war path, bearing in his mandibles an insect of the coleopterous tribe. He seemed to know his own whereabouts although I had found no water for his breathing habitation. Leisurely I followed on his tail, big with expectation, as to where he would lead, and what I should see, when suddenly he was lost. He had disappeared in his own path, as if by encha[n]tment. Carefully laying wide the rank herbage, I found he had disappeared in a well or cistern, about 10 or 12 in deep + 1/2 wide. Here was his own pool—his aqueous breathing house, dug out by himself. His instincts leading him to this method of preserving water for his pulmonary purposes, during the long summer droughts. Following up other trails I invariably found them terminating in similar pools. In many more, Acastus was quietly enjoying himself at home. In others he was on his morning foray.

R. P. Stearns, letter to Spencer F. Baird, Smithsonian Institution Archives, Record Unit 52, 2/12/1853 to 12/9/1853. Transcribed by Betsy Mendelsohn.

The interesting fact of making holes in wet prairies subject to drought was confirmed by a Mr. Mattson of Ill. a very careful observer who had discovered it many years ago.

I did not preserve any specimens, to examine if this Acastus was a new species, or the A. Bartonii under a new phase of habitation. I then expected to return with a bottle of spirits for collection, but sickness prevented.

Yours truly R. P. Stearns

P. S. Another season I hope to confirm these observations + make a collection

Southern Cook County and History of Blue Island Before the Civil War
Ferdinand Schapper

࿔ ࿔ ࿔ Sturgeon came up the Calumet River in the spring as far as the "dam" which obstructed their passage further up stream. They would jump out of the water while at play, and it was a strange sight to see these immense fish, 8 or 10 feet in length, and weighing 200 pounds, disporting themselves in the water. They were caught in numbers, the small boys always being around and begging for the "sturgeon nose," a gristly substance that they used instead of chewing gum.

Report of the Fisheries of the Great Lakes
James W. Milner

INVESTMENT IN THE FISHERIES
࿔ ࿔ ࿔ The fisheries of the lakes are an industrial interest of large extent and considerable commercial value, of which little is known except among those directly interested.

Ferdinand Schapper, "Southern Cook County and History of Blue Island Before the Civil War," vol. 1 (1917), typed manuscript at the Chicago History Museum.

James W. Milner, "Report of the Fisheries of the Great Lakes: The Result of Inquiries Prosecuted in 1871 and 1872," in *Report of the U.S. Commissioner of Fisheries for 1871 and 1872* (Washington, DC: Government Printing Office, 1874), 3, 7–9, 38–39.

Back from the lakes the very prevalent idea is that fishing is an unprofit-able employment for an irregular class of men who eke out a meager subsis-tence from year to year by this pursuit. Though the risks and uncertainties of this vocation make the yearly income very variable, the investments of fisher-men in their stocks are quite respectable sums, and compare favorably with the farming-communities, being all the way from three hundred to twenty thousand dollars, their sales reaching in some instances as high as $7,000 from their own nets. This refers to those men only who actually superintend their own fisheries. A few dealers who furnish the nets on shares sell five or six times as much in a year. Nor is there any truth in the aspersion on the class of men, who are industrious, hard-working citizens, and, considering the hard-ships and exposures incident to their calling, singularly free from the habit of hard drinking.

The fishing-stocks are necessarily a less stable investment than farming-lands, liable to frequent loss and injury, and as the success of a fishing-season depends much on the character of the weather, there is of course uncertainty in the yearly income.

The same as in other vocations, the alternation of abundance and scar-city does not develop the provident faculty that accumulates property, for though as a class not given to dissipation, they spend their money freely for comfort and good living when the fishing is prosperous. In spite of all these unfavorable conditions many attain comfortable circumstances.

The investment of fishermen and net-owners by itself is not inconsider-able. Under your instructions last year I visited nearly all the fishing-ports on Lake Michigan, and made an accurate count of the fishing-stocks owned and used on the lake.

IN LAKE MICHIGAN

As the fisheries of Lake Michigan were worked up in detail in the year 1871, a description of the character of the fishing in different localities may be valu-able for the light it may afford as to the necessity of discrimination in legislat-ing for different regions.

Beginning at South Chicago, near the head of the lake, there were ten pound-nets, distributed along about eleven miles of coast, lying three miles to the north and eight miles to the south of the Calumet River. Unlike pound-net men in other portions of the lake, they here seek to catch every variety of fish, finding sale to the peddlers of everything but the dog fish, *Amia calva* Lin.

At Chicago there were six boats fishing with trot lines off the mouth of the river; their catch being almost entirely the perch, *Perca flavescens.* One man is employed during the season at Milwaukee catching bait, shipping tubs full of minnows daily.

There has been no net-fishing here for years, the few experiments made proving failures. It is quite possible, now the filthy current of the river no longer flows into the lake, that there may be some success with nets.

At Evanston, the pound-net fishing was of very much the same character as at Calumet. At both points they have a spring and fall season, taking out the nets in hot weather, when the fish leave the shoal waters.

From Lake Forest and Waukegau to the Wisconsin line were twenty-seven pound-nets, fishing for both the fresh and salt fish markets. In this region comparatively few fishes are taken other than the white-fish. One proprietor has built a smoke-house, preparing and boxing the smaller white-fishes for the Chicago market, where they are sold as smoked herring. If there were no other objection to the capture of the small white-fish, than their useless destruction, this could be easily remedied by disposing of them in this way, as they find a quick and profitable sale, the demand being far in excess of the supply.

The season here is different from most other parts of the lake. Instead of a spring and fall season, with an interim of two months, in which the nets are taken out, the fishing, beginning late in May, lasts until the first week in September; the fisheries having their greatest run during the months in which the least fishing is done at most points on the lake.

It will be observed that in Illinois's share of the lake-shore no fishing is done, save with pound-nets. It is not likely that gill-net fishing would be undertaken here if pound-nets were prohibited, as it would be too hazardous of life and property. Gill-net fishing is adapted only to a coast with good boat-harbors, or at any rate favorable lees, as in high winds, driving heavy seas on the shore, there is great difficulty in landing, and often when there is not sufficient sea outside to prevent taking up the nets it is very difficult to launch a boat that would experience no inconvenience when once fairly out from shore; so that nets from a shore like this often remain out for days, while a few miles off from a harbor the boats run out and take up every day. Frequently they are caught in a gale when outside, and are obliged to run for harbors twenty of thirty miles to the north or south because of the danger they would incur in beaching. The large number of deserted fishing-shanties along the beach, on the east shore of the lake, attests the impracticability of gill-net fishing from a lee shore.

In this extent of shore there is no spawning-ground known for either the trout or the white-fish, though the fishing is not carried on sufficiently late in the season to find the species named upon their spawning beds.

From Kenosha to Sheboygan are some thirty boats, working the "large gill-net rigs," having from five to six gangs to the boat, and from twenty to thirty nets to the gang, making the string of nets a mile or more in length, and requiring a crew of five men. They run out from shore from eight to twenty miles, according to the range of the white-fish and trout at different seasons. The boats used in this region are principally what are called the "square stern."...

THE SALMON OR MACKINAW TROUT, *SALMO NAMAYCUSH* PENN

The trout of the great lakes is one of the three most numerous fishes, and, except the sturgeon, attains the greatest weight of any of the lake-fishes. It is captured almost exclusively by the gill-nets, the pound-nets in some portions of the lakes taking them during the spawning-season. In winter a great many are caught in the bays, through holes cut in the ice. They are found in all of the great lakes and in a few inland lakes in their vicinity.

As compared with the white-fish, their merits as a fresh fish are relative to taste, though the greater number would decide in favor of the latter. Salted trout bring a lower price in the market than white-fish, as they are inferior to them as a salt fish.

Their migrations, as far as observations have been made, are confined to the spawning-season. They do not ascend the rivers, and although they are known to be in a few inland lakes connected to the main lake by rapids, there seems to be no knowledge of their ever having been seen or taken in the outlets.

Their range of depths at other seasons than the spawning-period is in deep water. A few stragglers occasionally approach the shore, and are taken in the pound-nets, or with the hook, from the piers extending into the lake. In the northern portions of Lake Michigan they are taken in depths of fifteen fathoms, in small numbers, by the gill-nets, and more plentifully through the ice in the winter time, though a depth of over thirty fathoms is the most favorable ground for their capture....

The lake-trout is a ravenous feeder. The fishermen say of him that "he always bites best when he is the fullest."

In Lake Michigan, where the investigation of the character of their food was carefully made, it was found to be principally the cisco, *Argyrosomus hoyi*

Gill, Mss. The prevailing notion that they feed largely upon the white fish was not confirmed by these observations. Although it was continually asserted by the fishermen that the stomachs of the trout were found full of young white-fish, there was no instance under my observation where it was so. During 1871 no opportunities were omitted to observe the stomach-contents of the trout, when they were sufficiently undigested to determine the species, and often, when, to confirm the repeated assertions, a fisherman would throw out the contents of a stomach, to show me the young white-fish, the head and mouth invariably indicated the genus *Argyrosomus* Agass., and he would readily admit his mistake.

Questioning fishermen closely, who asserted that they found the young white-fish to be the principal food of the trout, they generally assented that they had not given close enough attention to decide positively between young white-fish and the cisco, though many gave testimony of finding un-mistakable white-fish, of mature size, in the stomachs of the overgrown trout taken in portions of the lakes.

Stragglers into the shoal waters, and the trout migrating into shallow places, to find their spawning-grounds, would undoubtedly prey upon the smaller white-fish as readily as they would upon any other species; but during the larger part of the year they make their home in deeper water than the young fish are found in.

An instance was related, in 1871, of a large trout having swallowed a smaller one, which the fisherman removed from its stomach in a good state of preservation.

It is not an unusual thing for a trout to swallow a fish too large for the capacity of his stomach, and the tail protrudes from his mouth until the forward part is digested. A trout measuring twenty-three and one-half inches was brought ashore at Two Rivers, Wisconsin, from the mouth of which some three inches of the tail of a fish, *Lota maculosa*, projected. The "lawyer," when taken from the trout, measured fourteen inches without the head, which had been digested.

Their exceeding voracity induces them to fill their maws with singular articles of food in the bill of fare of a fish. Where the steamers or vessels pass, the refuse from the table is eagerly seized upon, and I have taken from the stomach a raw peeled potato and a piece of sliced liver, and it is not unusual to find pieces of corn-cobs, in the green-corn season, and in one instance I heard of a fragment of a ham-bone.

They are readily taken with a hook baited with pieces of fish. They are a sluggish fish to pull in, taking hold of the bait with a tug at the line and then allowing themselves to be pulled to the surface, with no more vibration in the line than if a heavy sinker was the weight at the end. Parties going out with the fishermen often take a large number while the nets are being lifted, and in some localities the largest of the trout are taken in this way. While becalmed near Summer Island, in Lake Michigan, in 1871, two of us, in about one hour's time, took in fifty pounds of trout, in seventeen fathoms of water.

〰〰〰 DR. BROOKS BURR OF SOUTHERN ILLINOIS UNIVERSITY HAS provided me with the only three things that I know about the life of James Wood Milner: he was born in 1841, he died in 1879, and he engaged in fish culture. Beyond that, Dr. Burr also brought to my attention that Milner wrote the original descriptions of two species of deepwater ciscoes: the bloater (*Coregonus hoyi*) and the black-finned cisco (*C. nigripinnis*). There is still a commercial fishery for the bloater (sold as chubs or "smoked fish" in delis), but the black-finned cisco is now probably extinct, not unlike the traces of the man who described it.

Milner refers to two types of fishing. One relied on the pound net, a long net that stretched perpendicular from shore to deeper waters. Along its length were breaks that led to small holding areas. Any fish swimming along the shore would encounter the net, follow it downslope, and wind up in the "pounds," where they stayed alive until retrieved by the fisher. This approach allowed for safe and easy access to the nets, as well as fresh product.

The second type of fishing relied on gill nets. These could be placed at any depth or distance offshore, extending hundreds of feet. But during bad weather, fishers had to balance the loss of product with risks to their own lives. Indeed, the historical record is replete with accounts of men who failed to return.

Lake trout generated the most revenue of any Lake Michigan commercial fishery from 1890 through the 1940s. But overfishing, and, most particularly, predation by the newly arrived sea lamprey triggered a dramatic drop in lake trout numbers, eventually eliminating the species from the lake. This has been remedied more recently by massive reintroductions, so there is now a burgeoning lake trout sport fishery in Lake Michigan.

Sources: Norman Baldwin et al., *Commercial Fish Production in the Great Lakes 1867–1977.* Technical Report No. 3 (Ann Arbor, MI: Great Lakes Fishery Commission, 1979); Carl Hubbs, "History of Ichthyology in the United States After 1850," *Copeia*, no. 1 (March 1964): 42–60.

Sea and River Fishing: Fish in Season in November

S. C. Clarke

᭯᭯᭯ In many Western lakes and rivers may be found a peculiar fish, known to local anglers as the [bowfin], dogfish, mudfish, or lawyer, the latter name being probably given it from its rapacious habits. It is in general aspect something like a large chub or sucker, but has a truculent and savage expression of countenance, and devours everything it can master. . . .

This fish, *Amia calva*, is in soles, fins, and the force of tail allied to the extinct fishes of an older world. . . .

Well, this disreputable branch of an old family comes in the spring with shallow waters to spawn, and one day, many years ago, being with a friend on the drowned prairies to the south of Chicago in pursuit of snipe, which we found not, we found hundreds of these dogfish on the submerged prairie, and opening fire upon them we soon had quite a number. Being hungry, we thought we would try to eat them, in spite [of] their evil name. So we took them to dry land, built a fire, roasted them in the ashes, and really gave them a trial. I must add that even with the sauce of hunger we found it impossible to make much of a dinner.

Letter [Fish Pot Hunters]

Pickerel

᭯᭯᭯ I beg to call the attention of the proper authorities to the fact that parties are using seines in the Des Plaines River between Summit and Willow Springs. The game pot hunters have been exposed and held up to public contempt until they are well nigh routed, but not so with the fish pot hunters, who not satisfied with taking sufficient fish with the seine for private use, feed the surplus to their hogs, even fine pickerel.

S. C. Clarke, "Sea and River Fishing: Fish in Season in November," *Forest and Stream* 1, no. 15 (November 20, 1873): 236.

Pickerel, "Letter [Fish Pot Hunters]," *American Field*, May 22, 1886, 486.

Black Bass Fishing in Fox River

Fred J. Wells

ᔓ ᔓ ᔓ Few, if any, waters of Northern Illinois can compare with the Fox River for black bass fishing. The Fox has its source in Southern Wisconsin and flows through the northeastern portion of Illinois to its confluence with the Illinois River at Ottawa.

The water is clear and cold, and the bed of the river, rock, or hard sand or gravel. In the spring large numbers of fish of different varieties go up into the Fox River to spawn and many thousands are caught at the Government dam at Dayton, which is the first dam from the mouth. People go there from miles around and camp out for days, catching fish and salting them down for future use. The majority of the black bass caught in the river are of the small-mouthed species.

With my friend Love, I have spent many pleasant hours along the banks of the Fox and we have caught many fine strings of bass. One day we drove down the river to the small village of Oswego. From this place to Yorkville, a distance of some six or seven miles, there is excellent fishing.

On the bank near the village is located a cider-mill and the refuse of the press is thrown into the river. We stopped here and after putting on our waders, we took with one haul of the minnow seine enough minnows to fill our buckets. Chubs and shiners are abundant. They apparently live on the refuse from the mill and the small fry can be found here in the Fall when it is impossible to catch them elsewhere in the river.

In addition to the minnow buckets we had brought grasshopper buckets made of wire screen with a cover of tin and a hole in the cover through which to put the grasshoppers when catching them. The black bass is a very peculiar fish as regards what it will eat, and one should always be prepared to use different kinds of bait. We did not find any difficulty in catching a number of large grasshoppers among some small willows on the bank.

We entered the water and waded to what is known as the "swimming-hole." Close to the shore the water is very deep, and we found it necessary to fish from the weeds opposite the bank. We cautiously approached until we could throw over the weeds into the clear water. I put on a black chub, and

Fred J. Wells, "Black Bass Fishing in Fox River," *Outing Magazine* 28, no. 2 (May 1896): 130–34.

cast so the minnow would go down the current close to the weeds. Several times I tried, but without succeeding in obtaining a strike. I then put on a shiner but without any better success. Love had also failed to obtain a strike with minnows, and saying he was going to try a 'hopper, put on a large yellow one, and cast far over toward the shore.

The grasshopper never touched the water, for as it fell, out jumped a bass and took the bait. With a quick movement of the wrist, Love set the hook, and then followed as pretty a play with a fish as one could wish to see.

Down the river the bass went, and was turned just at the edge of the weeds. With a swirl he started up the current, the line cutting the water and going faster than Love could reel in. Twice the fish went up into the air and tried to shake the hook out of his mouth, but it was firmly fastened, and he evidently gave up the attempt, as he did not again rise to the surface. After making the run up and down the clear water several times, he at last weakened and was secured.

He proved to be what we call a tiger black bass. Upon Love's pocket-scales he registered three pounds four ounces. I told Love this was a pretty good one to start with, but that I would go him one better before we got through.

We both tried grasshoppers for some time longer in this place, but failed to catch another bass. The struggles of the first one had scattered the others into the weeds. We crossed to the other side of the river, and in a short distance we found a channel between an island and the shore that was free from weeds. The clear water was from four to five feet deep. About it the weeds were nearly a solid mass, with here and there an opening, up to the channel. Love took one side and I the other, using grasshoppers for bait. I tried several casts, letting the grasshopper float down the current, but without a strike. Telling Love I believed the fish were down on the bottom, I put on a sinker. This time I was successful, and landed a fine bass. I gave him but very little slack, drawing him into the weeds so as not to disturb the water any more than I could help. Love also leaded his line, and we fished down the channel, each having all he could attend to.

The chubs bothered us considerably by stealing our bait, so that we used a good many grasshoppers. A fisherman can always tell whether it is a chub or a bass that is after his bait. In fishing with grasshoppers where the chubs are numerous, one will wish they were in Guinea or some other far away country. They take hold of a leg or a wing and pull it off, and soon you will be obliged to put on another grasshopper. A bass takes the bait entire, and when he shuts his mouth and starts off is the time to strike.

When bass are taking grasshoppers in deep water it does not make any difference whether or not the grasshoppers are alive. Dead ones answer just as well. If you are using minnows for bait and are fishing in deep water, you must not be in any hurry to set the hook. A bass takes the head of a minnow first and will frequently make a long run before he attempts to swallow the bait. It is a nice question to know just when to strike and one can only learn from experience, and even then will not always be successful.

After we had used all our grasshoppers we tried minnows, but the bass would not take them and we emptied our buckets into the river. On counting our catch we found we had taken out of this channel fourteen black bass, all small-mouthed and ranging in weight from a pound and a half to three pounds four ounces.

We fastened our strings or rather chains of fish securely. I have never found a more convenient stringer than a light chain, about four feet long, with a bar on one end to hold the fish from slipping off and a heavy piece of wire on the other end, sharpened so that it can easily be passed from the outside through the lower jaw of a fish. A fish strung in this manner will live a great deal longer than one strung through the gills as is the common practice.

Taking our pails we spent about an hour in catching grasshoppers. It was now near noon and very warm. We secured our bait in the weeds and grass, and as we wore waders, when we got through there was not a dry thread on us. Catching grasshoppers may appear very simple, but if those who have never tried to catch a bucket of the large green or brown fellows will spend an hour at it on a hot day, without a net, they will change their opinion. Love thought we ought to try our channel again, so we went up the river and fished down again. This time we only succeeded in catching three, but all good ones.

The inner man now began to make his wants known, and as we could see our wagon down the river about three-quarters of a mile, we concluded to fish to it and have our dinner. It was some time before we found another place where the water was clear and this was in a narrow channel between two islands. The water was shallow in the channel, but ran into a deep hole just where the weeds commenced.

I let the current take the line some twenty-five yards down to the deep water. As the hook swung into the hole, there was a rush and a snap, and my reel started spinning. Quickly putting on the thumb-drag, I slowed the fish down and set the hook. I soon brought him up to where the current was strong and the water shallow, and such a splashing and a racket as the fish then made!

He darted first one way then the other, jumped into the air, and if I had not been very careful and lowered the tip of my rod each time he made a jump, I should certainly have lost him. I was using a nine ounce split bamboo rod and the strain made it creak more than once. By careful and patient work, however, I finally reeled him in. The bass weighed two pounds and ten ounces, but one twice the size in deep water could not have made a stronger fight.

A great deal is written about catching large-mouthed black bass in the Northern lakes. I have tried it, fishing from a boat, using frogs for bait, and have caught bass until I was tired of reeling them in, but I would rather catch one good-sized small-mouthed black bass in running water where one has to fight for every inch of line, than a whole boat load of lake bass. It means more genuine sport, more skill in the use of the rod and the excitement is a hundred times more intense.

Love let his hook float down into the hole, but without obtaining a strike. He tried it a second time and succeeded in landing a bass weighing a trifle more than a pound. We tried several more casts here but if there were any more fish in the hole they would not bite and so we went on down the river. We caught two more before reaching our wagon, making a string of nineteen for the forenoon.

There are a number of springs along the banks of the Fox, the water in which is clear and cold. Beside one of them under a shady oak we pitched our camp. After removing our waders and refreshing ourselves with a drink of the sparkling water, we started to prepare our dinner.

We first made a stove by taking two stones and setting them on edge and a third one on top and gathering some drift wood and dead branches soon had a rousing fire. While Love dressed several of the bass, I took from the wagon our "cook-box," as we call it, which contains a coffee pot, frying pan, tin plates, and, in fact, all the necessary utensils and dishes for preparing a meal.

Love fried the fish and attended to the coffee, and taking our box for a table I placed on it the dishes and some rye bread, onions, butter, sugar and salt. This does not seem very much, but with fried black bass and hot coffee it makes an excellent meal. Fish are never so nice as when taken from the water and immediately cooked, and one never enjoys them more than when tired and hungry from a tramp.

After we had put away our cooking outfit we took a quiet siesta for an hour or two, enjoying our pipes and dozing in the shade. About four o'clock we prepared for the water again. We fished every clear channel we could find, catching a bass here and there, but not finding a place where we could reel in any number.

As the sun commenced to go down we concluded we had enjoyed enough sport for one day, and started for home. On counting our catch we found we had thirty-two, making as fine a string of bass as was ever taken out of the Fox. There were only three fish that weighed less than one-and-a-half pounds.

◆◆◆◆ FRED J. WELLS (1859–1925) WOULD HAVE ELUDED MY EF-forts at trying to find out something about him except that his name appeared precisely once in a ProQuest search of the *Chicago Tribune*. The only mention is that he contributed to a charity, but his home was listed as Aurora. It turns out he was well known enough that his death merited a front-page obituary in the *Aurora Daily Beacon* and a file in the Aurora Public Library, whose staff very kindly sent me copies of what they had.

Born in Iowa, Wells arrived in Aurora in 1890 to work as "chief clerk in the office as superintendent at the local Burlington shops." After thirteen years he changed jobs, becoming an officer of the Aurora Foundry Company, where he stayed for the rest of his career. Wells was also active in a host of other organizations, some of which I presume had interesting hand signals: the Union League Club; Aurora Elks Lodge No. 705; Aurora Lodge No. 254 A.F. & A.M; Aurora Commandery, No. 22, Knights Templar; Consistory of Freeport, Tebala Shrine of Rockford; and the Hamilton Club of Chicago.

But what Fred Wells evidently enjoyed most was being outdoors. According to his obituary, "in his early days he spent two years hunting and trapping for the government and made a thorough study of birds." It was his love of fishing, though, that he pursued most diligently as a resident of Aurora. Local columnist Justus Johnson credits him with catching the largest smallmouthed bass in "northern waters": a seven-pound, four-ounce lunker pulled from the Fox River near the dam at North Aurora about 1890. Wells was also one of the founders and secretary of the Fox River Bait Casting Club, the precursor of the county chapter of the Izaak Walton League. Its 115 members encouraged fish stocking and other measures to enhance angling in the river.

What would an angler find today in comparison to what Wells described? It seems likely that heavy loads of sediment created by development in the watershed would have long since covered the clean bed of rock, sand, or gravel that Wells observed. And I would bet that it is no longer prudent to quench one's thirst by partaking of the springs along the river's bank. But to get a fuller picture, I contacted Steve Pescetelli and Robert Rung, biologists with the Illinois Department of Natural Resources who are authorities on the Fox River. They obligingly read Wells's article and noted some differences that exist today. The fish are probably fewer now and of smaller size, although the Fox still remains "a pretty good smallmouth river compared to others in the state."

But the single biggest change that has occurred over the century is the decrease in aquatic vegetation. This growth is "absent today due to turbidity caused by large populations of suspended/motile algae and bacteria. These, in turn, reflect high levels of nutrients and numerous instream impoundments."

Preventing the degradation of our local streams in an increasingly urbanized environment remains one of our greatest conservation challenges.

Sources: "Fred J. Wells, 66, Dies After a Long Illness," Aurora Daily Beacon, December 31, 1925, 1; Justus Johnson, "Now and Then," Aurora Daily Beacon, October 17, 1948 ("Fox River Fishing"), and October 24, 1948 ("Numbers Interested in Rod and Reel").

Gleanings from Nature

W. S. Blatchley

TWO FOPS AMONG THE FISHES

THE RAINBOW DARTER

"Little fishy in the brook."

᳆᳆᳆᳆᳆᳆ Not the one "daddy caught with a hook," but another, too small for the hook, too small for the frying-pan, too small for aught else but beauty, and gracefulness of form; and yet not the young of a larger fish, but full grown of himself. In every brook in the State he may be found, yea, even in the rill, no more than a foot in width, which leads away from the old spring-house on the hillside. You will not find him swimming about like the minnows in the still deep water of the stream, but where the clear cold water is rushing rapidly over the stones of a ripple he makes his home. There he rests quietly on the bottom, waiting patiently for his food, the larvae or young of gnats, mosquitoes and other such insects, to float by.

If you attempt to catch him, or your shadow suddenly frightens him, with a sweep of his broad pectoral or breast fins, he moves quicker than a flash

W. S. Blatchley, Gleanings from Nature (Indianapolis: Nature Publishing, 1899), 19–26.

a few feet farther up the stream and then as suddenly comes to a stop, and resumes his quiet "thoughtful" attitude. If you persist in your attempt to capture him he will dart under a small stone or submerged leaf, where, like the foolish ostrich which when pursued hides her head under her wing, no longer seeing you, he thinks himself secure.

On account of the shape of his body as well as on account of his rapid movements he has received the surname "darter." Belonging to the group which bear this surname, there are, in the eastern half of the United States, about 47 species or kinds, the largest of which, when full grown, measures only about six inches in length, while the smallest species never reaches a length of more than an inch and a half. They all have the same habits, and at least 29 kinds of them are found in Indiana; but the one of which I am writing, *Etheostoma coeruleum* Storer, is much the more common. He is from two to two and a half inches in length, and, like the other members of his family, has two fins on his back; "dorsal" fins they are called by naturalists, the front one of which contains 10 short spines. During eight months of the year, the males and females dress alike in a suit of brownish olive which is striped on the sides with 10 or 12 narrow, black cross-bars, and more or less blotched on the back with darker spots. But on the first warm days of spring when the breezes blow up from the gulf, awakening the gypsy in our blood, the little male fish feels, too, their influence, and in him there arises an irresistible desire to "a courting go." Like most other beings of his sex he thinks his every-day suit too plain for the important business before him. It will, in his opinion, ne'er catch the eye of his lady love. So he dons one of gaudy colors and from it takes his name—the rainbow darter—for in it he is best known, as it not only attracts the attention of his chosen one, but often also that of the wandering naturalist who happens along the stream.

The blackish bars of other seasons are changed to indigo blue, while the space between them assumes a hue of the brightest orange. The fins are broadly edged with blue and have the bases orange, or orange and scarlet, while the cheeks assume the blue and the breast becomes an orange. Clad in this suit he ventures forth on his mission, and if successful, as he almost always is, the two construct a nest of tiny stones in which the eggs of the mother fish are laid and watched over with jealous care by both parents until in time there issue forth sons destined some day to wear a coat of many colors, and "darters" to be attracted by those coats, as was their mother by the one their father wore.

Although so abundant and so brilliant in the springtime, the rainbow darter is known to few but naturalists. The fishes in which the average country boy is interested, are the larger ones—such as the goggle eye, the sucker, chub and sunfish—those which, when caught, will fill up the string and tickle the palate.

But there are, let us hope, among our farmers' sons and daughters, some who are learning to take an interest in the objects of nature which are beautiful, as well as those which are useful. To them I will say, if you wish to see something really pretty, make a seine from an old coffee sack or a piece of mosquito netting, and any day in spring drag two or three ripples of the branch which flows through the wood's pasture, and ten chances to one you will get some "rainbows." By placing them in a fruit jar three-fourths full of clear, cold water, and renewing the water every few hours, they can be kept for several days; but they can not bear the confinement long, accustomed as they are to the free running stream from which they were taken.

By taking the rainbow as the type of the darter and studying closely its habits, both in captivity and in the streams, much can be learned about a group which, in the words of Dr. S.A. Forbes, "are the mountaineers among fishes. Forced from the populous and fertile valleys of the river beds and lake bottoms, they have taken refuge from their enemies in the rocky highlands where the free waters play in ceaseless torrents, and there they have wrested from stubborn nature a meager living. Although diminished in size by their continual struggle with the elements, they have developed an activity and hardihood, a vigor of life and a glow of high color almost unknown among the easier livers of the lower lands."

THE LONG-EARED SUNFISH

Among the most brightly colored of all the fresh water members of the finny tribe is the long-eared sunfish, *Lepomis megalotis* (Raf.). When full grown its length is about eight inches and the breadth one half as much. The color is then a brilliant blue and orange, the former predominating above; the orange on the sides in spots, the blue in wavy, vertical streaks. The cheeks are orange with bright blue stripes; the fins with the membranes orange and the rays blue. Extending back from the hind margin of each cheek is a conspicuous blackish membrane termed an "earflap," which in this species is longer than in any other of the sunfish family, whence the specific name, *megalotis*, from two Greek words meaning "great" and "ear."

Within the placid pools of the brooks and larger streams of the State this sunfish has its favorite haunts. Mid-summer is the time when its habits can be best observed. On a recent August morn I sat for an hour or longer on the banks of a stream, which flows through a wooded blue-grass pasture, and watched the denizens of its waters. A peaceful calm existed, the water being without a ripple and with scarce the semblance of a flow—the air without the shadow of a breeze. Dragon flies lazily winged their way across the pool, now resting daintily upon a blade of sedge or swamp grass, now dipping the tips of their abdomens beneath the surface of the water while depositing their eggs. The only sounds of nature were the buzz of a bumble-bee feeding among the flowers of the [P]runella [Self-heal] at my side, and an occasional drawl of a dog-day locust from the branches of the sycamore which threw a grateful shade about me.

The sunfish "hung motionless" in the water, their heads towards me, holding their position only by a slow flapping of their dorsal and pectoral fins. Their nesting time over, their season's labor ended, it was with them, as with many other beings, a time of languor.

These long-eared fishes are the lords and ladies of the respective pools wherein they abide. When they move other smaller fry clear the way. If a worm or gnat, falling upon the surface, tempts them, it is theirs. A leaf falls near them and is seemingly unnoticed—a fly, and how quickly their dormant energy is put into motion. With a dart and gulp the insect is swallowed and a new stage of waiting expectancy is ushered in.

How admirably fitted their form for cleaving the water! They often seem to glide rather than propel themselves through its depths. Again, how swiftly the caudal fin moves when with straight unerring motion they dart upon their prey. At times one turns his body sideways and, with a slow, upward-gliding motion, moves toward some object on the surface which is doubtfully "good to eat." He even takes it into his mouth and then, not having faith in his power to properly digest it, ejects it with force, and turning quickly darts back to the friendly shadow of a bowlder beneath whose sides he has, in time of threatened danger, a safe retreat.

I throw a grasshopper into the pool. Like a flash six of the sunfish are after it. One reaches it a tenth of a second in advance of the others, and with a lightning-like gulp, which disturbs the serenity of the surface of the pool, swallows the kicking prey. The energy of the sun's heat and light, stored in grass, transmitted to move muscles in gigantic leaps, will, in a short time, wag a caudal tin and propel the owner through these watery depths.

Years are thus doubtless spent by these long-eared sunfish in a dreamy sort of existence, their energies quickened by the vernal season and growing duller on the approach of winter. Excepting the times when they are tempted by a wriggling worm on some boy's hook, theirs is a life exempt from danger. A kingfisher glancing down from his perch on the bent sycamore limb may, at times, discern them and lessen their ranks; but, methinks, the chub minnows, with fewer spines in their dorsal fins, are more agreeable to the kingfisher's palate. With all the tints of the rainbow gleaming from their sides they move to and fro, the brilliant rulers of those quiet pools.

The king or monarch of those noted was most gorgeously arrayed. In addition to the hues above described, a streak of emerald bordered his dorsal and caudal fins and was bent around the edge of his upper lip—a green mustache, as it were. By tolling them with occasional bits of food I drew him and his retinue close into shore. There, for some time they rested, watching eagerly for additional morsels. As I was leaving I plucked from my sleeve an ant and threw it towards them. A dart, a gurgle, a gulp—the leader had leaped half his length from the water, and the ant was forever gone. The ripples receded and finally disappeared, and the last scene in this tragedy of nature was at an end.

〰〰〰 WILLIS STANLEY BLATCHLEY (1859–1940) IS CERTAINLY among the greatest naturalists Indiana has ever produced. And the label of naturalist is one he would have worn proudly, for he once proclaimed: "There are, in my opinion, too many specialists and too few naturalists in the world today."

Blatchley was raised in Putnam County, Indiana, and had the unusual experience of attending the school where his father was teacher. Later he worked his way through normal school selling picture enlargements, maps, and notions. He taught for several years before entering Indiana University, an experience that changed his life. Two faculty members, giants of their time, guided him into the natural sciences, which he spent the rest of his life exploring with great adeptness. His two mentors were the geologist John C. Branner and, especially, the ichthyologist David Starr Jordan, whose name appears in many of these biographical profiles. While an undergraduate, Blatchley published nine scientific papers.

After graduating in 1887, Blatchley served as the chairman of the science department of Terre Haute High School for seven years. But it's clear that his duties did not satiate the full scope of his interests, for during his tenure at the school he accompanied Dr. Branner on a geological survey of Arkansas, participated in a field study of Mexico's Mount Orizaba, and collected fish in Ohio and Indiana on behalf of the U.S. Fish Commission. He also managed to find time to earn a master's degree from his alma mater.

In 1894 he began a new chapter in his career by being elected Indiana's state geologist, a post he would hold for four four-year terms. As state geologist, he traveled the length and breadth of Indiana, studying the dunes on the north to the Ohio River on the south. In addition to the annual reports he had to publish as part of his job, he produced twenty-seven papers and books, the best known of which is probably his classic *An Illustrative and Descriptive Catalogue of the Coleoptera or Beetles (Exclusive of the Rhynchophora) Known to Occur in Indiana*. Although published in 1910, it is still being used today.

When he left the office in 1910, he and his wife began spending winters in Florida. Despite the languorous temperatures and greater autonomy that comes with retirement in the South, Blatchley continued his labors, completing additional entomological treatises on such groups as the Rynchopora (weevils), Orthoptera (which include grasshoppers, crickets, and cockroaches), and Heteroptera (bugs). In recognition of his accomplishments, Indiana University awarded him an honorary doctorate in 1921.

The selection included in the anthology is from *Gleanings from Nature*, a book Blatchley wrote for the purpose of providing country youths with "better knowledge of nature": "With such a knowledge generally diffused there would be less dissatisfaction with country life and fewer farmers' sons and daughters would flock to the cities.... With a knowledge of some of nature's objects and a desire to ferret out for themselves some of her secrets, they would have something of which to talk and think besides crops, stock, work, neighborhood gossip, and local politics, and the attractions of the city would seldom excel those to be found on the old homestead."

Sources: W. S. Blatchley, "Some Interesting Birds I Have Known," *Indiana Audubon Bulletin* (1930): 7–17; "Willis S. Blatchley, 1859–1940," *Proceedings of the Indiana Academy of Science* 50 (1941): 1–2.

~~~~~~

The Fish Market of Chicago

Illinois Board of Fish Commissioners

֍ ֍ ֍ The City of Chicago is unique in many ways, but it stands alone in one thing, viz., the Jewish fish market, where the Bohemians, Poles, and Hebrews go to buy their fish. There is nothing like it in the United States. South Jefferson street, from Twelfth to Maxwell street, on one side of the street is lined with small, dingy buildings one story high, in which the retail fish merchant displays his stock of fresh water fish. On the outside is a stand on which the fish of all kinds are piled and mixed indiscriminately—black bass, black

---

Illinois Board of Fish Commissioners, "The Fish Market of Chicago," *Report of Illinois Board of Fish Commissioners* (Springfield, 1898–1900), 21–22.

fins, bullheads, crappies, herring, mullet, pike, pickerel, perch, rock bass, suckers, sheepsheads, sunfish, trout, whitefish, and white bass. On the inside of the store are tanks, where the live buffalo, carp, and dogfish are displayed, and the customer selects the fish, it is caught in a net, then wrapped in newspapers (a large pile being always on hand), and given to the customer, who takes the live carp and walks out of the store feeling sure of its being fresh. After the fish is wrapped in the paper it is perfectly quiet, although just previously it may prove highly pugnacious in its efforts to escape from its captor with the net. Possibly its new surroundings may numb its faculties or maybe it is smothered in the close wrapping of the paper. This is all the more strange as the carp will live longer out of the water than any other fish. Its tenacity is proverbial. The buyers are of all kinds, from the poor woman that takes two or three small suckers, to the prosperous merchant's wife, who brings her basket for a mess of black bass at 14 cents a pound or live carp at from 8 to 10 cents.

All go through the same routine of buying, selecting the live fish or taking up the dead fish in their fingers, examining it, smelling it to test its freshness, then handing over to the salesman the quantity of fish selected, who wraps it up in old newspapers and hands the fish over to the customer, taking pay for the same invariably in silver. The whole transaction is free from wrangling, for the customer selects his own fish and is to blame if it is not satisfactory.

Many stands are located on the curb along the street, where only dead fish are sold. The side streets in the neighborhood have also their stands. The fish are usually brought from the West Side Fish Company at a uniform price and sold for one price by all the dealers. The small dealers mix up several kinds of fish in one lot and cry out the price—three, four, or five cents a pound, as the case may be. The dealer that has the strongest voice usually sells the most fish. The stores handle the best quality, some of them selling live fish, the quality being graded down to the poorest as the small dealer is reached, who only has a pushcart to sell from.

In the season some of the retail merchants handle cured herrings, including Scotch and Holland. They display the fish in a barrel on the sidewalk, as in fresh fish. They are also subject to examination. Thousands of barrels are sold annually.

There are more fresh water fish sold in this market than in any other place in the United States. During the holidays it is not unusual to see ten thousand people buying fish. They always pay cash and take their purchases with them. Every one interested in the fishing industry should visit this market. It is one of the sights of Chicago.

Animal Communities in Temperate North America

Victor Shelford

ANIMAL COMMUNITIES OF STREAMS

I. Introduction

༄༄༄ The conditions in streams from headwaters to mouth have many features in common with lakes, like Lake Michigan. It is therefore appropriate that they follow the discussion of such a lake. The streams belong to two drainage systems—the Mississippi and the Saint Lawrence. All are tributary either to Lake Michigan or to the Illinois River. The principal tributaries of the lake near Chicago are the Chicago River, the Calumet River, Trail Creek, the Galien River, the St. Joseph River, and the Black River. The principal tributaries of the Illinois River, with which we are concerned, are the Fox River, the DesPlaines River, the DuPage River, the Kankakee River, Salt Creek (Ill.), Hickory Creek.

The factors of greatest importance in governing the distribution of animals in streams are current and kind of bottom. They influence carbon dioxide, light, oxygen content, vegetation, etc.

These factors are controlled by age (physiographic), length of stream, and elevation of source above the mouth, all of which are physiographic. The typical stream begins as a gully and works its way into the land. The importance of some of the factors is greater in some stream stages than in others. For example, in the younger stages (*a*) material eroded, (*b*) relation to ground water, and (*c*) slope of stream bed play a more important rôle than they do in later stages.

II. Communities of Streams

1. CLASSIFICATION The classification of stream communities is based upon physiographic history and physiographic conditions. In the early stages of stream development there are two types to be distinguished: (*a*) the communities of intermittent streams, and (*b*) spring-fed streams. As soon as the intermittent stream cuts below the ground-water level, it becomes

Victor Shelford, *Animal Communities in Temperate North America* (Chicago: University of Chicago Press, 1913, 1937), 86–105.

much like the spring-fed stream. Permanent streams are divided into brooks, swift and moderate, and rivers, sluggish and moderate, with communities named accordingly. We undertake a discussion, first, of the history of the communities of streams developing in materials easily weathered and eroded, containing bowlders, gravel, and occasional strata of hard rock.

2. THE INTERMITTENT STREAM COMMUNITIES There are two types of these—intermittent rapids and pool communities.

a) Temporary rapids consocies—Small gullies in which water runs only when it is raining do not have any aquatic residents. As soon as such a gully has cut a channel deep enough to stand below ground-water level during a few days or weeks of the rainy season, aquatic insects make their appearance. The species which is usually found in the smallest trickle of water is the larva of the black fly, *Simulium* As the stream grows a little larger, and perhaps even at such a young stage also, we sometimes find the nymphs of May-flies. Such streams have, however, no permanent aquatic residents. These aquatic forms are not aquatic during their entire lives. They require water only during their early stages. If the water is running at the time the female is ready to deposit eggs and if she is properly stimulated by the conditions, she deposits them without regard to future conditions. If the wet weather continues long enough, the larvae will mature and the other adults will appear, otherwise they die. This type of animals continues after the stream becomes large enough to have permanent pools. At such a stage the number of species is increased, but no two collections are alike. Clinging to the upper surface of the stones are black-fly larvae, caddis-worms (*Rhyacophilidae*); under stones, May-fly nymphs, those collected at different times often belonging to different species. On some occasions there are great numbers of unidentifiable dipterous larvae and caddis-worms without gills or cases. Such a stream may possess any or all of these on one occasion, and none or only a few of them on another.

b) Temporary pool consocies.—As a young stream grows deeper it often reaches some depression or marsh at its headwaters of which it forms the outlet in the early spring. It is now permanent for a longer period each season of normal rainfall, and small pools usually alternate with the rapids just described. In these pools aquatic insects, crustaceans, and snails which belong primarily to stagnant ponds make their appearance. The first resident species are the crayfishes. They are found in the pools in the early spring when the water is high. The drying of the stream calls forth behavior suited to the conditions, and in summer their *burrows* are common in the stream bed. They come out at night and are preyed upon by raccoons, the tracks of which are commonly seen.

c) The horned dace, or permanent pool communities.—The first permanent parts are permanent pools. In these, conditions such as current, sediment, oxygen content, etc., are intermittent or spasmodic. The current in the rapids is distinctly spasmodic and conditions in these rapids are similar to those in the stream before even temporary pools were developed. Streams with permanent pools are represented in the Chicago region by many which enter the lake where high bluffs are present. County Line Creek has been studied as an illustration of this type.

The larger pools possess a practically permanent fauna. The characteristic forms are the crayfishes (*Cambarus virilis* and *propinquus*). The young are to be found in the pools at all seasons of the year. Water-striders, back-swimmers, and water-boatmen are common. Occasionally one finds dragon-fly nymphs (*Aeshna constricta* and *Cordulegaster obliquus*), dytiscid beetles (*Hydroporus* and *Agabus*), crane-fly larvae, the brook amphipod (*Gammarus fasciatus*), and the brook *mores* of the sowbug (*Asellus communis*). These are common among the lodged leaves. They move against water current.

The species of fish which is most commonly found in the smallest streams and nearest the headwaters of the larger streams is the horned dace or creek chub (*Semotilus atromaculatus*). It possesses certain noteworthy physiological characters. Like many other species of fish, it goes farthest upstream for breeding. Its nest is made of pebbles. Often after the breeding season is over, and the adults have gone downstream, the water lowers so that young fishes are left in large numbers in small drying pools. Here they swim about, with their mouths at the top of the water, which is constantly being stirred up by the many tails, and which often contains much blackened, oxygen-consuming excreta and decaying plant materials. This would cause death to less hardy fishes. [W. C.] Allee found very little oxygen in the waters of such pools. As it is, the pools often dry up, and the fish die. The second fish to enter a small stream appears to have many of the characters of the first. It is usually the red-bellied dace (*Chrosomus erythrogaster*), which breeds on sandy or gravelly bottom but tolerates standing water, being found also in some of the stagnant ponds at the south end of Lake Michigan. In some streams, the black-nosed dace (*Rhinichthys atronasus*) is second from the source. These fishes go against the current, but avoid the places where it is most violent. This one also breeds on gravel bottom, and can withstand the stagnant conditions of the summer pools.

As the stream lowers its bed, this type of formation passes gradually into a later one. The beginning of the succeeding formation is heralded by the coming of the Johnny darter (*Boleosoma nigrum*), the common sucker (*Catostomus commersonii*), and the blunt-nosed minnow (*Pimephales notatus*).

d) *Characters of the communities.*—The intermittent-stream communities are made up of animals which are dependent upon water during only a part of their lives and which possess a means of attachment and move against current (positive rheotaxis). The pool communities are made up of animals tolerating great extremes of conditions and being also positively rheotactic. The fish are able to meet the current and to withstand the conditions of the stagnant pools. The crayfishes live in the water in the spring and burrow in the dry weather; adults of the aquatic insects creep into moist places when the stream dries. Allee has found that isopods are positively rheotactic and that they can be acclimated to extreme conditions.

3. SPRING BROOK COMMUNITIES In glaciated areas many of the streams are fed by springs which have not been produced by erosion, but are the result of porous and impervious layers of till arranged as in regions possessing artesian wells. The presence or absence and numbers of animals in a spring depend largely upon the chemical content of its water. Spring waters commonly have insufficient oxygen to support animals and at the same time may contain sufficient nitrogen and carbon dioxide to be detrimental if not fatal to animals. The mineral matter in solution may be large in quantity and in some cases poisonous also. As the water flows away from the spring it becomes aerated and diluted with surface water so that the animals of the spring brook can live in it. *Spring consocies* differ in different springs because of variations in the character of the water.

In an area where there are springs, they are usually numerous. The little brooks unite to form larger streams. Typically, such streams may not be larger than intermittent streams, but a nearly constant flow at all times of the year is one of the characteristic conditions. Pools and riffles are not so well defined, but contain some small fishes. The watercress grows abundantly at the sides of the stream and affords a lodging-place for aquatic animals not furnished so abundantly by young streams of other types. The water is colder in summer and warmer in winter than in other streams.

Spring brook associations.—Among the watercress are the amphipods (*Gammarus fasciatus*), the larvae of *Simulium* attached to the leaves, beetles, dragonfly nymphs, and young crayfishes. Here are also found occasional snails (*Physa gyrina*). The species of the cress association are nearly all found under stones or on stones in the riffles. On the stones are *Simulium* larvae and *Hydropsyche*, the net-building caddis-worm. Under the stones are the nymphs of the May-fly (*Baetis* and *Heptagenia*), the larvae of flies and midges (*Chironomus, Dixa,* and *Tanypus*), the brook beetles (*Elmis fastiditus*), and occasional amphipods and crayfishes.

4. THE SWIFT-STREAM COMMUNITIES As the spring brooks and the intermittent streams continue to erode their beds, they increase the extent of their drainage systems and become larger streams. Springs tend to disappear in connection with the spring brook and the intermittent stream reaches the ground-water level and becomes permanent. The two sets of conditions converge toward the *larger swift stream*. While the conditions in these are like those of the spring brook, the watercress is absent and there are few rooted plants. Pools and riffles are well developed and the flow of water is constant, but fluctuates in volume. These streams differ in size, but the formation *mores* are practically the same, although larger species commonly inhabit the larger stream.

a) Pelagic sub-formation is very poorly developed in the smaller streams and will be discussed in connection with sluggish streams.

b) Hydropsyche or rapids formations.—These are usually due to the presence of coarse material or an outcrop of rock. They are typical in streams with large bowlders and stones of all sizes. Here current is probably the control-ling factor. In these streams, we find the best expression of the riffle forma-tion, which we have seen is poorly developed in the smaller streams. This formation includes three ecologically equivalent modes of life, each meeting the current in a different way. These are (i) clinging to stones in the current, (ii) avoiding the current by creeping under stones, (iii) self-maintenance by strong swimming powers.

Upper surface of stones (stratum 1): Here again we find the black-fly lar-vae, particularly in the smaller streams. They are provided at the posterior end of the body with a sucker surrounded with hooks. The salivary glands are, as is common in insects, modified into silk glands and the silk is of such a nature that when it is brought into contact with a stone it adheres. The animals are usually found attached to the rock by the sucker, with the head downstream. The fans are extended and serve to catch diatoms and other floating algae. If for any reason the sucker gives way, the animal starts to float downstream. If the mouth can be brought into contact with a stone, the silk is exuded and the animal is held until it can make the sucker fast again. The pupae of this fly are also attached to the stones. They are surrounded with a cocoon. We have re-moved them from the stream and have found that they *cannot make this cocoon in the absence of the current, but make a shapeless tangle* instead. The adults deposit their eggs at the sides of the streams.

On the tops of stones caddis-worms (*Hydropsyche* sp.) usually have cases made of pebbles stuck together with silk. They also have a net for catching floating food. The net faces the current (usually upstream). The river snail (*Goniobasis livescens*) is common on the upper surfaces of the larger rocks and is distinguished by a strong adhesive foot. These snails are usually headed upstream. When placed in a long piece of eave-trough into which the tap water was running at one end, they nearly all made their way to the upper end within a short time. *They are ecologically equivalent to the caddis-worms and the black-fly larvae.*

Among the stones (stratum 2): Of the animals living among stones, the darters are most important. Of these the banded darter (*Etheostoma zonale*), the fan-tailed darter (*E. flabellare*), and the rainbow darter (*E. coeruleum*) live among and under the stones or in the algae which cover the rocks (especially the fantail). With them are sometimes found the Johnny darter (*Boleosoma nigrum*), the black-sided darter (*Hadropterus aspro*), and the small bullhead or stonecat (*Schilbeodes exilis*). These fish are all positively rheotactic. They apparently orient because of unequal pressure on the two sides of the body when it is not parallel with the direction of the current.

Under the stones (stratum 3): There are many more forms living under and among the stones than on the tops of them. Here are the May-fly nymphs, the flattened *Heptageninae*, and the very awkward damselfly nymph, *Argia*, evidently succeeding well together. This fact makes the value of the flattening as an adaptation appear nil. There are also the larvae of midges (*Chironomus* sp.) and of horse-flies (*Tabanus*). The adults of the latter deposit their eggs in great masses on the tops of the stones which protrude from the water. The stone-fly nymphs, similar to the *Heptageninae* May-fly nymphs in form and appearance, are found here also. Perhaps the most bizarre of all are the water-pennies. These are round flat objects adhering to the under sides of stones, and not looking like animals at all. They are the larvae of a parnid beetle (*Psephenus*).... The old larval back becomes the cover for the pupa. The adults live under the stones also.... Sessile or attached animals are common in the brooks, but their numbers vary greatly from year to year. On one occasion the surface of the rocks and stones in Thorn Creek was almost covered with sponge, and while some sponge is always to be found, we have not seen it so abundant again. Polyzoa are usually present under the stones. Such animals depend upon foods in solution and small floating plants and animals.

In addition to those rapids which have large rocks, are those in which the bottom is of coarse sand and gravel, with only a few small stones. Here we find the caddis-worm (*Helicopsyche*), which has a spiral case made of sand grains. These are most abundant where some sand and swift current are both found. There is from time to time some vegetation in such situations and on it we find the brook damsel-fly nymph (*Calopteryx maculata*), the adult of which is the black-winged damsel-fly.

Characters of the formation: The swift-stream formation has a striking behavior character, namely, strong positive rheotaxis. Other physiological characters, such as the toleration of only low temperatures and high oxygen content, and the necessity for current for the successful carrying-on of their building operations, are probably common to the animals. So far as the fishes of the rapids are known, they breed on coarse gravel bottom or under stones. The *mores* of the formation are, then, current resisting and current requiring, dependent upon large stones or rock bottom for holdfast and building materials.

c) Sandy and gravelly bottom formation (pools).—The pools of streams with characteristic formations are usually 2 or 3 to 10 feet deep, depending upon the size of the stream. The bottom is sand or coarse gravel. In these we find conditions very different from those in the rapids. The pools are the home of the rock bass (*Ambloplites rupestris*), the small-mouthed black bass (*Micropterus dolomieu*), the sunfishes (*Lepomis pallidus* and *megalotis*), and the perch (*Perca flavescens*)....

With these are also the mussels, frequently as many as nine or ten species, among which are *Lampsilis luteola, ventricosa*, and *ligamentina*, the little *Alasmidonta calceola*, and *Anodontoides ferussacianus*, the last-named being perhaps the most characteristic of them all. They are often found beneath the roots of willows along the sides of the pools. Mr. Isely found that mussels migrate to shallow water during flood time. Mussels are dependent upon fish for a part of their lives. The young are carried by the adult until ready to attach to the body of the fish. When they leave the fish they are able to take care of themselves. Burrowing in the gravel are bloodworms (*Chironomus* sp.), the burrowing dragon-fly nymph (*Gomphus exilis*), a burrowing May-fly, a caddis-worm, and occasionally snails, *Campeloma* and *Pleurocera*. There are a few plants that grow on the sandy bottom in such places, and among these one finds the snail (*Amnicola limosa*), May-fly nymphs (*Siphlurus*), and hair-worms (*Gordius*). In some localities bivalved mollusks (*Sphaeridae*) and leeches are numerous.

Under primeval conditions beavers are associated with the pool formation. They build dams which contribute to the deepening of the water of the
pools.... An old beaver dam is supposed to have turned the waters of the
DesPlaines out of the Chicago River and down the Chicago outlet.

Characters of the formation: The *mores* of the pool formation are distinctly
those of partially burying the body just beneath the surface of the fine gravel
and moving against the current. The few animals that make cases usually use
gravel or sand grains. A single caddis-worm makes its case from small sticks
such as commonly lodge in eddies. Some of the fishes breeding in these situations cover their eggs. Some fishes orient the body and swim upstream as a
result of seeing the bottom apparently move forward below as the fish floats
down. They behave the same if put into a trough with a glass bottom and the
trough drawn forward. Some orient also when their bodies rub against the
bottom when floating downstream.

〰〰〰 VICTOR SHELFORD (1877–1968) ENTERED THE UNIVERSITY
of Chicago in 1901, having transferred from West Virginia University when
one of his mentors made the jump first. At the time he was firmly in the camp
of the traditional zoologists, who favored laboratory work over that of field
studies, the bailiwick of those embracing the still nascent discipline of ecology. But in the words of his biographer, Robert Croker, "Shelford was won
over to ecology by the marvelous zoological and botanical field trips of Charles
Child and ... Henry Chandler Cowles and by Cowles's infectious enthusiasm
for the dynamic story of wind, water, and sand carving the ancient, plant-
covered shores of Lake Michigan" (xvi).

Shelford's first love was the family Cicindelidae, the multicolored predators known as tiger beetles. He would devote a dozen years to studying them,
and his 1907 dissertation addressed the life histories and larval biology of
twelve species. As he absorbed the ideas of Cowles, Stephen Forbes at the
University of Illinois, and others, the beetles would be his entrée into the
work that led to his classic *Animal Communities in Temperate North America*. It
did for animals what Cowles had done for plants. Underlying the effort was
this principle: "Ecologically comparable animals living under similar conditions possess certain similarities of physiology, behavior, habitats, and modes
of life." And like plant communities, animal communities also evolved in suc-
cessional stages. Succession was accelerated by biological processes, which
occurred more rapidly than physiographic changes. Shelford went so far as
to say that "each set of coexisting species made the environment less fit for

itself and more fit for its successors" (Croker, 27). Apart from its larger significance as a pioneering investigation of ecological systems and processes, the
book in my mind is also of great value in being the premier survey of all the
animal taxa in the region. The sheer effort of examining every possible type
of habitat to find all the animals that are present and then identifying those
creatures down to species is a remarkable achievement.

Shelford left Chicago in 1914 for the University of Illinois, where he would
spend the rest of his career. Like Cowles, his contributions extended beyond
the academy. The conservation movement continued to grow, and Shelford
took an active role. He became president of the Ecological Society of America
(ESA), an organization that had as part of its mission to involve professional
ecologists in preserving natural areas. Later, when ESA truncated its work on
conservation issues, Shelford and his allies formed the Ecologists Union (EU)
in 1946. It grew rapidly and began attracting like-minded people who were
not necessarily credentialed ecologists. In 1950, to reflect its new membership, the EU changed its name to The Nature Conservancy.

Source: Robert Croker, *Pioneer Ecologist: The Life and Work of Victor Ernest Shelford* (Washington,
DC: Smithsonian Institution Press, 1991).

The Nesting Habits of Certain Sunfishes as Observed in a Park Lagoon in Chicago

Carl L. Hubbs

᭜᭜᭜ Just back of the present building of the Field Museum of Natural History, in Jackson Park, Chicago, there lies one basin of a series of pretty
lagoons, which are connected with one another and with Lake Michigan by
means of narrow channels. These lagoons are well supplied with fish life. The
writer has records of more than forty species, of which those belonging to the
sunfish family (Centrarchidæ) are in many ways the most interesting to him, as
well as to the numerous small boys who delight in dodging the park police to
catch these little fishes from the shore. As this locality is so readily accessible to
the writer, he was able during the spring of this year to make daily observations
here on the nesting habits of the sunfishes. Although many, and perhaps all, of

Carl L. Hubbs, "The Nesting Habits of Certain Sunfishes as Observed in a Park Lagoon in Chicago,"
Aquatic Life 4, no. 11 (July 1919): 143–44.

the facts determined have already been recorded or are generally well-known, nevertheless these notes may be of interest to the readers of *Aquatic Life*.

The species which was first observed nesting is one which is not popularly associated with the sunfishes, though belonging to the same family and having similar habits, namely the large-mouthed black bass (*Micropterus salmoides*). The nests of this species, found only during the latter half of May and the first half of June, were all circles of exposed stony or gravelly bottom, surrounded by the finer bottom material—first a ring of sand and then one of silt—thrown outward by the bass in forming the nest. They were usually located in a small, cleared area among pond weeds (*Potamogeton*). The diameter of the central or gravelly portion of the nest, throughout which the eggs were found concealed, varied between one and two feet; the extreme diameter, from one and one-half to three feet. Most of the bass nests were in depths greater than two feet, and at distances from the shore greater than ten feet.

The single warmouth bass (*Chaenobryttus gulosus*) observed breeding in the lagoon was found on June 11, over its nest, about fifteen feet from shore. The nest, which resembled that of the black bass, was located at a depth of about three feet in a cleared area in a thick growth of *Potamogeton*. The fish was affected with fungus, and died two days later.

A green sunfish (*Lepomis cyanellus*) was found guarding its eggs on June 20. No nest whatever had been constructed, the eggs being attached to willow rootlets, which here projected thickly into the water at the very edge of the lagoon. The guardian fish, presumably the male, was very brilliantly colored; back and sides metallic green, rather indistinctly barred, grading into coppery below, each scale margined with darker; the cheeks with emerald spots and streaks, more interrupted than those of the pumpkin-seed; the opercular flap greenish black, margined with coppery; soft dorsal and anal fins each with a black spot at base of the last rays, the former fin with a narrow, the latter with a wide margin of orange. This fish was surprisingly "tame," repeatedly taking an earthworm from one's fingers, permitting itself to be touched, and rising to one's hand on the surface of the water, and only gently biting at one's fingers, like a dog at play, when the eggs were being examined in the rootlets. Occasionally it circled off to a distance of about two feet, but returned at once; even after being caught and examined a moment, it came back in about two minutes from the deeper water in which it had temporarily taken refuge. On the next two days, however, the fish darted off immediately upon approach.

After a few warm days, the last of the month of May, the two commoner species of sunfishes, the pumpkin-seed (*Lepomis gibbosus*) and the blue-gill

(*Lepomis incisor*) began breeding at approximately the same time in scattered localities throughout the lagoon. The number of nests gradually became more numerous, the breeding season for each species reaching its height between June 15 and June 20. A few days of hot weather then terminated their breeding; careful search from shore and boat disclosed no new or occupied nests during the last few days of June or the first few of July.

That sunfishes do not always construct their own nests is proved by the following instance. A blue-gill was observed guarding a certain nest in the lagoon on June 15, 16 and 17; on June 18 it was not in evidence, but a pumpkin-seed was swimming about the nest, though not actually resting on it; on the morning of June 19 apparently the same pumpkin-seed was guarding a mass of eggs in the nest, but by noon of the same day most of the eggs had been removed (by small boys?), and in the evening the male was observed spawning again with a female on the same nest.

〰〰〰 ALTHOUGH CARL LEAVITT HUBBS (1894–1979) WAS ONE OF the twentieth century's most eminent ichthyologists, he was also a throwback to the earlier all-around naturalist. Over the course of his long career, he published an astounding 707 titles, which dealt with an amazing array of subjects including mammalogy, ornithology, herpetology, archaeology, and climate. He was popular among other scientists, who acknowledged his generosity and wide-ranging interests by naming after him "various fishes, lichens, algae, mollusks, cave arthropods, insects, a crab, a bird, a whale, and a dry lake" (Miller and Shor, 374).

Hubbs was born in Arizona but grew up in California. He received both his BA and MA degrees from Stanford University. It was here that he also met Laura Clark, a mathematics major who was to become his wife and crucial collaborator for the rest of his life. In reading of Hubbs's background, I am struck by how much his family was involved. Laura, Carl, and their three children would go on collecting trips together, and to keep the offspring interested, Carl would offer financial rewards for their participation—"five cents for each species collected, one dollar for a new species or subspecies, and five dollars for new genera or subgenera.... Fortunately for them he was a taxonomic splitter" (Miller and Shor, 370). On one such trip, his daughter Frances met her future husband, the young ichthyologist Robert Miller. And one of Hubbs's sons, Clark, became a distinguished ichthyologist in his own right.

Hubbs's first job out of graduate school was at the Field Museum, where he stayed two years as an assistant curator. The rest of his career was spent at two institutions: the University of Michigan from 1920 to 1944 (it was here that he received his doctorate) and the Scripps Institution of Oceanography, University of California, San Diego, from 1944 until his death.

One aspect of Hubbs's work that has also been noted are the vast collections he made. Due largely to his labors, the University of Michigan fish holdings grew from 5,000 to 1,705,197 specimens. Over the course of one excursion to Indonesia and Japan, he brought back five tons of fish. Miller and Shor write that "during this trip Carl visited many fishery stations and ichthyological laboratories, repeatedly admiring particular specimens, an act that by Japanese tradition required that the specimens be donated to the admirer!" Later, under his direction, Scripps enlarged its collection to 1.2 million specimens, of which 350,000 are said to have been collected by Hubbs. His own personal library held 40,000 books and reprints. And the largest single contributor to the Scripps archives was Hubbs, whose ninety boxes of material contain everything from manuscripts and letters to receipts.

This guy truly deserves to be the subject of a book-length biography.

Source: Robert Rush Miller and Elizabeth N. Shor, "Carl L. Hubbs (1894–1979): Collection Builder Extraordinaire," in *Collection Building in Ichthyology and Herpetology*, ed. T. W. Pietsch and W. D. Anderson. Special. pub. no. 3 (American Society of Ichthyologists and Herpetologists, 1997), 367–76.

~~~~~~

A River Mussel Parasitic on a Salamander
Arthur D. Howard

☙☙☙ It is commonly known that fresh water mussels (Unionidae) are parasitic in the larval stage on fishes. Owing to the great demand for pearl mussels particularly by the pearl button industry, the United States Bureau of Fisheries at one time undertook the rearing of mussels. In doing this a great deal of study was necessary to work out this parastic relationship. It was found that certain of the desirable kinds of mussels required fishes of a certain species, genus or family, as the case might be. For instance, the yellow sand shell (*Lampsilis anodontoides*) was parasitic on the gars, and the [ebony shell] (*Fusconaia ebena*) on the blue herring; lake mucket (*Lampsilis siliquoidea*)

Arthur D. Howard, "A River Mussel Parasitic on a Salamander," *Natural History Miscellanea*, no. 77, Chicago Academy of Sciences (January 19, 1951): 1–6. Courtesy of the Chicago Academy of Sciences—Notebaert Nature Museum.

on the basses and perches; butterfly shell (*Plagiola lincolata*) on the sheepshead; [and] warty-backs (*Quadrula pustulosa*) on the channel catfish....

The glochidia will grow on their proper hosts. They will take hold on almost any fish that touches them in a manner to call forth their snapping reaction, but they will subsequently be shed or dropped off if not on an appropriate host.

Simpsoniconcha ambigua first came to our attention at Moline, Illinois, where we were investigating a problem with the United States Bureau of Fisheries. The washboard mussel, *Megalonaias gigantea*, an important shell to the button industry, was quite abundant there. To quote from our first account, "Nets placed near the mussel bed for the purpose of determining the host fishes of the washboard mussel yielded some material that presented quite another problem. With the fishes caught were a number of waterdogs,* *Necturus maculosus*. From a total of fifteen caught, twelve were infected with glochidia of mussels. Upon attempting to identify these, a few were found to be *Megalonaias gigantea*—not imbedded nor becoming so after an attachment for a known period of twenty-four hours or more. Evidently they were merely accidental infections upon an inappropriate host.

There was a great majority of infections caused by a glochidium then unknown. They were deeply imbedded in the external gills of the waterdog; that is, the tissue of the gill had grown completely over them. By keeping the animals alive all winter we succeeded in carrying the glochidia through to the juvenile stage, these being shed the last week in May, soon after which we obtained the young mussels from the bottom of the aquarium in which the *Necturus* were held.

The parasitic period is a long one, over seven months from the date the infected salamanders were captured, October 17th to the last of May. We have observed some summer breeding mussels which made a comparable development in one week. The higher temperature at this season would explain the difference, for heat hastens the development. However, the carrying through of the parasite is the test of appropriateness of the host.

These glochidia were different from any in our station collection, of which we supposedly had a complete faunal set with one or two exceptions which we knew did not answer the case in question. We looked up all the known Unionidae which might have a range to the Mississippi River in western Illinois and found that one species was given which we did not have. This was *Simpsoniconcha ambigua*. An inquiry at the Academy of Sciences, Davenport,

---

*Now known as mud puppy.

Iowa, disclosed the fact that there was one record of collection for Davenport, Iowa. Frank Collins Baker states that it is a small and characteristic species which is at once distinguished from other species found in this region [Illinois] by its fragile shell, delicate hinge armature and minute beak sculpture. Next to *Lampsilis parva* [*Carunculina parva*] it is the smallest species found in the region (less than two inches long). It lives "in rivers and creeks, under stones and other objects. It is also found in mud which is free from débris." The distribution is given as the "Great Lakes and Mississippi Valley."

Isaac Lea figures the glochidium of this species, but it is so small and so like that of other species that without dimensions a certain identification could not be made from it. As the likelihood of finding material in some collection seemed slight, we decided to look for gravid mussels and make a direct comparison of the glochidia.

The female mussel when gravid carries the young in the outer gill, the gill thus acting as a marsupium. From the literature and from such information as could be gathered from experienced collectors, this species has the peculiar habit of living under flat stones. This seemed to present a difficulty in collecting, for the water was deep at the point where the *Necturus* had been taken. We had dredged here considerably, but no examples of this species were in the hauls. Their habitat would seem to account for this failure to secure them as an ordinary boat dredge would be likely to miss them, protected as they are by the stone above. The best chance for success seemed to be to locate them in some small stream. Baker cites the collection of this species under such conditions. By correspondence with Mr. Baker we learned the names of the collectors who kindly gave us directions for finding locally the species which they had taken many years previously. The place they described was Hickory Creek, a branch of the Des Plaines River near Joliet, Illinois.

The first five specimens found were not gravid; upon examination with a microscope all proved to be males. Returning again to the stream we found nine gravid mussels out of a total of seventeen. The first of these contained glochidia which corresponded exactly with the glochidia found on the *Necturus* in the Mississippi, thus giving us the link desired to make out the life history.

The glochidia are clear white in color, of the triangular type, with well-developed hooks contrary to the description given by Lea who suggested, however, the possibility of hooks in more mature specimens. The dimensions are as follows: height 0.265 to 0.274 mm., length 0.247 to .255 mm. All of the adult individuals were found under flat stones of the flagstone type characteristic of the limestone in the region. Beneath a single rock we found four.

While exploring the under surface of these we felt and seized a wriggling animal which proved to be a waterdog. The finding of the *Necturus* under rocks with gravid mussels suggests that the salamander may become inoculated in this way. Glochidia released by the mussels in such a location would not have a rapid dispersal by currents so that the waterdog commonly seeking such a shelter would run the chance of a heavy infection.

We suspect that *Necturus* eats the adult mussel and in seeking food visits one rock after another. In satisfying its appetite it becomes infected with the mussel glochidia, nourishing them, and when they have matured serves as a transporting and distributing agent for the young mussels.

Mussel-eating fish such as the sheephead (*Aplodinotus grunniens*) are known to be infected particularly by the thin-shelled species which they eat.

The finding of a mussel parasitic upon a salamander as the appropriate host instead of a fish was the first instance we have known to be recorded for an American mussel. [Viktor] Faussek, in St. Petersburg, experimented with Amphibia artificially infected with glochidia of *Anodonta*. He reported successful infection upon the larval form, axolotl, of the tiger salamander (*Ambystoma*) and the Austrian cave salamander (*Protcus*).

R.V. Seshaiya, in India, reports finding tadpoles infected with glochidia.

〰〰〰OF ALL THE WELL-CREDENTIALED MODERN SCIENTISTS whose works contribute to this anthology, Arthur Day Howard (1874–1960) is in a select group with Donald Lowrie as having left behind the fewest details of their lives. Howard is of particular interest to me because he has strong local connections, beginning with his birth and, presumably, childhood in Glencoe, Cook County. He left the area to attend undergraduate school at Amherst College but returned in 1898. Back on his home turf, Howard taught at Englewood High School in Chicago and earned a master's degree from Northwestern University. (The title of his thesis was "Development of the Optic Nerve in the Lower Vertebrate.") He then went to Harvard, where he received his doctorate.

His first university job was at Westminster College in Pennsylvania, and it lasted from 1906 to 1908. He relocated to the West Coast, where he taught at the University of Washington (1908–11) and the University of California (1922–25). No later positions are indicated.

Much of his early research focused on mussel reproduction and the identification of specific fish that hosted the glochidia of specific mussels. But in the partial bibliography provided by the one source, there is a gap in his publications that extends from 1922 to 1951. The 1951 article used in this anthology identifies itself as "a contribution from the Laboratories of the Allan Hancock Foundation, University of Southern California, Los Angeles," but the university archivist I spoke with could provide no information on Howard. The only other tidbit I found as to what he did later in life is his stint as president of the Conchological Club of Southern California from 1951 to 1952. Lindsey Groves of the Natural History Museum of Los Angeles, himself a former president of the club, had never heard of Howard nor had anyone he asked. However, Dr. Groves was kind enough to go through the club's minutes and came up with the year of Howard's death.

Sources: R. T. Abbott and M. E. Young, eds., *American Malacologists: A National Register of Collectors and Biographies of Early American Mollusk Workers Born Between 1618 and 1900* (Falls Church, VA: American Malacologists; and Philadelphia: Consolidated/Drake Press, 1973); website based on Abbott and Young: http://www.inhs.uiuc.edu~ksc/Malacologists/HowardA.D.html.

A Survey of Fishes in an Illinois Stream
Loren P. Woods

〜〜〜 Running down the west side and around the foot of Lake Michigan lies a continental divide, the Valparaiso Moraine, left there by the retreat of the last (Wisconsin) glacier. This is a true divide, for water falling on the east face of it flows through the Great Lakes and out the St. Lawrence, while that falling on the west and south eventually reaches the Mississippi River and Gulf of Mexico. Actually this divide is scarcely noticeable when driving across it for it is of low relief and the streams flowing from its summit are usually sluggish. In fact, many of the streams have one or more of their source tributaries rising from the overflow of a marsh held between the low glacial-drift hills that make up the divide. Such a stream is Hickory Creek, rising in southwest Cook County, Illinois, flowing southwest, and emptying into the Des Plaines River at Joliet. Hickory Creek and its principal tributary, Marley Creek, resemble most of the smaller streams of northeastern Illinois in characteristics of low gradient, shallow valley, and banks alternately following along cultivated fields, pastures, woodlands, or through towns. Hickory Creek's fauna and flora are largely duplicated in the majority of other small streams making up our local drainage pattern.

Loren P. Woods, "A Survey of Fishes in an Illinois Stream," *Chicago Natural History Museum Bulletin* (January 1959): 6–7.

Hickory Creek is approximately 21 miles long from its farthest east tributary to its mouth, and it drains an area of about 100 square miles. During most of the year the water is quite turbid so that only by sampling with seine, dredge, or dip net is it possible to learn anything about what lies beneath the surface. This turbidity, characteristic of most of our streams, results from run-off of fields, eroding banks, and livestock wading in the channel. These factors introduce a large amount of exceedingly fine silts and clays, of which the soils of the surrounding country have a high percentage, into the streams.

STREAM WATERS CLEAR

In late winter while the surface water is still held frozen on the land the only water entering the stream proper is ground water seepage from springs. At this time the water is free from silt and it is possible to see the stream bed along its entire length. The invertebrates that have survived the winter, the fishes, and their nesting areas all are visible. On February 3 one year, although the water was clear, in the shallow headwater portions of Marley Creek a great deal of anchor ice was present and no fishes were seen. Anchor ice is formed where the current is too swift for the formation of surface ice. The turbulent water is cooled by the air below the freezing point but it does not freeze because of its motion. Near the bottom or wherever the current is sufficiently retarded the supercooled water freezes and the ice attaches to stones, frequently to such an extent that the whole bottom may be covered by ice.

Supercooling on clear cold nights when the air temperature is below zero degrees Fahrenheit will often cause the formation of innumerable free crystals of slush ice (sometimes called frazil). The ice crystals may be sufficiently abundant to make the water milky. Slush ice and anchor ice scouring the bed and polishing the rocks or covering up the bed may greatly reduce the numbers of all kinds of animals in the creek. Some kinds of fishes survive living in the mouths of springs and some in riffles too swift for the attachment of anchor ice, but most kinds migrate downstream where they find deep holes.

The thawing of the ice held on land and along the edges of the stream along with spring rains often swells the volume to flood stage. It is well known that during such periods of rising waters many kinds of fishes migrate upstream.

The white sucker and creek chub move upstream and spawn on gravel beds that may be covered with 12 to 24 inches of water only during flood periods, that is, within the intermittent portion of the stream. The young creek chubs remain in this part of the stream and sometimes perish if the water falls too rapidly. Other species that migrate upstream as far as they can are the stone roller, the little green sunfish, and golden shiner. The last two species even enter tiles draining fields and may work their way up to a break so they come out in a flooded field or perhaps in a suburban garden.

Lampreys and carp also migrate upstream to spawn, the lamprey very early in spring (April) and the carp a little later (May and June). The non-parasitic brook lampreys spawn on gravel riffles where the water is not more than two feet deep. With their sucker mouths they carry stones until they have constructed a shallow depression about 12 to 24 inches in diameter. The pair then attach themselves to a large stone at the upstream edge of the nest to spawn. The freshly laid eggs stick so firmly to the stones of the nest that any attempt at dislodging destroys them, but after a day or two the eggs are washed off and lie loose among the pebbles. Carp seek a shallow weedy area for their spawning—a marsh or even a flooded pasture. Usually one female is attended by several males and with much splashing the eggs are scattered widely. These eggs are adhesive and cling to plant surfaces. Many eggs are lost, but carp are very prolific—one female will produce 300,000 to 700,000 eggs in a season but not more than 400 to 500 are deposited at one time.

Altogether 38 species of fishes have been collected from Hickory Creek. Since this is a small stream, nearly all the thousands of individuals taken have also been small (less than 10 inches). They are principally of kinds that are most often found inhabiting creeks although some kinds also live in larger streams or lakes where they grow to larger sizes. In Hickory Creek there are 4 kinds of suckers, 14 species of minnows, 4 species of catfishes, 5 different sunfishes, 8 kinds of the dwarf perches (called darters) and the mud-minnow, the sculpin, and the black-striped top minnow. It is unusual to find such a diversified lot of fishes living in a stream of this size. The present inhabitants of the creek are almost completely isolated from other streams by pollution at the stream mouth. Studies on Hickory and Marley Creeks have been carried on intermittently by many people during the past 50 years and a survey made within the last five years revealed the same kinds as those in Museum collections gathered 50 years ago. The fact that until recently the watershed and stream have remained relatively unchanged is no doubt responsible for this stability.

During the recent survey no little pickerel were collected, although these are still abundant in adjacent streams and were reported from Hickory Creek by the early collectors. Very likely this one species has disappeared from this stream and it is the only one of which there is a record of extinction. Two exotic species, carp and goldfish, have appeared in the stream either by invasion or introduction since the early survey was made around 1905.

Looking at the stream from the marshes at the head to the sludge-laden mouth, similar habitats of pools, gravelly riffles, broad meandering mud-bottom stretches, and weed patches occur again and again. Many of the species have very definite habitat preferences and no species is found distributed throughout the stream in a random or uniform manner. The rock bass can always be found in the deep holes under bridges while the catfishes and suckers live in the deep, mud-bottom holes of meanders or where a tributary joins the main stream. The sunfishes and several kinds of minnows prefer deep stretches with gravel or hard bottom and the darters and sculpin live on the riffles or parts of the stream where the current is most rapid. In one stretch there is a forested section and the stream here has more than usual gradient, cutting into dense clay. Here, in submerged holes in the banks, lives the translucent madtom catfish. The lower portion of the stream where broad weed patches grow in summer is the habitat of smallmouth bass 8 to 10 inches long.

The two most important conditions that determine how animals are distributed in streams are current and kind of bottom. These two conditions influence the vegetation, light, and carbon dioxide and oxygen content. On the other hand the conditions of current and kind of bottom are determined by the physiography, the length of the stream, and elevation of the source above the mouth.

The habitat preferences change as the fish grows. The very young fry of most species seek protection in shallows where the grass or rushes grow dense and the battle against the current is least. As they grow and change their feeding habits they move to other parts of the stream.

Since fishes are sufficiently motile they are able to seek out the place along the stream that suits them best. Once established, the individuals tend to remain throughout the season. Although they may be temporarily dislodged by a summer flood, the majority return to the same spot and never wander very far from it.

During the summer the stream population is fairly sedentary. The principal movements and migrations occur in late fall and early winter when most kinds of fishes cease to feed and seek the protection of deep holes where they crowd together. In the spring, even before the ice has completely melted from all parts of the stream, some kinds—for example, the suckers and sculpins—begin their migrations to suitable spawning areas. The other kinds—sunfishes, minnows, and catfishes—disperse later. Several kinds of sunfishes remain on the spawning beds all summer, holding a territory against others of the same species, guarding eggs or caring for successive broods.

As the population of the city and suburbs grows, few streams in our area remain unchanged. Siltation and domestic and industrial pollution reduce the streams to conditions far from suitable for most kinds of fishes. Subdivisions and country homes along the valley usually destroy the very natural beauty that made the site desirable. The Hickory Creek fauna survived the establishment of farms and pastures, but within the past five years many sections have changed because of growth of villages in the watershed and building along the valley. Dredging and straightening have begun. I predict that a survey 50 years hence will be so unproductive that no biologist is likely to be interested in making it.

Observations on Lake Michigan Fishes at Chicago

Loren P. Woods

᭡᭡᭡ Extensive reaches of the shores along the southern part of Lake Michigan consist of sandy beaches. Here the waves constantly shift and sort the sand at shallow depths—depositing it, transporting it a little distance, and redepositing it. The sand is ground to incredible fineness, and because it is mixed with silt it resembles mud. It does not act like mud, however (that is, it does not pack firmly), and so it is possible for the deep-reaching waves of storms to stir the bottom in water at least 25 feet deep. This is one of the reasons why the cribs where Chicago's drinking water is gathered are located two to five miles offshore in depths of 32 to 37 feet. The sand may remain close to the bottom, but the turbulence of the waves draws the silt up where it remains in suspension, making the water cloudy for several days after storms and waves subside.

Such a habitat of shifting sand is an extremely difficult place to live. The smothering effect of sand plus poor light resulting from the frequently turbid waters are very unfavorable conditions for both plants and animals and few can maintain themselves here. So, generally speaking, the vast areas of sandy beaches extending considerable distances offshore are barren "deserts." A careful search has revealed *nothing* large enough to be seen living on beaches in depths of less than 5 feet. Occasionally schools of small fishes (perch) may visit during periods of calm, and occasional snails or plants such as *Elodea*

Loren P. Woods, "Observations on Lake Michigan Fishes at Chicago," *Chicago Natural History Museum Bulletin* (June 1957): 5.

or *Myriophyllum* are drifted in, but there exists no permanent flora or fauna.

The majority of kinds of plants and animals live a pelagic existence in the upper open lake-waters well off the bottom, in the quieter waters of the lagoons, or on the rocky reefs where they are raised above the constantly shifting sands. There are several natural rocky outcrops along the shores of the Chicago area. Bottom-living plants and animals are the most abundant where the rocky substrate permits attachment of algae and the holdfast organs of animals. Snails (*Goniobasis*) may live here after being washed in from deeper waters but they are probably not permanent residents. Lake trout formerly spawned on rocky reefs off Lincoln Park.

MAN-MADE REEFS

Jetties, breakwaters, and sea-walls with their extensive protective foundations of large rocks have created a vast series of artificial reefs along the city waterfront and greatly increased the living space for many reef-dwelling forms, of which the most conspicuous are crayfish and log perch.

Almost no investigations have been made of the biology of either the natural or artificial reefs here, probably because they are rather inaccessible, relatively barren, and not very interesting.

One day last summer when I was swimming off the sea-wall near the Planetarium the water was unusually calm and clear. Although it was late afternoon, visibility was good to 15 feet. By diving with face mask and swim fins I made observations of the pilings and rocks on the bottom. On the surface a school of whirligig beetles swam aimlessly, as they might on any quiet pond, scattering with every nearby disturbance. The wooden pilings were covered with short golden-green threads of algae that moved with the slight motion of the water. Among the pilings and algae were hundreds of small yearling perch, 2 or 3 inches long, too small to be interested in the shiners baiting the hooks of the fishermen lining the promontory. Although none of these perch were seen feeding, they probably ate occasional sidekickers (amphipods) and other small invertebrates that live in the tangle of filamentous algae.

Several medium-sized rock bass, probably 6 or 7 inches long, were hiding in the spaces between the pilings. Around them was an area clear of small perch. Neither perch nor rock bass appeared to be frightened by me; but if they were approached, they quickly swam just out of reach and returned as soon as I passed.

CRAYFISH ON GUARD

The bottom here close to the wall is about 12 feet below the surface and consists mostly of angular rocks 1 to 2 feet in diameter. The exposed surfaces of the rocks were clean, probably kept this way by the strong currents of waves deflected by the wall. Between the angular rocks were crevices of various widths and lengths. In many were wedged bottles, bottle caps, and beer cans, and among them, living here in great abundance, were large crayfish. They did not retreat as did the other animals, but when my hand passed above one it would rear up in its most threatening manner with its pincers spread, ready to do battle.

On the smooth rock surfaces were a few large (3 to 5 inches) log perch. These beautifully shaped, zebra-patterned fish skipped rapidly from one to another rocky prominence with a smooth darting motion. The light yellow of their backs and sides matched the color of the rocks, and their narrow black bands actually made them inconspicuous until they settled on the rocks. Log perch are members of the subfamily of darters in the perch family. These darters are small (1 to 3 inches over-all length with the exception of the log perch). They have a fusiform body and enlarged pectoral fins, but they do not have a swim bladder and are thus heavier than water. As soon as they stop swimming, they quickly sink to the bottom where they brace themselves by their pectoral fins. Most kinds of darters are stream fishes living in the swift waters in the rapids of streams of our area. Log perch, although adapted to large swift streams, are also at home in the strong currents that may move along the face of a promontory sea-wall. They are quite active during the winter and not torpid as many fishes are. Their food is small crustaceans and insect larvae.

Diving in other places along the lake front or at other times of the year would certainly add to these sketchy observations, but so far I have not had the opportunity to repeat the experience under such favorable conditions.

◆◆◆◆ LOREN P. WOODS (1913–1979) ANNOUNCED EARLY IN LIFE that he wanted to be an explorer. It seems that this desire to discover was directed toward zoology by Stella Boyle, a teacher at Poseyville High School in Indiana. Woods continued sending her journals with his articles for many years. (I would think that teachers could feel few other professional pleasures more profound than being told by students that they helped provide the inspiration to accomplish great things.)

From Poseyville, where Woods was born into the family of a local physician, he entered Earlham College at Richmond, Indiana, where his interest in fish was fixed. Upon graduating, he went to Northwestern University, from which he received his master's degree. In 1938 he joined the staff of the Field Museum to work on its public school programs. He stayed with the museum until his retirement in 1978, serving as curator of fishes for thirty-four of those years. Woods authored fifty-one articles, plus two books for children.

While Woods stayed put with respect to his employer, the position enabled him to travel extensively. He enlisted in the navy, which put his expertise to good use by first having him teach commercial fishing to officers at the military school at Princeton University. Then he was dispatched to Kyushu to head the fisheries program for the military government there. A few years later, he was part of a team studying the effects that nuclear testing in the South Pacific had on local fish. On another trip he found himself trapped inside a coral formation and would have drowned had he not been able to batter his way out using his air tanks.

One memorable expedition was to western Mexico, where he spent three months with his wife and two children. They wound up collecting ten thousand specimens, many of which were caught by the use of the poison rotenone. He ran into difficulties one day at Acapulco, when a glass-bottom boat operator arrived with a full load of customers. Expecting to see a rainbow of sprightly jewels, they were met instead by the horror of dead and dying fish everywhere. The captain knew the culprit and raced back to the dock, where he gathered a few police and other boat operators. When they angrily confronted Woods, he explained that the mortality was justified for it was being done on behalf of science and thus of benefit to everyone. As for the poison, they needn't worry since it would become inactive within a day. His words were convincing, for the once-hostile crowd left content.

The final thing I want to say about Loren Woods is that he was an early advocate of protecting the health of aquatic ecosystems. He wrote and lectured on the changing fisheries of Lake Michigan, focusing on the impacts of pollution, fishing, and the colonization by such alien species as sea lamprey and alewives. And despite his travels through much of the world, his article on Hickory Creek shows he also cared about less exotic territory. Unfortunately, few others have shared that concern, for with the urbanization of its watershed, the creek continues to lose biota. Still, its condition is not as bad as Woods had predicted it would be by now, and the rushing waters still harbor sensitive creatures worthy of all the preservation efforts that can be mustered.

Sources: James Bland, "Hickory Creek Revisited," *Field Museum Bulletin* (January 1976): 8–10, 18–21; "Field Briefs: Loren P. Woods, 1913–1979," *Field Museum of Natural History Bulletin* 50, no. 7 (July/August 1979): 3; John Keasler, "Catches Fish by the Thousands," *St. Louis Post-Dispatch*, August 7, 1955; "L.P. Woods Rites Set; Curator at Field Museum," undated and uncredited obituary in archives of Fish Department, Field Museum; "Son of Local Physician Is Curator and Well Known Scientist," *Poseyville News* 73, no. 47 (October 21, 1955): 1.

Mindscapes

Miscellany [Wolf Hunts]
H.

~~~~~~

January 10, [1834]: I have been here [in Chicago] more than ten days without fulfilling the promise given in my last letter. It has been so cold, indeed, as almost to render writing impracticable in a place so comfortless. The houses were built with such rapidity during the summers as to be mere shells; and the thermometer having ranged as low as 29 below zero during several days, it has been almost impossible, notwithstanding the large fires kept up by an attentive landlord, to prevent the ink from freezing while using it, and one's fingers become so numb in a very few moments when thus exercised that after vainly trying to write in gloves, I have thrown by my pen, and joined the group composed of all the household around the barroom fire. This room, which is an old log cabin aside of the main house, is one of the most comfortable places in town, and is of course, much frequented, business being, so far as one can judge from the concourse that throng it, nearly at a standstill. Several persons have been severely frost-bitten in passing from door to door; and, not to mention the quantity of poultry and pigs that have been frozen, I was just told a horse has perished from cold in the streets at noonday.... [T]he wolves, driven in by the deep snows, which proceeded this severe weather, troop through the town after nightfall, and may be heard howling continually in the midst of it.

The situation of Chicago, on the edge of the Grand Prairie, with the whole expanse of Lake Michigan before it, gives the freezing winds from the Rocky Mountains prodigious effect, and renders a degree of temperature which, in sheltered situations, is but little felt, almost painful here.

January 13: [The writer continued his letter on January 13, relating how on January 11 he was invited as a passenger in a race between two horse-drawn "carioles." During that exercise, a wolf was roused, an event that prompted a hunt for the following day. That hunt produced two "prairie wolves" and one "grey wolf," the latter which the writer killed with his own knife after dogs had

H., "Miscellany," *Chicago Democrat* (April 23, 1834). [Letters originally appeared in the *New York American*, no. 18.]

brought it to bay. With little else to do, I suppose, a goodly number of residents planned yet another hunt for January 13.]

It was a fine bracing morning, with the sun shining cheerily through the still cold atmosphere far over the snow-covered prairie, when the party assembled in front of my lodgings to the number of ten horsemen, all well mounted and eager for the sport. The hunt was divided into two squads; one of which was to follow the windings of the river on the ice, and the other to make a circuit on the prairie. A pack of dogs, consisting of a gray hound or two for running the game, with several of a heavier and fiercer breed for pulling it down, accompanied each party. I was attached to that which took the river and it was a beautiful sight, as our friends trotted off in the prairie, to see their different colored ca- potes and gaily equipped horses contrasted with the bright carpet of spotless white over which they rode, while the sound of their voices was soon lost to our ears, as we descended to the channel of the river, and their lessening fig- ures were hid from our view by the low brush which in some places skirted its banks. The brisk trot into which we now broke, brought us rapidly to the place of meeting; where, to the disappointment of each party, it was found that nei- ther had startled any game. We now spread ourselves into a broad line, about gunshot a part from each other, and began thus advancing into the prairie. We had not swept more than a mile, when a shout on the extreme right, with the accelerated pace of the two furthermost riders in that direction, told that they had roused a wolf. "The devil take the hindermost" was now the motto of the company, and each one spurred for the spot with all eagerness. Unhappily, however, the land along the bank on the right was so broken by ravines, choked up with snow, that it was impossible for us who were a half a mile from the chase when it started, to come up at all with the two or three horsemen who led the pursuit. Our horses sunk to their cruppers in the deep snow drift. Some were repeatedly thrown; and one or two breaking their saddle girths from the prodigious struggles of their horses made in the snow-banks, were compelled to abandon the chase entirely. My stout roan carried me bravely through all; but when I emerged from the last ravine on to the open plain, the horsemen who led the charge, from some inequality in the surface of the prairie, were not visible; while the third, a fleet rider, whose tall figure and Indian headdress had hitherto guided me in the chase, had been just unhorsed, and abandoning the game afoot, was now wheeling off apparently with some other object in view. Following on the same course, we soon encountered a couple of officers in a train, who were just coming from a mission of charity in visiting the half- starved orphans of a poor woman, who was frozen to death on the prairie, a day or two since—the wolves having already picked her bones before her fate became known. One by one, our whole party collected around to make their inquiries about the poor children, and the two fortunate hunters soon after joined us with a large prairie wolf hanging to the saddle bow of one of them.

It was now about eleven o'clock; we were only twelve miles from Chicago; and though we had kept up a pretty round pace considering the depth of the snow, in coursing backward and forward since eight, our horses generally were yet in good condition, and we scattered once more over the prairie with the hope of rousing more game. Not ten minutes elapsed before a wolf, breaking from the dead weeds which shooting eight or ten feet above the level of the snow indicated the banks of a steep ravine, dashed off into the prairie pursued by a horseman on the right. He made instantly for the deep banks of the river, one of whose windings was within a few hundred yards. He had a bold rider behind him. The precipitous bank of the stream did not retard this hunter for a moment, but dashing down to the bed of the river, he was hard upon the wolf before he could ascend the elevation on the opposite side. Four of us only reached the open prairie beyond in time to take part in the chase. Nothing could be more beautiful. There was not an obstacle to oppose us in the open plain; and all of our dogs having long since given out, nothing remained but to drive the wolf to death on horseback. Away, then, we went, shouting on his track; the hotly pursued beast gaining on us whenever the crust of a deep snow drift gave him an advantage over the horse; and we in our turn nearly riding over him when he came to ground comparatively bare. The sagacious animal became at last aware this course was soon up at this rate, and turning rapidly in his tracks as we were rapidly scattered over the prairie, he passed through our line and made at once again for the river. He was cut off and turned in a moment by a horseman on the left, who happened to be a little behind the rest; and now would come the keenest part of the sport. The wolf would double upon his own tracks while each horseman in succession would make a dash at and turn him in a different direction. Twice I was near enough to strike him with a horse whip, and once he was under my horse's feet, while so ferociously did each rider push at him, that as we brushed by each other and confronted horse to horse while riding from different quarters at full speed, it required one somewhat used to turn and wind a very Pegasus to maintain his seat at all. The rascal who now and then looked over his shoulders and gnashed his teeth, seemed at last as if he was about to succumb—when after running a few hundred yards in an oblique direction from the river, he suddenly veered his course when everyone thought his strength was spent; and, gaining the bank before he could be turned, he disappeared below it in an instant. The rider nearest to his heels became entangled in the low boughs of a tree, which grew near the spot; while I, who followed next, was thrown out sufficiently to give the wolf time to get out of view, by my horse bolting as he reached the sudden edge of the river. The rest of the hunters were consequently at fault when they came up to us; and, after trying in vain to track our lost quarry over the smooth ice for half an hour, we were most vexatiously compelled to abandon the pursuit as fruitless, and return to the village with only one scalp as the reward of our morning's labors.

Six Months with the Indians

Darius B. Cook

⌘ ⌘ ⌘ It was a clear, cold morning in December. Breakfast was over; the huge log fire was burning briskly as we sat smoking our pipes and discussing whether we would go to our distant traps or stay around our cabin. The owls during the night had disturbed us exceedingly. It seemed as if all in the forest had gathered with the wolves to give us one grand serenade. They were attracted by the venison which hung upon poles outside. In fact the owls had feasted upon it during the night, and the tracks of wolves were numerous around the lodge. We finally resolved to make war upon the owls and for two hours only. Both could imitate them to perfection. We started out in opposite directions and for some time we could shoot them without calling, being very numerous. When the call began they came from all directions. As fast as we could load and fire they would fall. We were not over eighty rods apart, judging from rifle reports, and we were striving to see who would bring in the greatest number. We met promptly on time, Rhodes having twenty-two to our eighteen. He could beat us in loading and firing.

It was exceedingly cold and we concluded to spend the day and pick off the thick matted feathers from the birds, make each a pillow, for as yet our cases were only stuffed with fine hemlock boughs. There seemed to be over two bushels of them, but on drying them a few days there were not enough for two pillows but by mixing them with the boughs they were all any one could desire.

Darius B. Cook, *Six Months with the Indians* (Niles, MI: Niles Mirror Office, 1889; facsimile reprint, Berrien Springs, MI: Hardscrabble Books, 1974), 26–28, 54–58, 65–66.

At noon our picking was finished and our appetites were appeased by our usual home dinner. Our dishes were washed according to our custom by turning them bottom side up. We had just shouldered our rifles to visit our wolf traps when in came Pe-make-wan. He loosened his belt giving us to understand he was very hungry. We had plenty of cooked venison and cold roasted potatoes left and he cleaned it out in quick time and started with us for the traps. The snow was over a foot in depth. When we reached there we found one wolf dancing on the end of a pole, caught by the fore paw. He was soon dispatched, the trap re-set and our owls were left for bait. The wolf was left in a tree and we proceeded to a few traps on the bank of the river, about thirty-five rods distant. Reaching the river, brush were heard to crack on the opposite side. "Ke-wob-em," exclaimed Pe-make-wan, that is "you see?" In a moment a deer leaped into the river almost opposite us, and dashed up the bank not more than thirty yards from us. "Hold! hold!!" said Rhodes, don't shoot the deer!" No sooner had he spoken than two immense timber wolves dashed into the stream in the same place and were quickly over. As they showed their heads above the bank, one of them fell by the deadly aim of Rhodes' rifle. The other received a bullet from our rifle, but made a leap for us. The second leap his back broke in the centre, the bullet cutting it half in two. The Indian discharged his rifle, cutting through the ear close into the head. With a broken back he struggled desperately for us. A bullet in his head from our revolver, put an end to him before a rifle could be loaded. These were soon skinned and their carcasses thrown into the river. On reaching our cabin, near evening, we found our only near neighbor, Mr. Chambers, there with a supper all ready, roast venison, etc., with delicious wheat bread, which Mrs. C. frequently furnished, and perhaps we four did not enjoy a hearty meal and a pleasant evening. Mr. Chambers left for his home about 9 o'clock, but soon returned for a torch to protect himself from wolves....

"Captain, there are fresh deer tracks between here and the spring," said I early one morning as we returned with a pail of water, "and I propose to get one before breakfast," seizing our rifle. "Yes, but I'll have breakfast in half an hour or so," said the Captain. "I'll be here on time, and if you hear me shoot come out and help me drag one in," said I, jokingly. "Yes, yes," said the Captain laughing.

Out we went, and in less than thirty rods we saw a fine doe lying down, and she was shot dead in her bed. But a few steps off a fine buck jumped up, and ere we could load was out of reach. By a cautious pursuit we got in a shot at long range, but drew blood freely and followed on, but knowing we would be late to breakfast we retraced our steps for the doe, and much to our surprise there was nothing left but a small piece of skin which we took back for a cushion to the shaving horse. The wolves had carried it all off. This could not be believed until the Captain examined the ground.

Breakfast being over we resolved to pursue the wounded buck. Taking the track, the blood showed he had a serious wound, and we had not gone far ere he sprang up and dashed off, getting a second, but it seems not a fatal shot. Thus we pursued this wounded buck far off from our lodge until the shadows of night began to set in, and we saw there was danger of not getting to our lodge. The day was cloudy and we had no idea how far we were from our camp. Our compass told us which way to steer our course to strike Rabbit river, and we hastened on as speedy as possible. But darker and darker it grew. Old Jim set up his great bass howl in the same place he always did, and we knew we must be a long distance off for he was between us and the lodge. If we could reach Rabbit river before dark we were safe for we knew of an Indian encampment up that river, but the distance to it seemed too great. Coming to a tamarack swamp we made up our mind our only salvation was to strike a fire, for the wolves were on our track and when darkness fairly set in an attack was certain. We gathered a lot of dry tamarack poles and kindlings but to our sorrow we could not find a match. Every pocket was searched in vain. We tried rubbing dry sticks together, but could not succeed in getting any thing but sparks. We determined to discharge our rifle and load with powder and tow, which we had in our game bag, which would set the tow on fire. After doing so and taking the tow out of the bag a solitary match dropped from it into the snow which we seized with the utmost care. Preparing well for a fire we lit the match, set the tow on fire and the dry bark and sticks were soon going, and as darkness fairly set in we had a fire which illuminated the wilderness for a long distance. We listened for the signal gun in vain, we were beyond hearing it. Every short time we would fire our own gun, but it was useless. The wolves surrounded us in large numbers, but the fire was our protector. Sometimes when it became a little dim they would approach nearer. Their howls and growls were terrific. They would often have a fight among themselves, and their clear voices would ring for miles around. When their eyes were turned towards us they would glisten by the light of the fire, and occasionally we would shoot as near as possible between them. The noise of a rifle would still them but for a moment when a louder and more terrific howling would be set up. Thus all the long night we worked to keep up the fire, and dry tamarack near us was getting scarce. To venture out too far was certain food for them. On one occasion, one wolf more daring than the rest, while we were procuring a dead tree four or five rods from the fire, came so near we heard him snuff. Turning around we saw his glaring eyes not over three rods off. Dropping our pole we took good aim at his eyes and he fell dead, but we did not know it then. He was apparently crouched for a spring when we shot. We only knew the eyes disappeared. It was a cold and dreary night. We had fixed a place to sit down, and in a minute we fell asleep and fell off. This awakened us and we dare not sit down again. A few frozen roasted potatoes were found in our pocket, which we thawed out and eat, which were refreshing. Daylight dawned at last, and as it grew lighter the pack drew off and their noise was hushed. We went to see the effect of our shots and found the venturesome wolf dead in the snow. Those farther off had been hit, as seen by blood, but not fatally. All around and within eight rods and less the snow was completely tread down, and here and there blood and hair, caused either by their fights or bullets, by bullets we imagined, for we sent not less than ten in their direction.

It was a night of terror, long and dreary. Almost ready to surrender to fatigue we pursued our northern course slowly and sadly, for even then we began to think we must perish alone in the forest. At last we struck the river and took new courage. Here there was a half beaten Indian trail and our steps were quickened. Onward we pressed and at last we beheld one of the most beautiful pictures the eye could imagine. It was smoke curling up among the trees. It was an Indian encampment and we were greeted with a hearty welcome. Here were those whom we had chastised for robbery. We told them our story, and two young bucks started for the wolf with ponies, and in less than an hour he was brought in.

They feasted us on their best—boiled muskrat, corn bread and potatoes. They brought up their best ponies and one strapped the wolf on, took the lead, and in an hour we reached our lodge. It was about 10 a. m. and the Captain had taken our track, but a shot from our rifle was answered near a mile off and he speedily returned and it was a joyful meeting for he never expected to see us alive. We remained in camp three days before we recovered sufficiently to be out. . . .

Porcupines were so numerous and so useless to us that we seldom killed one. Go in any direction and there they were. The good old Indian Gosa wanted a lot of them, and he had been so kind to us we volunteered to kill them. He called by appointment at an early hour in the morning with his pony and a sack made of deer skin large enough to hold forty or more. There was no trouble in finding them for they were quite plenty, and we were not over four hours killing twenty or more.

On our return to camp we came upon a party of Indians who were cutting a bee tree. It was a dead pine tree about a foot through. It was a mere shell. The bees went in about fifteen feet high. In cutting they found old candied honey at the butt. It was full of solid honey over thirty feet high and did not contain less than five hundred pounds. They were several days in carrying it off. They presented us with three muhkuks full, which became very acceptable on pancakes. This honey had to be strongly barricaded with heavy trees to keep the wolves off for they were fond of everything sweet. It was with maple sugar Gosa caught a wolf for us that hung around his wigwam.

On our return from camp we found our door open and tread cautiously to catch a thief. On looking in, a large black bear had found the potatoes in the hole under our bed and was devouring them as speedily as possible. We closed the door upon him, and had him fast. This done we climbed upon the top and had a fine view of the black monster. He made a desperate leap against the door and then at us, and as he did that his hind feet got into live coals which made him desperate, and he raved and tore around the lodge. It was our desire to rope him and secure him alive, but we found it an impossibility, and he was finally dispatched by the Captain with two shots in the head. Indian Gosa and Su-na-gun skinned him and took the carcass excepting one quarter which we desired for our worthy neighbor Chamber's use.

Every thing in our lodge was now in great confusion. Our bed, our cooking utensils, were scattered all over the lodge. Our plates, were overturned and had to be washed up. An hour or more was spent in cleaning after bruin before we could get a meal.

In the mean time we had a call from Mr. Chambers, who brought us potatoes, having heard by an Indian of our loss. But notwithstanding the bear had devoured many, still we had plenty and he enjoyed a venison supper with us.

～～～ JONATHAN WUEPPER IS AN AVID BIRDER AND NATURALIST IN Berrien County who shares my passion for historical natural history. Over the course of fifteen years, he has examined most every newspaper ever published in the county in a search for faunal and floral references. (This exercise would be valuable in the rest of the region as well, but it requires great patience. Anyone interested in helping on such a project please contact me.)

He attributes *Six Months with the Indians* by Darius B. Cook (1815–1901) as the work that triggered his "interest in the old stuff." And I am indebted to him for introducing me to Cook. In an e-mail, Jon provided the following biography:

> At first I thought *Six Months* was exaggerated, but on doing quite a bit of research, I think it is mostly fact. All those people he mentions are buried up in Allegan County. I've read the 1840 U.S. Census and they all check out. The Rabbit River is still there.... I envision the wolves jumping over the creek deep in the woods.
>
> Cook was born in Litchfield, Connecticut. He worked in several East Coast newspapers and was a reporter in D.C. (he heard Henry Clay, J. Q. Adams, and other political giants). After a stint at the *Kalamazoo Gazetter*, he returned to Connecticut, where he married Jane Wadhams. He returned to Michigan in about 1840. In the spring of 1842, Cook took a printing press from Detroit on the back of a wagon, with the intent on moving to Chicago. He got as far as Niles, Michigan [Berrien County], where he stopped for the night. The townspeople met him and persuaded the Democratic-slanted Cook to stay in their village.
>
> An avid hunter, Cook became editor of the *Niles Republican* in [the] spring of 1842. After the Republican Party was founded in 1854 and became a national player, he inserted the word "Democratic" into the title banner: *Niles (Democratic) Republican*. He sold the paper twenty years later, and started the *Niles Weekly Mirror* in about 1870/71. That venture lasted until a year before his death due to kidney failure.

One note: Allegan is the second county north of Berrien.

Source: E-mail from Jonathan Wuepper, dated November 22, 2005.

Letter in Defense of Snakes

John Kennicott

ᐧᐧᐧ Well, what about snakes? A good deal might be said, and very appropriately and profitably too, were we in the mood for saying it pleasantly....

John Kennicott, corresponding editor, "Letter in Defense of Snakes," *Prairie Farmer* 17, no. 21 (May 21, 1857): 162.

Snakes are much abused animals. As supposed types of the first deceiver, a sort of religious dread has ever attached to them among Christian people, and a few of species being really venomous, and others possessing *imaginary* attributes, for transcending the actual posers of any of the class, it is not very wonderful that all the sons and daughters of Eve should inherit a hearty hatred of *snakes*. And we believe such is the case, with the exception of some savage nations, who according to their own notions, are not descendants of Grandmother Eve, and therefore have no family feud with the tribe of serpents.

But we had no intention of going back so far, when commencing this article. Our desire is to notice two or three points very briefly, that have been stated by our correspondents.

First, what are the venomous snakes? [Briefly discusses moccasin, copperhead, diamondback, and banded rattlesnakes—Ed.] But the *snake*, about cures for whose bites so much has been said in this paper, is quite a different customer—not a very agreeable intimate of one's house (though we have killed two found in ours) and quite sufficiently venomous for the snake's own purpose. Still, that our prairie rattlesnake has ever caused the death of a single human being—whether doctored or not—we have yet to learn.

And this brings up the second question. Is there any specific antidote for snake poison? Possibly. But who knows it? Not we, and we studied medicine, and *believed* in medicine for nearly thirty years.

...[A] residence of over twenty-one years in Illinois with as extensive a country practice as any other physician perhaps—and in a region and during a time, where and when prairie rattlesnakes abounded, no death from their bite has ever come to our knowledge. That is no death of man, woman, or child—a few small animals, usually bitten on the nose, have died; and death among large animals have been reported to us, but we never saw such a case. Of human subjects, we have treated many cases; and known many that had no treatment at all, or were treated in all ways, and the result was always the same—all recovered, though some suffered horribly, for a little while.

As to the popular remedies, or most of them, ... they are all well enough in their way, and have certainly *seemed* to give relief at first. Perhaps their operation is chiefly of the mind, and even if so, we need not tell our readers that FEAR is often more deadly than snake poison; and in *this* way, by lessening the horror, these nostrums may be considered as "specifics" for snake bites—as like ones in cholera....

But, after all, it is no longer a question of much moment, to prairie farmers. Our little snake is of no great account, any way; and it is of little consequence whether "Dr. Perry" or "Infant Sawbones" is in the right.... [U]ntil we are assured of some human death from the bite of our snake, we shall content ourselves with the Let Alone Treatment, in mild cases, and whiskey etc when needed.

We have but a word to add to this hasty dissertation on snakes. Let every farmer bear in mind that the whole tribe of serpents are insect eaters, and the benefactors of their human persecutors. Rip up the stomach of one, and you will find it stuffed with insects, or enlarged by the backs of meadow mice. Except in killing an occasional bird or frog, nearly all of our snakes are as useful to vegetation as they are harmless to mankind; and it is not only an act of wicked barbarity, but a species of suicidal folly to destroy them. For better aid in determining the species and their dissemination, as urged in the circular by Robert Kennicott, and show that you are above the vulgar prejudice against these persecuted creations of the Greater Author of all animal life, who made nothing without an object, and made them for our good.

〰〰〰 JOHN KENNICOTT (1802–1863) SEEMS TO HAVE BEEN ONE OF those rare people who not only was way ahead of his time, but was respected and influential nonetheless. He was born and raised in New York, where he taught school and studied medicine. At the age of twenty-seven, he moved to New Orleans. His productivity over the course of his seven-year stay was noteworthy in a number of respects: he headed the city's first public school system; treated patients; ran an orphanage; produced New Orleans's first magazine covering literature, science, and religion; wed May Ransom; and fathered two children, including the future naturalist Robert.

For reasons unknown to me, the Kennicotts took leave of New Orleans and moved to what became Glenview, Cook County, Illinois. Their homestead was called the Grove, which today includes a wonderful visitor center on eighty-five acres of woods. The property is protected as a holding of the Glenview Park District.

In northern Illinois, Kennicott devoted himself to three realms that he considered vital to the betterment of his fellow citizens: medicine, horticulture, and education. Being one of only three physicians around, Kennicott's medical practice flourished. Astride his brown pony, he visited his scattered and isolated patients on a route that today would have included the towns of Lincolnshire, Niles, Wilmette, and Highland Park.

Kennicott's innovation was to stress the value of good diet. He wrote that "in the western country, a great majority of persons either suffer from habitual constipation or the reverse. These conditions of the digestive organs are produced by faults of diet, especially eating too much pork, whiskey drinking, and the use of purgatives.... For all this wide spread misery, on which quacks fatten—this waste of health and life—we offer a cheap, safe, and efficient remedy, in *fresh garden vegetables* and *ripe fruits*."

Kennicott's belief in the salubrious qualities of fresh produce is linked with his activities in horticulture. In 1842 he started the Grove Nursery, which became one of the largest in this part of the state. Jack White includes an amusing quote describing Kennicott as he pursued his two trades: while making the circuit of his patients, Kennicott on his favorite pony would look like "either a man or a monkey in the saddle. Sometimes the strange creature would be carrying a tree whose top and branches hid it altogether—a tree on horseback." The good doctor emphasized that only plants suited to local circumstances should be used, and that the chances of their success could be improved by proper care and appropriate soil preparation. And as his essay on snakes demonstrates, his views on the relationship between agriculture and nature were quite progressive as well.

The third of Kennicott's great contributions lay in the field of agricultural education. Illinois was evidently the center of the successful movement to create a nationwide system of universities that would emphasize agriculture and mechanics in their curricula. Kennicott was a major player in this effort, working hard to convince Congress to approve the Land Grant Act, under which the federal government would provide states with the money or land to found such institutions. He traveled to Washington, D.C., to personally lobby legislators and used his position as an editor of the *Prairie Farmer* to garner support among farmers.

Source: Jack White, "Ecological and Cultural Significance of the MacArthur Foundation's Property at the Grove," unpublished report by Ecological Services, Urbana, Illinois, to the MacArthur Foundation, 1995.

A History of the County of Du Page, Illinois

C. W. *Richmond and* H. F. *Vallette*

ᔫᔫᔫ Until within a few years, this part of the county was infested with wolves, which were a source of great annoyance to the whole community. The farmers, however, were the principal sufferers of their depredations; for sometimes whole flocks were destroyed and scattered by them in a single night. To rid the country of these mischievous animals, it was the custom for all who were able to "bear arms," to rally once every year for a wolf hunt, which was usually a scene of much amusement, and often times of the most intense excitement. These expeditions were conducted in various ways. The general hunt, which was perhaps the most common, was conducted upon the following plan.

Notice of the time of starting, the extent of county to be traveled over, and of the place of meeting, which was usually at the common center of the circle of territory to be traversed, was first given to all the participants in the hunt. At an early hour on the morning of the day appointed, the hunters assembled and chose a captain for each company, whose duty it was to station members of the company at short intervals upon the circumference of the circle alluded to, and then the game was completely surrounded. At a given time, the line of hunters began their march, and when they had approached near enough to the centre to close in, captains advanced with their sharp shooters to ascertain whether any game had been surrounded. If an unlucky wolf or deer had been drawn in the snare, upon making his appearance before the lines, he was sure to be riddled by rifle balls. We have been informed by one who frequently participated in hunts of this kind, that he had known of 60 wolves and as many deer being killed in one day. This mode of hunting the deer seemed altogether too cruel and cowardly in the eyes of some, but no scruples were entertained in thus exterminating the mischievous, thieving wolves. To see the harmless deer penned up with no chance of escaping, darting about bewildered, with eyes almost started from their sockets, and

C. W. Richmond and H. F. Vallette, A *History of the County of Du Page, Illinois* (Chicago: Scripps, Bross, and Spears, 1857), 181.

then to see them slaughtered in the manner described, appeared cruel in the extreme.

The mode of hunting wolves adopted by the settlers of Downer's Grove was different from that described, and obviated the appearance of cruelty in slaying the deer. The wolf hunt was a source of amusement in this town for years, and whenever a wolf dared to show his head above the prairie grass, the boys were instantly in pursuit of him. The pursuers usually went on horseback, carrying in the hand a short club, and the captain of the company was the one who had the swiftest horse. The plan of action was to spread out in every direction and scour the prairies until the game was started, when by a peculiar yell, the whole company was called together and the chase commenced. Every horse was now put to his utmost speed, and with his rider, would go flying over the prairie like the wind. It is utterly impossible to describe the wild excitement that attended the wolf chase. Generally a race of from three to five miles would bring Mr. Wolf down; then the day's sport would be ended, and the party would return home in a sort of triumphal procession, bearing the fallen hero. Such reckless headlong riding was attended with much hazard, and although no serious accident ever happened to the riders, yet is surmised that the horses might have suffered from ring-bones and spavin induced by undue speed.

~~~~~~

A Day's Hunting in 1871 on the Club's Ground

South Shore Country Club

ᘒ ᘒ ᘒ When a person mentions the South Shore Country Club in 1870, a smile of incredulity is provoked and the listener is justified in raising his eyebrows and saying very audibly, "Squirrel food!" But the location of the South Shore Club in those days was just as good a spot for sport as it is today.

There were no buildings that can be recalled, but there were water hazards, bunkers, and "rough" in the greatest abundance. As for hunting, there was no place along the lake shore to compare with it for snipe and plover shooting; and speaking of duck and Canadian brant in the season for them,

---

"A Day's Hunting in 1871 on the Club's Ground," *South Shore Country Club Magazine: Golden Anniversary* 42, no. 2 (August 1956): 34.

the air was so full of them that the merest tyro could get as many as he cared for with a minimum of effort.

About the nearest habitation was the Hess House a mile or two south, and a little further on was the Kleinman house where hunters could put up for the night and Kleinman boys would pole them around the Calumet swamps with a guarantee that they would get their bag of game by evening or no pay would be expected.

Abe Kleinman was the champion wing shot of the United States about this period, although John H. Turrell and A. S. Bogardus also held the title for a number of years. But Abe Kleinman was supreme in the Calumet district and with Abe and one of his brothers—John, Henry, or George—for a guide, it was a certainty that the hunter would not come home empty-handed.

The writer well recalls a trip in the spring of 1871. There was no such place as South Chicago in those days, as the first station on the Lake Shore Road was Calumet. Arriving there, Henry Kleinman met us and announced that the ducks weren't flying, but the jacksnipe were good and plenty, and that there were a few wild pigeons going north.

Kleinman had brought along a setter dog and he flushed up so many jacksnipe that it was less than an hour before we had as many birds as the right kind hunters care to kill. We built a fire near the shore and ate roast snipe until the liquids we had brought along with which to wash them down were entirely exhausted.

Well on into the afternoon we started back and before we had got a mile on our way, Kleinman suddenly pulled the team up and ejaculated, "What's that?" Pointing to the southwest we saw what appeared to be a dark cloud moving rapidly towards us. "They're wild pigeons going north," exclaimed Milligan. "Drive fast George and we'll run right into them."

Guns were loaded hurriedly and the team went on at jump-through-the-sand leaps. But the luck of the chase was against us. When the birds sighted Lake Michigan, they made a sudden turn to the east and the nearest we got to them was a mile from shot gun distance. When we arrived at the Kleinman house we found Lycergus Laflin there and he had brought down a dozen or more from the fringe of the flock as they passed by the house.

To tell the size of a flock of wild pigeons in those days is a strain on the credulity of the listener. But when the sky is fairly black with them for an hour at a time, it takes algebra and the radical sign "made double" to express in figures the approximate number that passed over.

~~~~~~

Field, Cover, and Trap Shooting

Adam Bogardus

PIGEON-SHOOTING

᭜᭜᭜ I began to shoot pigeons in 1868, when I had been a field-shot for more than eighteen years. I had often been invited to go and witness contests of the kind, but cared nothing for them, and up to 1863 had never seen a pigeon-trap. The first public pigeon-shooting into which I entered was a series of sweepstakes at St. Louis. I had some success; so much, in fact, that R. M. Patchen, who was with me, forthwith made a match, in which I was to shoot against Gough Stanton of Detroit for $200 a side. Expenses were to be paid to whomever travelled to the other, and he came to Elkhart. The match was fifty birds each. He brought with him a plunge trap, the first I had ever seen of that character. However, I consented to the use of it, and won by killing forty-six to his forty. I was then just about as good a shot at pigeons as I am now, except that I was anxious about the money, and sometimes missed owing to that.

I next shot against Abraham Kleinman. John Thomson, a stockman of Elkhart, made the match on my part. It was for $200 a side, fifty birds each from a spring-trap. There was a dispute about the quantity of shot to be used, he contending that it was to be limited to an ounce. We made a sort of compromise, by which I was to pull my own trap, while he was allowed a man to pull for him. The match was trap and handle for each other. He had an old trapper named Farnsworth to do this on his part, while my man, as afterwards appeared, did not know an old bird from a young one. Before we began I offered to bet that I killed forty-six out of fifty. This wager was eagerly accepted by Farnsworth, who wanted to bet a larger sum on the point. Kleinman killed forty-nine and I killed forty-six. I told Kleinman that I could and would beat him before long, and went home to practise in the field. I challenged him for the championship of Illinois, and we shot for $200 a side, at fifty single birds and twenty-five pairs of double birds each—the single birds ground-trap, the doubles plunge-traps. Of the single birds I killed forty-three to Kleinman's forty-two. At the doubles we killed forty-three each. It was at Chicago in 1868. Soon after I shot with another man two or

---

Adam Bogardus, *Field, Cover, and Trap Shooting* (New York: J. B. Ford, 1874), 300–304.

three times, and won; but I shall not mention his name in this book, for sufficient reasons.

The next match I took up with Abraham Kleinman was rather singular in
character. It was at single and double birds. I was to shoot from a buggy at
twenty-one yards, the horse to be on a trot or run when the trap was pulled.
Kleinman shot from the ground at twenty-five yards. I won it. I afterwards
shot two other matches on these conditions, one with King at Springfield,
and one with Henry Conderman at Decatur. Of these I lost one, and won the
other. My shooting from a buggy at plover, grouse, and geese had made me
very quick and effective.

In the spring of 1869 R. M. Patchen made a match, in which I was backed
to kill five hundred pigeons in six hundred and forty-five minutes, with one
gun, at Chicago. I was to load my own gun, and the stakes were $1,000 a side.
There were heavy outside bets that I could not do it. I won the match, however, in eight hours forty-eight minutes, and thus had one hour fifty-seven
minutes to spare. In the third hundred pigeons I killed seventy-five in consecutive shots. In the last one hundred and five birds I scored one hundred;
and in the seventh hour killed ninety-five. I shot with a muzzle-loader. It was
twenty-one yards rise and fifty bounds. Before this match came off I had, in
practice, killed five hundred birds in five hours and seven minutes; but then I
used two guns, and had a man to clean them, though I loaded them myself. I
missed thirty-four out of the whole number shot at.

I was next matched to kill a hundred consecutive birds at Chicago in July,
1869; $1,000 to $100 that I could not do it, and three matches to be shot if I
failed in the first and second. In the first I had killed thirty when the lock of
my gun broke, and being obliged to borrow one which was a poor article, I
lost. On the 21st of the month I tried it again, and won. At Detroit in the same
season I undertook to kill forty birds in forty minutes, to load my own gun,
and gather my own birds. I killed fifty-three in twenty minutes forty seconds,
and won. In the fall of 1869 I shot a match for $1,000 a side against King at
Chicago. It was fifty single birds and fifty pairs of double birds, making one
hundred and fifty each, plunge-traps, twenty-one yards rise. I killed all my
single birds. Mr. King killed forty-one of his. I killed eighty-five of my double
birds, Mr. King seventy-five of his.

I shot and won a great many matches which I need not mention here.
In 1870, Mr. Nathan Doxie challenged any man in Illinois to go to his place
and shoot against him for $100 at twenty-five birds. I went there and killed
twenty-two to his twenty-one. At the Chicago tournament I killed ten

straight at twenty-one yards, as did several others. Under the conditions we went back to twenty-six yards to shoot the ties off at five birds each. Mr. G. K. Fayette, of Toledo, Ohio, and I tied four times more at this distance, killing all our birds. I then killed five more, making twenty-five consecutive birds at twenty-six yards. Mr. Fayette killed four of his last five, but missed the fifth, so I won. Later on I shot against Mr. J. J. Kleinman, of Chicago, at five traps, fifty birds, mine at twenty-eight yards rise, his at twenty-five. I won, and in the course of the match killed thirty-three consecutive birds.

～～～ COLONEL ADAM BOGARDUS (1833–1913) WAS FAMOUS DURING the last half of the nineteenth century as being the premier bird shooter of the time. He was born in Albany County, New York, and began shooting at the age of fifteen. He describes himself this way: "I was then a tall, strong lad, and have since grown into a large, powerful, sinewy, and muscular man." In New York he distinguished himself early for his ability to kill snipe and quail: "I used to stint myself in quail-shooting time to twenty-five brace a day." He eventually moved to central Illinois, where he settled in Elkhart, Logan County. During one three-month period hunting in nearby Christian County, Bogardus personally shot over six thousand birds, including prairie chickens, geese, ducks, whooping cranes, snipe, golden plovers, and Eskimo curlews, the last two of which made up the largest number. As for golden plover, he writes: "I have often seen as many as four hundred or five hundred... together, and they sometimes fly so close in the pack that a great many can be cut down with two barrels."

Nowhere can I find any clear statement that he fears that the game he has hunted for all those decades might be in danger of disappearing. But, then, hunting is how he earned his livelihood and upon which his fame was based. (Perhaps it is the same impulse that moves oil men in the twenty-first century to mock those who advocate energy conservation.) Bogardus does suggest small measures that would help the game species: setting back the season on quail by two weeks and prairie chicken by one month to enable the young of the year to become better fliers and for the cooler weather that would result in less spoilage of game; banning the spring hunting of upland sandpipers; and the cessation of spring burning in the prairie chicken range because it destroys eggs.

Eventually, however, states began outlawing the use of live birds for shooting contests. Bogardus showed ingenuity and adaptability by developing a glass ball to replace live targets. Clay pigeons then replaced the balls, as their movement through the air more closely approximated birds. In 1880 Bogardus joined another prominent shooter in a national tour to promote the newly introduced Ligowsky clay pigeon. Later Bogardus performed in circuses with some of his thirteen children, and when travel became burdensome, he opened a shooting gallery in Elkhart.

The section I have selected is his chapter on passenger pigeon shoots, many of which took place in Chicago. No animal has suffered a more vicious assault by human beings than this wondrous bird. Estimated to number from 3 to 5 billion birds in 1800, the last wild passenger pigeon was shot in Ohio in 1900. The very last living specimen died in the Cincinnati Zoo fourteen years later. The hunting contests of the kind in which Bogardus participated were only one aspect of the incredible carnage, but not trivial when one considers that almost forty-five thousand pigeons died in a single shoot held in New York in 1874 and fourteen thousand at another in Peoria. And many thousands more succumbed in the process of securing the live birds.

Sources: Adam Bogardus, *Field, Cover, and Trap Shooting* (New York: J. B. Ford, 1874), 300–313; Trapshooting Hall of Fame website, http://www.traphof.org.

~~~~~~

History of Kendall County

E. W. *Hicks*

⤳ ⤳ ⤳ Wolf hunts were common. A stake would be set up on the prairie, say beyond Lisbon. The settlers would be engaged and would come in a narrowing circle from miles in every direction, driving everything before them. As they neared the central point and the enclosed game came into view, the excitement became intense. The wolves and deer tried to run the blockade, but were beaten back from every point until they were nearly crazy with fright. Then the slaughter commenced and it was rarely that one escaped. After all was over an equitable distribution was made. In the hunt of 1835, 18 wolves and 24 deer were killed. These hunts, however, like every other amusement, soon degenerated. The settlers in some localities would privately agree to shoot their game on the way, and afterwards came in for a share in the common stock, thus defrauding their neighbors from other places. This cheating brought the hunts into disrepute.

E. W. Hicks, *History of Kendall County* (Aurora, IL: Knickerbocker and Hodder, 1877), 158–59.

Game Hunt

Anonymous

☙ ☙ ☙ Our sportsmen have chosen sides and made final arrangements for an old fashion game hunt on next Tuesday. The following gentlemen represent the two sides, with H. W. Ray and Ed. N. Hatch as captains. [Names of participants omitted.]

The following list will show how game will count:

Bear 300	Pigeon 10
Deer 200	Rice hen 10
Otter 100	King rail 25
Fox (black) 100	Rail (small) 3
Fox (red) 75	Mud hen 3
Fox (gray) 50	Blackbird 1
Mink 40	Turkey 75
Coon 40	Swan 75
Woodchuck 15	Goose 50
Porcupine 15	Brant 40
Skunk 10	Duck (mallard) 25
Rabbit 15	Duck (red-head) 25
Squirrel (fox) 15	Duck (canvasback) 30
Squirrel (grey) 10	Duck (wood) 20
Squirrel (black) 10	Duck (widgeon) 20
Squirrel (red) 4	Other ducks 15
Muskrat 5	Gulls (large) 8
Yellow leg plover 10	gulls (small) 4
Golden plover 10	Eagle 40
Other plover 5	Owl 10
Grass snipe 5	Hawk (big) 20
English snipe 6	Hawk (small) 10
Curlew 30	Crow 10
Woodcock 25	Yellow Ham'r 4
Partridge 40	Crane 25
Quail 15	Bittern 5

"Game Hunt," *St. Joseph Traveler [MI]*, October 26, 1878.

The plan agreed upon is to start out early on Tuesday, each sportsman to choose his own route, but it is understood that every one shall report with his game at the basement of the Lake View House before 10 o'clock p.m. on Tuesday. On Wednesday evening the party will be treated to a fine oyster supper at the Lake View, the losing side to bear the expense of the same.

Duck Shooting Around Fox Lake

Anonymous

᭡᭡᭡ Another source of danger—or will it eventually be a salvation to fall back upon—is the increasing quantity of mud hens [American coot]. Some think that they will in time crowd the ducks away by eating up their food. It has been suggested that clubs should organize and pay a premium of, say, two or three cents a head for their destruction, and if they were transferred to the feather dealer this would almost cover the premium. Certainly these birds are a great nuisance, and the lakes, rivers, and approaches are black with the pests, which neither afford food to eat, nor fun to shoot. As scouts for the ducks, however, they are invaluable. If you are jumping ducks, and you are pushing noiselessly around some sly corner, where perchance teals or mallards are enjoying a siesta, an acre of mud hens will commence moving away, and thus convey a strong suspicion of danger to those left behind. Another instant, and the acre of nincompoops is churning the water with the rushing noise of a powerful river, which is a signal for every duck to save his bacon.

Then and Now—the Extermination of Game

Lowther

᭡᭡᭡ A recent visit to the grand prairie of Illinois, which for many years long was my favorite autumn shooting ground, has brought home to me

"Duck Shooting Around Fox Lake," *American Field*, November 12, 1881, 313–14.

Lowther, "Then and Now—the Extermination of Game," *American Field*, September 22, 1883, 269.

a realizing sense of the fact so often proclaimed in your pages of the rapidly increasing scarcity of game in this once favored state, and the certainty of its extinction at no distant period.

When about 28 years ago the Illinois Central Railroad was first opened to the now populous city of Kankakee (then consisting of one shanty and a blacksmith's shop), the whole country south of that point was very sparsely settled, and was the undisturbed breeding place of myriads of grouse and quail. The slough and timber along the river bottoms were, in their season, alive with mallard, rabbits, and sometimes deer, and thousands of sandhill cranes (I once killed seven at one double shot) might be seen any field day tearing the corn shocks, hop dancing upon the stubble, or circling at noon, vociferous and sky high above the hunter, resting for refreshment in the free watermelon patch. Momence, Middlefork, Onarga, Ashkum, Chebanse, Clifton, all lay in the middle of a then unbroken prairie teeming with game, and the average gunner from any one of these villages could (and did, if not too lazy) within easy walk, late in an afternoon, and without dog, bag, on the stubble and corn fields, and bean patches, and fringes of slough grasses, a pot-contenting backload of 2 or 3 dozen prairie chickens that had tamely tumbled to his rusty shooting iron. It was no feat then to kill a hundred head a day to each gun, and when, as it sometimes occurred, that a hundred or more were slaughtered to satisfy some citizen ambitious to boast at home of "bag," the game could seldom all be given away, but had, in warm weather, to be fed to the hogs.

Probably no one living then imagined that the inevitable destruction of game, foreseen though it was, would be so speedy, near, and complete, and that this paradise of hunters would be lost to their sons mainly because of their own indifference or neglect to preserve it. As the lazy Bourbonnais said, all agreed: "*Laissez faire: apres nous le deluge.*" ["Leave it be: after us the flood."]

I think it might safely be presumed that in 20 years more the larger game of Illinois will have ceased to exist there, so swiftly and surely is it disappearing before the settlements of man with his lawless, reckless, and destructive habits.

It seems to be established that nothing in food for man increases naturally as fast as his demand for it, but grows scarcer and scarcer until it dies out, unless preserved and cultivated. To do this is a work of civilization; and to this phase of it we are of necessity fast coming in order to ensure the very existence in the future of many useful and beautiful species, both bird and beast. Already throughout the country the best prairies and marshes visited by waterfowl are secured by the few who can best pay for the privilege of shooting thereon.

Prairie Chicken Shooting and Trapping Fifty Years Ago

Anonymous

☙☙☙ Thousands of prairie chickens were sailing up from the unplowed and still unburned prairie, and coming together to this field of sod corn. Some dropped on the stakes or top rail of the fence. These were sentinels and their wary vigilance had to be avoided if I intended to have any success. But most of the evening flight curved up as they neared the corn, and dropped into it about 50 yards from the fence. Some of the birds flew directly over my head, but no one in those days ever thought of shooting on the wing. The chickens as soon as they touched the ground began feeding and joined in a song, much like the singing of tame hens. I then crawled up to the fence, and looking down the corn rows, until I saw 2 or 3 in range, that is in line so that I could kill them all at once, for after each shot I had to return to the house with my game and have my gun reloaded. Finding what I thought was a pretty good chance, I thrust the long gun through the opening between the rails, and sighting, pulled the trigger. The roar that followed my shot, occasioned by the beating of innumerable wings, was astonishing. So large were these flocks that came together at times in the late Autumn and early Winter that their sudden and simultaneous rising would make the earth tremble beneath the shooter's feet. Only three years ago I enjoyed this almost forgotten sensation while hunting on the Sullivan, now Sibley, tract of prairie in Ford county, Illinois. It was the first time in 30 years that I had seen and heard so many prairie chickens in one flight.

I can not remember killing more than five, that I was able to secure, at one time. A young acquaintance held high over me, by claiming to have killed ten at one shot. But he had baited the birds with some straw scattered over the snow, near which he lay in ambush. Quite a number of years after, when I had become familiar with the use of the rifle—a percussion lock then—I once killed three prairie chickens at one shot, and several times two, shooting all through the head or neck.

"Prairie Chicken Shooting and Trapping Fifty Years Ago," *American Field*, December 1, 1883, 807.

In Winter, when the snow was deep, and nearly covered the high grass on the prairie, the birds, after feeding in the morning, left the corn-fields and flew into neighboring trees, where they perched, basking in the sun, until three or four o'clock p.m., when they sought the corn field for their evening feed and thence to their roost in the high grass.

At this time, when the scattered corn in the fields was covered with snow, the birds were sometimes pinched for food. This was the time for trapping.

We would take an oblong box with slats across the top, and raising one end on a set of "Figure four" triggers, bait them with a nubbin of corn. Then strewing some oat straw about on the snow to attract their attention, and some corn to keep them there and occupied until some hungry or curious bird would pick at the nubbin or step on it and spring the trap. A single fall of a good-sized trap would often secure a dozen birds. Then would ensue a great fluttering and beating of wings against the slats, but if the trap was well constructed, their attempts would be in vain, and at the next visit of the trapper he would be well regarded for his cunning and ingenuity. When, as often happened, the number of birds caught were more than could be eaten, the breasts only of the birds were dressed and then salted and smoked like dried beef. These were very fine eating, the fiber being tender and still retaining the prairie chicken flavor.

To the modern sportsman the shooting of prairie chickens as above described, spotting them in flocks and trapping them by hundreds, no doubt seems cruel and unworthy to one who hunts for sport and exercise as we do now. But before passing final judgment we should consider the difference of circumstances which surrounded the shooters of that time and those of the present. Then there were many shooters but very few sportsmen, as we understand the word. Then almost every one was dependent upon the forest and prairie for the animal part of his food, and necessity, not pleasure, called him to the chase. The time thus occupied was so much labor taken from his fields, and he felt the need of securing game rapidly and as easily as possible. Pleasure, such as modern sportsmen enjoy so keenly, was, no doubt, an element of the satisfaction the shooter of that period felt, when, with his clumsy, uncertain, rusty, neglected gun, he bagged a half-dozen prairie chickens at one shot, or when on going to his trap he found a dozen or more thus easily, without waste of time or ammunition, procured for the sustenance of himself and family.

The destruction and diminution of the game was not for a moment considered by him, as its present great abundance made the thought of future scarcity seem absurd.

The fact that game was so plentiful and easily obtained did much for the settlement and development of Illinois, for had it not been so the imperfect guns of the period and the limited time the settlers had to devote to its pursuit would have made the scarcity of food a drawback to immigration.

While all true sportsmen cannot but regret the great destruction of game, they cannot blame their early settlers. They indeed utilized a far greater proportion of what they killed than do the sportsmen of our day, who travel hundreds of miles sometimes to shoot. The sportsmen of that time shot for food, while the sportsmen of today shoot chiefly for pleasure.

Plover Shooting on the Prairies

Rock Jr.

☙ ☙ ☙ I have seen from time to time in the columns of the AMERICAN FIELD communications from your Eastern correspondents on the subject of plover shooting. I would like to give a description of a plover shoot which I enjoyed, to show our Eastern friends the way we shoot plover in the prairie States.

One day in the latter part of April my friend, John. H., and myself, started after snipe and plover, taking with us about 30 hollow tin plover decoys, and my friend's retriever, old Sport. We found snipe very scarce, bagging only about ten, and soon turned our attention to the golden plover.

About 4 o'clock in the afternoon we selected the center of a big flat prairie for our scene of action, and set out our decoys. There was nothing to build a blind of, so we lay down on our stomachs and trusted the color of our hunting suits to screen us from observation.

About 4:30 the birds began to fly, and they seemed to all start at once. A flock of about 20 soon headed for the decoys, and when they were 30 yards we rose to our knees. The birds, taken completely by surprise, immediately scattered right and left; my friend killed two and I did likewise. I was going to let the dog recover the slain, but Johnny commences to call the plover, and those foolish birds turned right back toward the decoys (a thing they frequently do after being shot at), and 3 more of their number were added to our pile. We then went out to help Sport pick up the dead, and while standing there in full view I killed two single birds that came by.

Rock Jr., "Plover Shooting on the Prairies," *American Field*, December 8, 1883, 532.

Returning to our post we awaited further developments, nor were we kept long waiting; a flock of small waders known by the various names of May plover, gray plover and sandys paid us a visit, and left 6 of their number on the ground.

Golden plover now began to come to the decoys in double quick order, and the sport was royal. Sport distinguished himself by an act which showed his familiarity with the game we were hunting. A flock of birds came in from the north and passed over the decoys, leaving 5 of their numbers dead. Sport watched the flock as it passed on, and suddenly started off in the direction the birds had taken. We called and whistled and finally yelled after him, but his only response was a wag of his tail and an accelerated pace. He went about 400 yards from us and returned with a plover in his mouth, apparently very well satisfied with himself. We shot for an hour and a quarter and our bag was no mean one, as it consisted of 56 birds, mostly all golden plover.

They are, without exception, the best birds to decoy I ever saw, and will decoy well in any kind of weather. I would not hunt them now in any other manner. They are not as suspicious as ducks, and any kind of a blind will serve your purpose, and if you have none at all lie down in the field (your clothes being of the proper color) as we did, and you will experience no difficulty. Keep still till the birds are over your decoys, and then pick out single birds in the edge of the flock.

Recollections of Calumet

Blue Wing

⌘ ⌘ ⌘ Calumet! What a host of recollections this well-known name will recall to the memory of Chicago sportsmen, old as well as young!

Who has not been to Calumet hunting?

How many have stood on the river bank and heard the whiz and flutter of the blue-wing teals as they wheeled in in countless thousands on some beautiful, clear, crisp, frosty morning in the latter part of September, just arriving from their northern summer homes! How the blood fairly races through one's veins as just at daybreak the first flocks are seen skimming along the river! How your gun flies to your shoulder, and the bang! bang! followed by

Blue Wing, "Recollections of Calumet," *American Field*, April 5, 1884, 322.

the swash, swash, swash of from 2 to 6 of the little beauties assures you that during the long summer you have not forgotten how it is done! How fast they come, and how swiftly they skim along! They hardly mind you at all, and for an hour or two you scarcely have time to put in your shells, and when the flight is over, as it will be by 8 or 9 o'clock, he is a poor shot indeed who cannot count his 50 birds.

It is certain a royal sport, and happy is the lucky sportsman who is at Calumet on such a morning as this. Although the flight is over on the river, and the birds have settled on the lake, the "initiated" knows a thing or two, and, by getting in to the right spot in the marsh, Henry Kleinman, a year ago last September, gathered in 250 after 2 o'clock in the afternoon. The writer, not being one of the initiated, and thinking the flight over, had retired. He has been waiting for just such another chance ever since, but in vain, as there was, for some reason unknown, no regular flight last fall at Calumet.

It is not long after the blue-wings appear before the bluebills begin to show themselves, and then the sport commences in earnest, although bags of 250 are for the rest of the season somewhat scarce. Somehow or other the Calumet boys do not seem to hanker after bluebills as they do for teals, and so they never overload their boats with these feathery bipeds—well, hardly ever.

Along with the bluebill come the royal red-head and stately canvas-back, and it is about this time that the genuine sportsman begins to show the mettle that is in him. There is something to do now besides standing on the riverbank and firing into the flocks as they pass.

A Plea for the Wild Duck

Anonymous

⌘ ⌘ ⌘ Almost the only bird now left to the sportsman in Illinois is the wild duck. The wild geese and sandhill cranes are pretty much all gone, the quail, ruffed grouse and prairie chicken are fast disappearing; and the same may be said of the woodcock, snipe and wild pigeon. The ducks alone are the sole representatives, in any considerable numbers, of that vast wild feathered colony which, a few years ago, made the hunter's heart glad, and made the prairies of Illinois a paradise for wing-shooting. But with the murderous

---

"A Plea for the Wild Duck," *American Field*, May 10, 1884, 442.

practice of wholesale slaughter now in vogue the days of the ducks also are numbered, and a few years more of such practice will suffice to rid the state of every game bird from Lake Michigan to the Ohio River.

Anyone who, of late years, has watched the markets of the cities and large towns in the Spring of the year must have noticed the large quantities of wild ducks offered for sale, to be reckoned by the ton instead of by pairs or dozens; all of which, with the exception of some of those sold in Chicago, come from the rivers and lakes of Illinois. And it must also have been noticed that these birds, being generally poor in flesh, as they are apt to be in the Spring, are sold cheap, and barrels of them are utterly spoiled by being kept so long in the market.

Now, how is this extermination to be stopped, and the wild duck, almost the sole survivor of the vanished races of Illinois game birds, to be preserved? There would not be any difficulty about it if sportsmen would be satisfied with only a decent modicum of capture. The mallard, teal and others of the anatidae, which haunt our waters in the Spring, are so prolific they would survive ordinary shooting of sportsmen, and we should see each season a return to our rivers and lakes of flocks sufficient to satisfy all needs, and greeds even, if kept within any reasonable bounds. But the devilish frenzy (we cannot call it by any other name) that possesses so many people to perpetrate indiscriminate slaughter will not be stayed or stopped by any sense of propriety or decency; and when it comes to a question of reason or conscience, such people have no more of either than a Comanche or a Hottentot. All they care about is to shoot and kill from morning until night, and as long as they can stand up under the recoil of their guns.

Statement Read before the Central Illinois Sportsmen's Association by Its President

George Hayden

ᔕᔕᔕ The wild pigeons—where are they? Where are they that formerly darkened the sky with their numbers? If they are heard of in the utmost parts of the country, the trappers are there. If a breeding place is heard of, the

---

George Hayden, "Statement Read before the Central Illinois Sportsmen's Association by Its President," *American Field*, October 3, 1885, 316.

pigeon destroyers are there, and the squabs are taken from their nests. Is this a wise policy? Certainly not. Utter extermination now seems to be the fate of the wild pigeon unless some prompt action is taken in their behalf. Trapping and netting should be prohibited and breeding and roosting sites protected.

Letter from Princeton, Illinois [in Favor of Hunting Hawks]

Anonymous

᷎᷎᷎ A friend and I have been out looking for something to shoot at times when the weather would permit. Have succeeded in capturing one chicken-hawk, one bullet hawk, two red-tailed hawks, two ape-faced owls, and two large horned owls. The rabbit fur on the talons of most of them showed of what their fare consisted. It is no wonder that game birds are becoming scarce in this locality for man is allowed part of the year to kill them, while hawks by day and owls and vermin in the shape of minks, coons, skunks, and foxes by night give them no rest at all. While man is fined for killing birds out of season, the pests can kill them anytime, rob their nests or drag them off from their nests and destroy their young ones. They destroy more in the closed season than the breech loader does in the open season. What benefit do the people of this state derive from a law which fines a man for shooting a game bird in the closed season, and lets the birds of prey and vermin destroy it at any and all times without placing a bounty on each and every one.

Letter [Wasteful Hunting on the Des Plaines River]

J. G. Nattrass

᷎᷎᷎ I was traveling in a wagon lately, within twelve miles of State Street Chicago and being in the vicinity of the river, I could plainly hear the

---

"Letter from Princeton, Illinois [in Favor of Hunting Hawks]," *American Field*, March 21, 1885, 269.

J. G. Nattrass, "Letter," *American Field*, April 3, 1886, 315.

shooting that was being done along the banks. One gun in particular, I could hear booming louder than the others. It could not have had in it less than ten or twelve drams of powder and two oz of shot, and I reckoned it to be about four-bore. The river was so high and [the] ground so flooded that half of the teal shot could not be recovered, as the current was so swift that a boat was useless and a dog could not make any headway upstream. I could see a flock fly over the point where the shooter was stationed, and then I would hear his thunder, and see a lot fall into the middle of the river, most of which he could not gather. It is a shame.

Our Hearts Were Young and Gay

Edward R. Ford

᭰᭰᭰ Teen age bird students in the early nineties had no family motor cars in which to make their excursions. To reach a new field, sixty miles from the city, three of us laid our plans. I had gone to work and had a weekly stipend and so was able to go on ahead and get together the needed supplies. The others were, first, to take the cable-car to 63d Street and then, burdened only with shot-guns and blankets, to *walk* the I. C. tracks to Kankakee. It was spring vacation for the two school-boys; I had leave, in the cold April weather, to take half of my annual two weeks holiday.

My part done, I waited patiently in the Kankakee depot for my friends to appear. Night came on; I became sleepy and ready to quit my vigil and go to the hotel across the way. Hour after hour passed, but still I hoped and fixed my eyes on the dim-lit street crossing down the track. At length two gun barrels shone wanly in the distance and presently two footsore and taciturn lads came wearily up to the station. They spoke only to ask if I had bought the grub. I showed them what I had provided, the load was divided and we set out on the road east, away from the town.

Some two miles, stumbling along the dark pike, brought us to a hayshed— the kind of which the upper part only is enclosed—and there we piled in. I had had supper in town and they had eaten something en route and it was sleep rather than food we craved. But they were dead with fatigue and I was brightly awake, and it was I alone who heard the cattle munching the hay

Edward R. Ford, "Our Hearts Were Young and Gay," *Audubon Bulletin*, no. 46 (June 1943): 13–15.

and rubbing their sides against the uprights, and it was I alone who heard the mice scurrying and squeaking all through the night.

To get across the river we had to borrow a boat and, to make return of it, one was chosen by lot to re-cross the stream, walk three miles to the bridge and three more to rejoin his companions. Meanwhile there was shelter to be thought of. In a woodlot near the river lay piled cord-wood. Three pairs of hands re-arranged these to make a three-sided shelter. Poles, covered with the weather-matted top layer of a hay-stack, shut out the sky, and more hay made a bed.

With all this we were not too busy to preclude the finding of a horned owl's nest and to bring the newly-fledged young into camp. Supper done we were ready, literally, to hit the hay; but hardly had we done so when there burst from overhead a challenging and soul-terrifying "Hoo-Hoo-Ah-h!"—a sound which continued for most of the night. (We liberated the young owls the next day.) While this was going on the night grew so cold that we were compelled to build a big fire before our shelter, standing watch, turn about, to keep it going.

Morning sunshine set all aright and, what with bread (without butter), fried salt pork and coffee, we made a rapturous meal and joyfully discussed our position. To use our one rubber blanket to cover all three, each rolled in his own woolen one, was folly. One might have known that to use the rubber one to close the opening of the shelter would be the most effective. And so it proved.

Prairie chickens were booming, meadowlarks singing, red-tails calling, chewinks scuttling in the dry leaves; the marsh hawk quartered over the dead slough grass, and boys with full bellies and an endless week of days ahead were lusty and loud in their bantering and horse-play.

Skins had to be made, eggs blown, guns cleaned, rabbits shot and dressed, partly at least—but what's a hair or two when one is eating hare anyway? Water had to be brought from the brook (not so much for washing, either), there was wood to gather and fresh hay to be carried in for bedding. (Oh, you modern campers-out, see what you miss!)

The osage orange hedges are gone, alas! The shrikes must adapt them-selves to less formidable chevaux-de-frise and less convenient meat-hooks. And the little quail, who could feed for miles without leaving their shelter, now must dart ever more furtively beneath a less protective cover. And is there a Cooper's hawk left there—that swift-striking "blue darter" whose destruc-tiveness of poultry must, to his great disadvantage, be set over against the thrill of his bold, predaceous presence. And the owls are gone, and the wolves whose cubs the old Frenchman took each year from their cave and claimed the bounty and let the old ones go free that they might, another year, give him more revenue. But still on the pastures and prairies of the Illinois country are born, each year, a new generation of Bartramian sandpipers, shore larks, yellow-winged sparrows and grass finches—the names by which we knew them in a day long gone.

Iroquois—Kankakee: stirring and historic names. Where the two streams join there was, in that day, a great pasture and down to its green levels and spring pools came the flocks of splendid golden plover. Mazed by their number and beauty I shot when they were, to all seeming, out of range, and brought down one. The following flock stooped to the wounded bird and then were away.

Were I an Indian I'm sure I should hope that the Happy Hunting Grounds would be as the region of the Iroquois and the Kankakee appeared to me in that vibrant spring of more than fifty years ago.

Skokie Memories

Edward R. Ford

Golden-throated warbler tell me—
Or you bittern, croak and tell—
Skokie memories compel me,
With desire ineffable,
To make question if its creatures
Fare as when I spoke farewell.

Then, oh, rufous-coated sparrow,
Then, oh, joyous-throated wren,
There were watery ways and narrow,
Where the silent water hen,
Sought her ancient sanctuary.
Is she hidden now as then?

Edward R. Ford, "Skokie Memories," *Audubon Bulletin* (Spring 1922): 25. The Skokie referred to is the Skokie Marsh, an inter-morainal wetland that extended roughly from Winnetka to Highland Park. In the 1930s most of it was converted into the Skokie Lagoons. Decades later the one surviving portion became the Chicago Botanic Gardens.

Does the swallow like an arrow,
Skim the Skokie's marge,
Iris blue and white with yarrow,
Does the bobolink enlarge,
As of old themes hedonistic?
Comes sir redwing to the charge.

If such Junes as once befell me,
Skokie wanderers yet may know,
It were very kind to tell me.
Do the winds of Skokie blow,
Spicy with the smell of flagroot?
Give me answer—I must go.

The Naturalist
Edward R. Ford

Over and over the night comes in,
Over and over the mornings rise,
Again and again is the wonder wrought
And the marvel enacted before his eyes.

Forever and ever the secret is hid;
They babble of mystery, bee and bird:
He listens entranced to their jargoning
But listens in vain for a clearer word.

Forever to seek and forever to ask—
What if his labor is never done?
Labor entrancing is never a task:
Other reward, let it be, there is none.

Edward R. Ford, "The Naturalist," in *Birds of the Chicago Region*, special publication no. 12 (Chicago: Chicago Academy of Sciences, 1956).

➤➤➤➤ EDWARD RUSSELL FORD (1875–1951) ENTERED THE PUBLISH-
ing business at the age of fifteen and continued until retiring in 1929 as secre-
tary-treasurer of Periodical Publishing Company of Grand Rapids, Michigan.
Of even longer duration was his great interest in birds, which spanned most
of his life.

Ford was closely associated with the Chicago Academy of Sciences,
where he served as an honorary curator, first in zoology and then in ornithology. Through his extensive contacts with other ornithologists, he orches-
trated trades that significantly increased the academy's collection of avian
material. And it was the academy that published the work for which he is
best known, *Birds of the Chicago Region*. He submitted a draft manuscript to
the academy just six months before he died.

In addition to the academy, Ford was active in a number of other bird and
conservation organizations such as the Illinois Audubon Society and Inland
Bird Banding Association. He also published articles and/or poems in the
Auk, *Bird-Lore*, the *Audubon Bulletin* (where most of the writings used here
originated), and the *Chicago Tribune*.

Beginning in the late 1930s, he largely abandoned Chicago for a summer
home in Michigan and a winter home in Florida. But on his way to and from
those destinations, he would stop and tarry in his old haunts, particularly at the
Chicago Academy of Sciences, where the "staff looked forward in pleasurable
anticipation to 'Ed Ford's semi-annual visits.'" Howard Gloyd, then director of
the academy, concludes his biography of Ford with these words: "Mr. Ford will
be remembered by his friends for his kindly personality; for his gentle whimsical
philosophy; for his quiet but stimulating conversation … ; and for his modest,
untiring efforts to add to our knowledge of the natural history of birds."

Sources: Howard K. Gloyd, "Edward R. Ford—An Appreciation," in Edward R. Ford's *Birds of
the Chicago Region* (Chicago: Chicago Academy of Sciences, 1956); Alfred Lewy, "In Memoriam:
Edward R. Ford," *Audubon Bulletin*, no. 78 (June 1951): 14–15.

Who Kills the Birds?
Mary Drummond

> Who kills the birds?
> "I," said the Woman, "although 'tis inhuman,

Mary Drummond, "Who Kills the Birds?" *By the Wayside* 4, no. 7 (November 1901): 1.

I must have dead birds."
Who sees them die?
"I," said the Man, "whenever I can,
For my sport they must die."
Who tolls the bell?
"I," said the Boy, "I love to destroy.
I toll the bell."
Who digs their graves?
"I," said the Girl; "for a feather's neat curl
I'd dig all their graves."

* * * * * * * * * * *

So the men and the boys by the woodland and streams,
And the women and girls, with their hats like (bad) dreams,
Are robbing the earth of its bird life and song,
With never a thought of their rights, and our wrong.
But, isn't it strange, if their hearts have no pity
For the poor little birds in the country and city,
They never remember that some summer day,
Not a bird can be found that a human can slay.
Why, what will become of the boys and the men,
Who can't shoot at birds, for there'll be no birds then?
And as for the women and girls of that day,
With their featherless bonnets and hats in array,
'Tis dreadful to think what their sorrow will be,
And yet it is something I'd much like to see.
For it's certainly true, and the truth must be said
If we kill all the birds, all the birds will be dead.

➤➤➤ ONE MAJOR THEATER OF THE CONSERVATION WAR OF THE early twentieth century targeted the wanton killing of birds for the millinery trade, which claimed over 5 million individuals a year in this country alone. Egrets, with their long nuptial plumes, roseate spoonbills, and terns probably suffered the most, but all kinds of birds (or parts thereof) wound up adorning ladies' headwear. Ornithologist Frank Chapman, originator of the Christmas bird count, did a different survey one day in Manhattan: he found that at least some portion of 20 different species adorned 542 out of the 700 hats he encountered. (Apparently even turkey vulture feathers wound up as hat decorations.)

Pitched battles in this struggle were waged state by state. In 1897 the Illinois Audubon Society came into existence with one of its major goals being to eliminate the trade in bird feathers. Three years later it was the largest of the state Audubon societies, enjoying a membership of over 10,000, of whom 800 were adults and the rest children. Among those 800 were such pillars of the scientific and philanthropic worlds as Stephen Forbes (University of Illinois), Frank Baker (Chicago Academy of Sciences), Mrs. C. H. McCormick (wife of *Chicago Tribune* publisher), and M. A. Ryerson (of the steel company).

The Illinois Audubon Society helped Illinois enact a protective statute that was deemed to be the strongest in the country. Of equal importance was trying to change consumer tastes. To convince adults, the society publicized the brutal way feathers were obtained, and to enlist children, they distributed a variety of educational materials (see the Vrablik letter on p. 384).

Mary Drummond (1846–1926), who lived in Wheaton and then Lake Forest, was secretary and treasurer of the society. She also appeared before other organizations, such as the Illinois Humane Society, to solicit support for bird preservation. I know nothing else about her. The poem presented here appeared in *By the Wayside*, "official organ of the Wisconsin and Illinois Audubon Societies" published by the Wisconsin group. (At one time the magazine also represented the Michigan Audubon Society.) By 1916 Illinois had its own journal.

Sources: "Work of the Humane Society," *Chicago Tribune*, May 8, 1898, 11; "Work of the Bird Savers," *Chicago Tribune*, April 1900, 8.

Colonel Isaac Washington Brown: The Bird and Bee Man

Henry A. Pershing

᭟᭟᭟ Indiana has been the home of many well known men—poets, writers, travelers, orators, artists, president of the United States, but it had only one Colonel Isaac Washington Brown, known all over the country during his life time as "The Bird and Bee Man" of Indiana.

Henry Pershing, "Colonel Isaac Washington Brown: The Bird and Bee Man," *Indiana Audubon Bulletin* (1930): 22–26.

He was a unique character. For some twenty years, he traveled over the United States, sometimes on foot, sometimes by horse and buggy, sometimes by train to deliver his message. And the burden of his cry was, "Don't kill the birds!"

He was born in Carroll County, Indiana, May 27, 1848, the son of John D. and Rebecca Brown. He was brought up on the farm, assisting his father and at the same time getting well acquainted with birds and bees as would any normal boy. He attended the ungraded schools nearby and secured as much of the rudiments of an education as was available in the poorly taught country schools of that early day, but he learned to love nature, roaming the woods and meadows listening to the quail, the meadowlark, the robin and wren.

At the age of sixteen, he could not resist the call of volunteers in the Civil War. As he was rather young to secure enlistment at home, boy-like he ran away and enlisted in Co. O. 135th Infantry.

After his return from the war he continued to work on the farm. Later his father removed to another farm near Lake Manitou and in 1870 the young man came to live in Rochester, Indiana, and lived there until his death. In 1872 he married Emma Strong. A short time later he took up the study of law but the detail-ridden life of a small-town lawyer was distasteful to a man who loved action.

Later he became a secretary for Hutchison, the board of trade manipulator. In the employ of "Old Hutch" he traveled over the country assisting in tabulating the crops. Eventually Brown became quite an operator on the Chicago Board of Trade and many are the stories told of his flush times. The capital he acquired there gave him the opportunity to carry out business plans he had evolved. At one time he bought a meat market and advertised that everything would be on a cash basis and that he would do no delivering. His "cash and carry" market was a failure. He was twenty years ahead of his time. He bought a hotel where no meals were to be served. The traveling public were not ready for that either. The hotel investment was a loss, but nothing could dampen Brown's energy and enthusiasm. After getting some funds together he went back to Chicago. Again he was successful on the Board of Trade. When he returned to Rochester with the profits of his latest investments, he hunted up every old soldier he could find around Rochester, and presented him with "a ham of beef."

But Brown's good luck could not continue forever. In an attempt to corner the wheat market, Brown and his preceptor, Hutchinson, lost everything. He returned home shorn but not whipped, spurred to action by the necessity for money.

He had always been a good talker, quick to think and with a decided ease of expression. He was red-headed, big and strong, with the forceful personality that makes so great a public appeal from the platform. He conceived the idea of making a special study of the birds he had always loved and talking to the school children about them. He was familiar with the habits of birds and wild life and this task would be easy and fraught with great good.

He had a friend, Sherman Gibbons, to whom he explained his intentions and who gave him a number of books on birds with the result that he soon had prepared a twenty minute talk which he felt sure would be well received. In order to become proficient in his delivery he would go into the woods, learn bird-calls, and at the same time practice his speech until he felt that he was master of his theme. If the birds would sit on the fence rails and listen to him, surely he could interest the children. But he must find some audience on which to try his first address.

Everybody liked Brown and he had many friends. Among them were two country school teachers in a two-story brick school house in the village of Talma, nine miles from Rochester. Arthur Deamer was the Principal, and Roy Jones, later county school superintendent, was his assistant. Consent to deliver his bird talk at Talma was readily granted. Brown brushed up his speech and walked the nine miles along the beautiful Tippecanoe River, little dreaming that he was beginning a career which would take him from one end of the continent to the other, carrying the gospel of being kind to the birds with the battle cry, "Don't kill the birds!"

In the northwest corner of the big second-story room in the old brick school house, Brown delivered his first bird talk. The fascinated children listened to every word and gave him a storm of applause. His reception at Talma decided his course of action. If he could interest one school room full of children, he could interest a thousand school rooms.

As he was poor, with no horse to carry him from school house to school house, he went on foot. On one of his trips, he walked to Mexico, Indiana, a distance of some twenty miles. Sometimes he gave notice that he would be at a certain school house in the evening to talk to the farmers and their friends. He would be greeted with good audiences and then, having given notice to that effect, would take up a collection. Some would give money and others would give him a few lengths of sausage or a ham or maybe potatoes. He seldom failed to get sufficient funds to pay his expenses with something over for the good wife, Emma.

He was a near neighbor to Mr. Henry A. Barnhart, former Congressman from the thirteenth district, who appreciated the "Colonel," as he was commonly known. It was a frequent occurrence of Brown to come across the lawn, shouting as he approached, "Well, Henry, hitch up the old mare. I want you to take me to a lot of school houses. You can talk for five minutes at every school house on the American flag and I'll talk one hour on birds."

The work with the birds was a serious matter to the Colonel and so earnest was he in his presentation that his fame began to go abroad, his name became familiar to the readers of the papers and he received calls from different places to come and make talks about the birds. In fact, it may be truly stated that he was the pioneer speaker in the country on that subject. His manner of presentation was known to almost every one and he was always listened to with attention. He talked in all the Rochester schools and was a frequent speaker at the Indiana State Normal School. He did not always wait for an invitation. He used to appear at school houses in small towns, walk into the room, introduce himself in his free and easy manner, "How do you do, Mr. Teacher. I was wondering if you would like to have me talk to your children about birds." Of course the teacher had always heard of him and his eccentricities and was invariably more than glad to halt his classes for the Colonel's talk.

One morning he appeared on the grounds of the Winona Lake Assembly and at an early hour gathered a crowd about him while he talked of Purple Martins and how they destroyed mosquitoes with which the grounds were well supplied. He repeated this for two or three mornings. The management noticed that the early meetings in the Auditorium were very poorly attended. On learning the reason, they invited Brown to become one of their regular speakers. He was on the program at Winona Assembly for nine consecutive seasons, from 1902 until 1910. In that time he was instrumental in having bird houses placed all over the grounds. Thousands of children were brought to a realization of the beauty and value of birds.

At Winona Lake in 1906, Dr. W. W. White, then editor of "The Bible Record," heard Colonel Brown and wa[s] so impressed with his interesting talks, that he wrote to Miss Helen Gould of Tarrytown, New York, concerning the effectiveness of Brown's work. She wired the Colonel to come to Tarrytown at once. At their first meeting she told him that she wanted him to take her house guests on a bird walk the next morning.

"All right," said the Colonel. "Meet me on the lawn at four o'clock."

"But, Colonel Brown, my guests cannot meet you at such an early hour."

"Very well. There will be no bird walk."

The next morning at the unheard of hour of four, most of the guests were there and Brown took them on a long walk, telling of the birds they saw, giving their names, songs, habits, and usefulness. These walks were taken often, for they were new experiences to these people and their eyes were opened to the beauties of nature. From that time for two years, Brown was in the employ of Miss Gould who gave him $5000.00 a year, with a pass over all the Erie lines, all expenses paid, with Dr. Chas. Gould as his companion on all his various trips. He traveled in twenty-four states. In Texas, he applied his knowledge of birds to the fight against the Boll Weevil.

After his contract with Miss Gould ended, he decided that he would like to be at home oftener. He was a man of great affections, and he loved his wife and children and the thousands of friends of his native state. He arranged for talks all over the state and was in constant demand.

Many anecdotes are told about this amazing old man. He was open and generous hearted, outspoken and impulsive. Once when talking to the inmates of Longcliff Hospital, Brown stated that the plover would fly from Indiana to the Gulf of Mexico in eleven hours. Fred Landis, who was listening to the address, interrupted him, "Don't you think that's a pretty long journey to make in eleven hours?" Brown turned on him with "Now look here, Landis, I don't want any interruptions from a novice," and went on talking.

Hon. John W. Kern was on the train once when Brown entered and sat beside him. Kern asked him if he knew Congressman Barnhart, at Rochester. "Know him!" exclaimed Brown, "Well, I should say. These were his pants I'm wearing right now."

He and Barnhart were very good friends and while Barnhart was in Congress the two of them were instrumental in the introduction and passing of the first bird laws ever passed by the national government. Brown became very popular and more than one bird talk did he give before congressmen and senators. Brown had no use for President Roosevelt. When asked whether he would call on the president before he left, Brown replied, "No, sir. I have given strict orders that the president is not to speak to me."

He once met William Watson Woolen of Indianapolis, who was a close student of nature and knew birds scientifically. Mr. Woolen was reported as saying that he did not think Brown knew very much about birds. When Brown was told this, he exclaimed, "Woolen's old style. I'm up to date."

Mrs. Barnhart, the mother of Henry, ventured to suggest to the Colonel as he was passing the house one day, "I should think, Colonel, that you'd ask the Lord to help you in your good work."

"I'm getting along very well, Mother," he said and passed on. Then turning, he called back, "Needn't tell the Lord anything about it. I'll try to figure it out somehow."

When Governor Thomas R. Marshall lived at Indianapolis he gave Brown the following letter of introduction:

Executive Department,
State of Indiana,

January 11, 1911

To the Superintendents of the several State Educational Institutions:

The bearer of this letter is Colonel Isaac Brown, a native born Indianian, veteran of the Civil War, commonly known as the "Bird and Bee Man." He is deeply interested in the preservation of the birds of Indiana and his lectures upon this question are not only valuable from the standpoint of conservation of our natural resources but they are also valuable in training young people along right ideals.

If consistent with the conditions of your institution, you might arrange for the Colonel to address your pupils. I am very sure you would never regret having done so.

Very respectfully yours,

Thomas R. Marshall, Governor

Colonel Brown was not only a speaker, but a frequent contributor to magazines, writing tersely and to the point. He loved children and they understood him. In the Chicago Examiner of 1907 is an account of his speaking to a crowd of little children in Jackson Park:

"An old man with the light of kindness in his eyes stood bareheaded yesterday afternoon on Wooded Island in Jackson Park and spoke to several children who stood in a sheltered group on the veranda of the Japanese pagoda. It was a strange plea that came from the bearded lips of the patriarchal speaker. 'Be good to the little birds, my children,' he said over and over again. This wonderful old stranger was the 'Bird Man' agent of Helen Gould, and a strange, picturesque old man he is, too—large of body as he is of soul with a merry smile for the children. He seemed an ideal harbinger of mercy and love. He stepped to the bushes and showed them how the birds lived and builded. He has studied the birds and knows their language, he trilled like a thrush, chirped like a sparrow, and cawed like a crow. It was sundown when the old man took his way back to the city but he left with the children his kindly gospel of bird love."

One afternoon in August, 1914, he came from Chicago, sick and trembling. As he walked in the door of his home, he said to his wife, whom he loved dearly, "Well, Emma, I've come home to die." He was suffering from acute indigestion. He lingered a day or two and died on the twenty-seventh of August after a life of great usefulness in spreading the doctrine of being kind to the birds. Congressman Barnhart, writing from Washington after his death, said of him:

"He was free-hearted, emotional, out-spoken, over-enthusiastic, industrious and picturesque. There was but one 'Ike Brown.' Some did not like his way, and some do not like yours or mine. But he had a mission in life and accomplished vastly more for posterity than many who sneered at his methods of attraction. Wherever he went—on the street corner, college, school, church, shop, or field, he worked for practical results. He talked birds in trains, cottages, hotels, parks and palaces, and his quaint but convincing logic made bird friends of everyone who listened to him. And nothing could be more in keeping than that the robins, the orioles, the turtle doves, and the bluebirds should warble a requiem for the eternal rest of their great benefactor."

He was a member of the church, a Knight of Pythias and an Odd Fellow, but, strange to say, no marble slab records his place of burial and although sixteen years have passed since his death, his lonely grave lies unmarked.

〰〰〰 IN 1882 HENRY PERSHING (1857–?) MOVED TO SOUTH BEND, Indiana, where he lived most of his life. He eventually relocated to the Masonic Home in Franklin, Indiana, where he celebrated his ninety-third birthday, if not still later ones. Pershing is best known for two books, *Poems: Sentiment, Nature, Humor* (1938) and, particularly, *Johnny Appleseed and His Times* (1931), which was favorably reviewed in the *New York Times*. The *South Bend Tribune* in its September 9, 1936, issue referred to him as an "author, humanitarian, and well known citizen." This is another instance of where a person of some prominence left a trail swept virtually clean by the breezes of time.

Source: Henry Pershing in "Journalism—Authors File," South Bend Public Library, Indiana.

When the Lotus Are in Bloom

Laura Buchannon

From Mineola Bay canoeing,
Sweethearts shy go coyly wooing,
Where the air is scented with that faint perfume,
When the dew is softly falling,
And the whippoorwills are calling,
O'er the waters when the Lotus are in bloom.

Chorus:
Good old Indian Fox Lake bow'rs,
Dear old Grass Lake's fragrant flowers,
Make one's heart rebound and boom,
When the Lotus are in bloom.

And when the sunlights brightly streaming,
And the Silver Bass are gleaming, in the waters of the
Nippersink Lagoon,
When Pistakee's skies are glowing,
And the current gently flowing,
In August when the Lotus are in bloom.

Then you can hear Rohema's laughter,
Sure and Blarney Islands chatter,
Where real mirth and joy and gladness chases gloom.
Where all nature prompts romancing,
As the waves go merr'ly dancing,
O'er the waters when the Lotus are in bloom.

So both in sun and moon and starlight,
And by evenings dim lit twilight,
We'll go floating thru that flow'ry landscape soon,

Laura Buchannon, *When the Lotus Are in Bloom* (words and music) (Chicago: J. T. Buchannon, 1917).

With pure love's eyes fondly smiling,

Days and nights and hours beguiling,

On the waters when the Lotus are in bloom.

〰〰〰STRADDLING THE BOUNDARIES OF LAKE AND MCHENRY Counties along the Fox River, the Chain O' Lakes had long been admired for their beauty and bountiful wildlife. One of its most popular destinations was Grass Lake, which harbored extensive beds of American lotus sprouting from its shallow waters. Local promoters claimed that such a sight, and the sweet "intoxicating" scent emanating from the golden flowers, was duplicated nowhere else in the world except China and Egypt. A large tourism industry developed, with hundreds of visitors arriving on trains from Chicago to board the vessels that would take them to their destination.

This song by Laura Buchannon (about whom I could find nothing) was written when this ecotourism was at its heyday. But over the decades, the acreage of lotus beds on Grass Lake plummeted due to high water, the commercial harvesting of flowers for ornaments, and heavy boat traffic. By 1959 aerial photos show no remaining acres of lotus beds. More recently, however, the plant has staged a modest comeback.

Letter to Illinois Audubon Society

Olga Vrablik

Seward School, Room 201

4600 South Hermitage Ave.

[Chicago]

June 1, 1922

To the Illinois Audubon Society,

Dear Friends:

I am glad to belong to the Audubon Society. We put our pennies in a box and when we had a dollar, we sent it to the Illinois Audubon Society. The Membership card is hanging up in our room. We thank you for it.

Olga Vrablik, letter to *Illinois Audubon Bulletin*, Fall 1922, 29.

We know sixteen trees. We have learned about trees by seeing the leaves and pictures. We know the shapes of the leaves and the different edges.

We have learned about wildflowers too. Our teacher, Miss Kelly, brings leaves and flowers to school. We love the song birds. We know 32 song birds from bird pictures. We love them for their beautiful songs. We love them for their beautiful colors. The birds contribute to our "Daily Bread" by eating bugs, worms and insects, and by eating weed seeds, too.

All the people must be very kind to birds. No boy in our room will ever harm a song bird.

Your little friend,

OLGA VRABLIK

Aged 9 years

▬▬▬ OLGA VRABLIK (1913–?) REPRESENTS ONE OF THE MANY youngsters who were so critical in helping promote the conservation cause in the early part of the last century. Given both the nature of her writing and likelihood that her surname would change relatively early in life, I did not hold much hope that I would be able to find anything about this young author. But, in fact, she (at least I think it is likely to be the same person) turns up in a *Chicago Tribune* article on January 26, 1947. It seems that the Slovak American Citizen League of Cook County held an annual "Night of Progress" and part of the festivities was a "popularity contest" in which a "Queen of Progress" was coronated. Ms. Vrablik was one of that year's nominees.

〰〰〰〰〰〰〰〰〰〰〰〰〰〰〰〰〰〰〰〰〰〰〰〰

Nature Recreation in Chicago
William G. Vinal

ৰ৶ ৰ৶ ৰ৶ As early as 1868 the State Natural History Society of Illinois assumed "the duty of supplying Natural History materials to the schools prepared to use them." Today in the Chicago district alone there are at least thirty-five agencies disseminating natural history in one form or another. These various bureaus and societies, like the Arabs, must have come silently

William G. Vinal, "Nature Recreation in Chicago," *Recreation* 29 (March 1936): 503–4.

in the night and set up their institutions as Chicagoans hardly know that they exist. When these facts are marshalled into a table they present a very potent power which contributes to the cultural life of Chicago.

One can readily find a list of "the tallest buildings" and every sidewalker can point out the world's largest hotel or largest stockyard or largest something else. Every loyal Chicagoan knows that the Navy Pier is one mile long. Recreation leaders can promptly say that there are 125 baseball diamonds in Park X, but asked for a picture of their nature activities they are silent!

Nevertheless, Chicago has had its Babe Ruths in Naturedom. There has been a succession of noted trainers in nature leadership commencing with H. H. Straight in 1883 who came to Cook County Training School from Oswego Normal. In 1889 Wilbur S. Jackman came from Pittsburgh. Both were coached and schooled by that teacher of teachers, Louis Agassiz, the Great. Then came Ira B. Meyers in about 1905 followed by Otis W. Caldwell. In 1911 the dynasty of Elliot R. Downing commenced and today O. D. Franks is the chief factum factotum. A noted legion dating back to Agassiz, yet the whole family tree hidden under a bushel!

It has taken considerable coaxing and maneuvering to bring this information to light and yet it ought to be useful not only in Chicago but to leaders in general who are trying to organize their own communities. It will not only provide source material and experienced people to contact but will indicate the trends of the time. The nature services and opportunities of any locality do not come about spontaneously but must be credited to enthusiastic leaders for having been born. That is not all: A favorable environment and nurturing is necessary. The recreation leaders who can visualize this picture most clearly will see an opportunity that parallels the other cultures—namely, drama, art and music. To them let it be a hint that they hang this "Bird's-Eye View of Nature Activities in Chicago" alongside of the old Farmer's Almanac and contemplate it now and then in planning the future.

ORGANIZATIONS CONDUCTING NATURE ACTIVITIES

Public and Semi-Civic Organizations

1. Adler Planetarium
2. Board of Education, Department of Education
3. Board of Education, Bureau of Recreation

4. Brookfield Zoo
5. Chicago Academy of Science
6. Chicago Public Library
7. Chicago Recreation Commission
8. Field Museum
9. Morton Arboretum
10. Museum of Science and Industry
11. Shedd Aquarium

Parks

12. Chicago Park District
13. Cook County Forest Preserves
14. Dunes Park of Indiana
15. Garfield
16. Humboldt
17. Lincoln
18. Washington

Schools

19. Northwestern University
20. University of Chicago

Clubs

21. Chicago Ornithological
22. Chicago Woman's Club, Forest and Garden Class
23. Conservation Council
24. Friends of Our Native Landscape
25. The Geographic Society of Chicago
26. Illinois Audubon Society
27. Izaak Walton League
28. Kennicott Club
29. Outdoor Art League
30. Prairie Club
31. Wild Flower Preservation Society

Social Organizations

32. Adult Education Council
33. Hull House
34. Outing and Recreation Bureau
35. South Chicago Neighborhood House
36. Y.M.C.A.

〰〰〰 WILLIAM GOULD VINAL (1881–1976) WAS BOTH AN ENGAG-
ing character and a brilliant teacher, necessary qualities for leading a nationwide innovation in education—the nature study movement. Indeed, he came
up with the term. According to the editor of *Nature Study*, Cap'n Bill, as he
liked to be called, "is perhaps the last of the 'second generation' of naturalist-
educators in North America. He and his contemporaries... followed close
upon the days of Louis Agassiz and Asa Gray. They were the great innovators
in the development of techniques to revolutionize the education of children
and the non-scientist with respect to science and human affairs."

Vinal eschewed lecture in preference of student participation. He believed that conservation education was best accomplished "by providing
seasonal, enriching education experiences in a natural environment." And
these activities were imbued with two of his fundamental tenets, as recalled
by a former student: "1. Learning is fun and... fun can be a truly educational
experience. 2. We may be born with the ability to see, hear and feel, but we
have to be taught to observe, listen, and interpret our senses."

Born in Norwell, Massachusetts, Vinal spent most of his life in New England, receiving graduate degrees from Harvard University and Brown University. (His dissertation at Brown was on the bivalve *Anomia simplex*, whose
structure and fragility made it, in fact, rather complex.) After stints at Marshall University, Western Reserve University, and the National Recreation
Association, he began his fourteen-year tenure at the University of Massachusetts in Amherst in 1937. Upon retiring, he returned to his place of birth,
where he resided until his death. The South Shore Natural Science Center in
Norwell has honored Vinal by naming their nature center after him.

During his two years with the National Recreation Association (1935–37),
Vinal's principal task was to identify the nature programs offered in various cities. The piece used here is a report of what he found in Chicago. It is
quite interesting to see what was available in 1936—many of the institutions
promoting natural history back then remain in the business, although others

have either faded into oblivion altogether or have turned their attention elsewhere. Vinal's report is even more heartening when one realizes that he overlooked local groups that were in the nature trenches. The Evanston North Shore Bird Club, for example, founded in 1919, is an active partisan on behalf of bird study and conservation. And there were bird clubs in other places, too, such as Barrington and Maywood, to name two. The only colleges Vinal mentions are Northwestern and the University of Chicago, but other local institutions of higher learning were also "conducting nature activities." As one example, there was the work of C. W. G. Eifrig at Concordia Teachers College.

Sources: J. A. G., "Editorial," *Nature Study* (Winter 1974–75); South Shore Natural Science Center, "Cap'n Bill Vinal Campfire Roast and Norwell Centennial Open House," May 15, 1988; William Gould Vinal, "The Growth of the Concept of Nature Recreation," *Nature Study* (Winter 1974–75): 1–8.

A Prairie Grove
Donald Culross Peattie

ᏬᏬᏬ These pages are those thoughts, a memory of what was gone before I came. I must remember not with my trivial faculties but with the blood that is in me. I must remember for my kind who came before me, and even for those who spoke languages that are alien to me. For they too went before me and stood where I stand, and they looked where I look, but they saw a grander and harsher world.

That world is this story, and I write it not literally but as a true legend. There are many kinds of story; we all like to hear and tell funny stories and Lincoln stories and wonder stories, and stories of the old times, lucky stories, war stories, and some of us esteem the worth of a bitter anecdote or some tale of a grand folly. We recognize that these are so many facets of the gem of truth; they give it its sparkle and its prismatic surfaces. Merely one surface is the relation called history.

The writing of history is for historians. I may not walk within their preserves. But they do not allow themselves to stray where I intend to go. My

Donald Culross Peattie, A *Prairie Grove* (New York: Literary Guild of America, 1938), 15–19, 118–24. Courtesy of Random House.

purpose keeps me to the theme of the island grove, the trees and the great grass, the wildfowl and the furred and antlered beasts, and tall men, very small, moving about in their roles beneath lofty boughs and across wide spaces.

We call those roles their history, and because they are finished, they seem inevitable now. But in their day these people lived as you and I do now, from moment to moment, the tense adventure of existence. Not knowing what would come, they ascribed, as you and I do, false reasons for the things they did. History sums up their story; it detects the great propulsive movements to which they were subject as are children. It cannot stop to listen to individual forgotten heart beats. But I shall listen, and I will claim the poet's right to say he knows what they think and feel who are too headlong in life to make a song of it.

The names of Father Gabriel Forreste and Father Pierre Prud'homme you will not find in all the seventy-two volumes of the *Jesuit Relations*. For one thing, they were not Jesuits. The work of the Recollet Franciscans has been neglected for the more articulate members of the Society of Jesus. I say you will not find the names of these characters of mine, yet they are there none the less. You will find these men in the *Relations*, unless you are blind, and you will find my Robert Du Gay in La Salle, Frontenac, Champlain. You will find the *coureurs de bois* everywhere in frontier history, and you will discover that they were the first men to reach the sources of the Mississippi, the first whites who ever gazed astounded upon the Rockies where the Big Horns jut out in the prairie province.

I guarantee, therefore, that no one in this story is wholly fictitious. It is not necessary to invent either character or detail; the gold and scarlet cloak of La Salle, the bourgeois bigotry of Father Hennepin, the humility and the death of Marquette, are matters of record. They are traditional folk airs in the great song of America.

You will not find on any map my island grove, yet it is here and I walk in it. It has had its history, predestined to it by its site between master rivers, crossed by a portage trail, and by the sort of men and women who there laid down first hearthstones, founding that honorable and half-awkward, half-lovely way of life that gave us Clemens, Altgeld, Grant, Logan, and Riley. My children are the sixth generation of their line in the grove; to go back ten generations, my people came westward by stages; they too are in my story, as they are in me. In old records, in county histories, in memoirs and letters I find the men and women who all over the state endured, exulted in, the privileges and pains of living in that day. I bring them to my story; I need invent little.

For the scene is all, the habitat group, man most significant in it. The drama is a biological drama; in the play we see how the white man came to the wilderness, and what happened then. The island is my stage; it would be falsifying, in effect, to name it; instead I hang out this sign: A Wood.

I walk beneath old trees that Du Gay might have known. I know my grove by winter and in summer; I know it through the night hours, when the vesper sparrows sing and the black-crowned herons are most active. My diaries tell me when the birds come back, when the thrush stops singing, when the first cicada praises heat. In my inventory are every one of the four hundred and fifty-three species of flowering plants that grow here. I have learned what pollinates them and what eats them and what nests among them, and what the Indians and pioneers used them for, and—have no fear—I shall not tell you much of this.

Humans have to take their place where it falls in the fauna. But there is no plot; this is not a novel, not a historical romance, not a popularization of history. I say that I am remembering, remembering for the trees and the great grass province and the passenger pigeons and the wild swans. I say that the coming of our species was an event, perhaps an impermanent one in the greater story. So my characters are transient, even shadowy. Individual character does not matter to Nature. In the end she absorbs all individualities; she knows only races and their rise and fall. But the ideas of our species are the human scent we leave upon the wild turf. They drift and linger on the airs after we have gone, and are the things most worth remembering about us....

It is an old miracle how spring will come back in spite of snow and sorrow and war. But in other lands than ours one cannot say much for autumn, except that there must be an end to all things. In temperate North America, in the hardwood forest belt, it is autumn that triumphs over drought and summer weariness; it sweeps in with the sense of freshening, of a new coming to life and an actual reawakening of the instincts.

There was in the wilderness days a whole great biological pulsation that was autumnal. The prairies then filled with grass herbage like a rising lake. Ankle-deep and starred with little wildflowers in the spring, it rose above a man's head in the fall. It was gorgeous with the purple spikes of blazing star and the gold of the sunflowers. The big grasses flowered in autumn, and when the green began to go the bronzes came, the tawnies like the wildcat's fur, the low-toned, burned-out vermilions like old war paint. And then at last, when the squaws had pulled the corn ears and the children, with the little melon bellies, had dutifully rolled the pumpkins to warm another side in softening sunlight, the prairie suddenly died. Color drained out of it like light from the top of the evening sky. And then the tumult in the island groves began.

Part of it was crows, a clan of birds that is shocked by everything, and part of it was the flight of the does, who do not come to heat so soon as the bucks. They crashed through the forest with eyes starting and exultant, knowing the rapture of flight from a pursuit that meant no harm to them. They say that a buck would follow one doe who would elude him by running among her sisters, crossing her scent with theirs; then he would drive the pack of them, rearing and bounding, divide it and quarter it and scatter it, trying to find his first desire and losing himself in a maze of scents, all of them female and all of them now dissipated upon the sudden fresh wash of the north wind through the softly applauding forest. Then his neck would swell with passion; he would throw up his head with his nostrils flaring and his antlers tossing the sumach leaves, the sur-royal tines clattering the dead black twigs of the hickory boughs. After weeks of such pursuit, his body wasted with running and fighting his rivals, this torment-ridden fury discovered all at once that the does were wraiths no longer. They were suddenly still in their tracks, fine legs trembling, dark dilating eyes turned back toward him; in the calm steeping sunlight the perfume of these gentling friends washed back to him, bathed him and promised kindness; the scarlet and the orange and the gold of the leaves rained through the antlers and drifted about the motionless black hooves.

For the autumn colors were part of the tumult. There is no other land in the world with autumns like ours. We pile the treasure of the year into a great burial fire. Tongues of flame go up to the sky, the garnet of black and red oaks, the leaping maples and the flickering aspens and out of the midst of it all one exulting spire of light where a cottonwood shakes primal yellow at the primal blue of the American sky. From the boughs pours down the glory of the vines—woodbine and corded grape and poison ivy. The thickets fill with the cymbal colors of the sumach—orange and scarlet and stain of wine; the leaning dwarf forest of the hawthorns begins to drop its shower of little pomes—ruby color overcast with purple bloom. They tumble in a circle, a wild harvest no less bounteous because only mice and children gather of it. Under the trees curls the violet breath of the asters. And still sometimes, where the cattle have not trampled, I find a lonely gentian hoarding blue. It keeps its corollas closed against the bee, dropping pollen from the linked brotherhood of the anthers upon the stigma, like some divinely descended royalty that must propagate within its own sacred circle.

The jays, blue crows that they are, have much to say in autumn, and they talk as though they remembered the clamor of the old abundance. For those were the days when upon the dwindled river marshes already crowded with wildfowl, the hordes from the north descended. The redwings devoured the wild rice; they rose in irritable black tempests when the Indians came among them, bending the freighted heads of the rice over their canoes, beating the grains into skins spread waiting on the boat bottoms. The yellowlegs and the plovers came back then, teetering, piping, foraging at a run upon the mud. The wild geese went over, high, too high for the upward rain of arrows. The great cranes felt the disconsolation of shortening days, and began to stream away—mere etched lines of gray on the soft gray plumage of a sky promising moisture and the break of drought.

They say the cougar mated in the fall, but he is more vanished now than the credible memory of the elk. King of the antlered kind, the elk in the great rutting season was a creature of terror. Nocturnal then, his fights were like the matching of Sioux strength with Iroquois. Rousing himself out of the mud wallow where he had retreated from the stings of the horseflies, the old master of the herd stumbled up, blew out his matted nostrils, and began to remember the number of his does. He rounded them together, perhaps a score of them, with a warning scream of his perpetually impending displeasure.

When they strayed, he struck at them unmercifully, for he smelled other bucks upon the air. He knew that some of the young does of the year had already eluded him and got to the gatherings of the males. Already there were minor passages at antlers, for the possession of these escapes. But these does were too few for the increasing herd of hunting males, and they followed up the wind to the old one's chivied wives.

So he must turn, and in the moonlight show in a hideous grin his hatred of those younger males that gathered in a waiting, wavering row, their hindquarters deep in the pool of forest darkness and the safety of retreat.

Then a challenger would step forward and bring his muzzle down to the luscious river of doe scent on the grass. The prongs of his antlers pointed then directly at his foe. They were twice as terrible as the buck deer's, branched like the snagged tree that tears the bottom out of a shooting canoe, and between the mighty arches of the shafts sprouted the two brow tines and the two bez tines. These were the weapons of the close attack, and the moonlight sharpened them dangerously.

The scream of the challenger was answered by the down thundering of the old Turk's charge. They met with a crash and a shock that sent the other rivals scattering, plunging, snorting away in the greater desire of escape. There was a sound of the snapping of the slenderer tines, and the harsh wrangling of the locked branches. They swayed and pushed and panted, and the does looked back from their cropping with the soft eyes of the enslaved. The old lord found some second strength; he began to thrust his enemy—his own son—back with a measured merciless science. The young buck reared back for the breakaway and found that his tines were snagged in his conqueror's. The fighters sensed their mutual danger; they rolled their heads in one frantic futile purpose. The fight staggered and crashed into the darkness, and the young bucks came back, and snorting and screaming they cut the harem this way and that, driving off the does by fours and fives and sixes, stopping to battle with each other and losing their favorites to a third. So under the eyes of a moon sagging in harvest orange toward the west, the wild irregular mating went on and was repeated on the next night, and the next, until the old lord and his challenger staggered dead in the forest and the young bucks, spent with their revels, sick of them, ravenous for grass, left the does some peace at last, left them to follow, chained now by the unrebellion of their new state.

Some Scenes Years Ago at the Dunes Recalled with Great Pleasure

Donald Culross Peattie

ৎৡ ৎৡ ৎৡ Two requests for information of a highly personal nature have come to me which I have failed to answer. Now I am twice again challenged, and though I blush something will have to be done about it. The substance of these communications is simply this: Who am I? Perhaps some of them mean, who do I think I am? Am I the son of my mother and father? Could I be the impossible child they seem to remember, playing on south side streets?

The answer in every case is yes. Without being yet aged and infirm, I remember what is now Rainbow beach when the rainbow bathing suits were not seen there, and there was, rather magnificently, nothing but the beach. I remember the dunes there, and the gracile wild oat and marram grasses growing on them. I remember the thickets of crab apple behind the dunes,

Donald Culross Peattie, "A Breath of Outdoors: Some Scenes Years Ago at the Dunes Recalled with Great Pleasure," *Chicago Daily News*, December 15, 1936.

where on the gnarled dwarfish stems, amid the ferocity of thorns, the delicate pink blooms burst so miraculously in the late, wind-smitten Chicago springs. I remember the marsh behind our back fence, where Lady ferns grew, and the big prairie, where old Jim Bushnell had his farm, and the shooting-stars sprang out of the sod like an earthborn miracle in May, and the unseen ventriloquist meadowlarks whistled eerily from beyond the sloughs flowering with wild blue flag. Gulls and sandpipers calling and wheedling on the beach; flickers in the black oaks; the lisping chatter of the great cottonwoods even on breathless days, and the lashing of the big swamp willows in the storm. I heard these things before our street became a roaring river of motors, before the fire engines had sirens, before the church in the next block had a bell.

No more than any other youngster did I suppose that Illinois nature—its wildflowers and its birds, its sounds and smells—was romantic. I had to go far away and see some of earth's wonders, and come back, to appreciate that there is a tender beauty about this marshy plain where Joliet portaged from the Checagou. Now I am one of its lovers. Will that suffice as my letter of credit?

〰〰〰 DONALD CULROSS PEATTIE (1898–1964) WAS A PROLIFIC writer who possessed both scientific training and the voice of a poet. Born in Chicago to parents who themselves were writers, the young Peattie endured a youth plagued by illnesses that forced him into the gentle province of reading. But a stay in the Great Smoky Mountains of North Carolina improved his health and forged a love of natural history.

Peattie entered the University of Chicago as a student of French but left to follow his parents to New York City in 1918. His interest in botany rooted while at the university, but a visit to the New York Botanical Garden changed the trajectory of his life. He transferred to Harvard University, graduating in 1922 with honors in the natural sciences. He also received the university's Witter Bynner Prize in poetry.

His first job out of school was as a botanist for the federal government, but he left after two years to pursue writing full-time. During this period, he married his high school sweetheart, Louise Redfield, who, besides being a writer herself, was the great-granddaughter of John Kennicott. Although he was getting some of his nonfiction published, Peattie wanted to concentrate on fiction. Joining a trend popular among American writers of the time, the Peattie family (which by now included children) moved to France in 1928 with the financial aid of Louise's mother. Both the Peatties produced novels during their five-year stay, but none of them were particularly remunerative. Eventually, they returned to the Kennicott family home at the Grove, where they stayed for three years. (Peattie said for the duration he was never gone for as much as overnight.)

Perhaps they should never have left their home state, for Peattie scored his first major success in 1935 with *An Almanac for Moderns*, a collection of his daily musings on natural history topics. It provided enough income for him to build a house in Santa Barbara, California, where the family relocated a few years later. He followed that book with two of his most popular, *Singing in the Wilderness* (about John Audubon) and *Green Laurels: The Lives and Achievements of the Great Naturalists*. These three books firmly established Peattie as one of the twentieth century's premier nature writers. Over the next twenty-five years, his output was prodigious, with over twenty books alone, all but one of which were nonfiction. Hal Borland wrote of Peattie's two volumes on the natural history of trees: "They combined his scientific knowledge and his lyric gift with words in one of the most eloquent and authoritative studies ever written of America's woodlands."

In my opinion, if Peattie's only book was *The Prairie Grove*, he would be deserving of lasting respect. (The *New York Times* obituary does not even mention it, and Ann Keene incorrectly refers to the book as a collection of essays.) This unique melding of fiction, history, and natural history tells the story of the Midwest by focusing on one wooded tract, modeled after Kennicott's Grove in Glenview. What makes it an extraordinary success is the power of Peattie's language and the depth and breadth of his knowledge. He is obviously steeped in the old texts that tell of cougars, martens, and bison, and the human beings who were their fellow tenants. But he was also a field biologist who understood the phenological changes that endear us to this great ecotone. Keene writes that most of his works are now forgotten. She may be right, but what a shame. As appreciation of the region's biodiversity grows, so perhaps it will be with *The Prairie Grove* and other of Peattie's heartfelt odes to the natural world.

Sources: Hal Borland, *Our Natural World* (Philadelphia: J. B. Lippincott Co., 1969); "Donald Culross Peattie Is Dead; a Leading Naturalist and Writer," *New York Times*, November 17, 1964, 41; Ann T. Keene, "Peattie, Donald Culross," *American National Biography Online* (Oxford: Oxford University Press, 2001).

Chicago Park District vs. Arthur Canfield

Illinois Supreme Court

᭐᭐᭐ The public are paying for their [parks] establishment and main-tenance, not primarily as means of travel but for the purpose of establishing places for recreation in surroundings pleasant to look upon. A householder who has spent a day in the office, store, or factory, does not take his family to the railroad or industrial yards for a picnic lunch or to rest his eyes on pleasant scenes, but shuns such places. He chooses a park, for there the family may enjoy rest, beauty, and quiet, or games and exercises if they choose. It is now recognized that in large cities where great multitudes of people live, and com-mercial pursuits find it necessary to combine into a comparatively small space much of dust and strain and noise, and to erect on land so used many large grim-walled buildings, the wear of business and industrial pursuits takes a large toll in health and life and clean enjoyment. It indubitably follows that to secure surcease from those strains, and a chance to develop or gratify a love of natural beauty, it is a proper function of government to provide places in such a community where commercialism, unpleasant noises and scenes are elimi-nated. Experience has shown the public parks have not only contributed to public welfare, generally, but have specifically and directly restored and im-proved the public health. It is generally well-known that the natural beauty, quiet, and opportunities for exercise and amusement to be found in the public parks, away from business, and the noise and bustle of industry and trade have been the tonic to better health and an inspiration to better living. Cer-tainly these things are not only worth while but are important.

◄◄◄◄ IN *CHICAGO PARK DISTRICT VS. ARTHUR CANFIELD* (370 ILL. 447), a 1939 case, the Illinois Supreme Court examined an ordinance enforced by the Chicago Park District that made it illegal for a vehicle bearing advertising to ride through the city parks because it contributes to "objectionable travel and traffic." (The Arthur Canfield referred to was head of the local soft drink company that bears his name.) Justice Stone, writing for the majority, held that although "governing authorities may exclude any use of public property which is inimical to the purpose for which the property is maintained," this

---

Chicago Park District vs. Arthur Canfield (370 Ill. 447), Illinois Supreme Court, 1939.

particular law was flawed because it was too indefinite and left too much discretion to the enforcers.

<hr />

Siftings

Jens Jensen

ⁿⁿⁿ ... When I first set foot on Illinois soil, the buffalo still roamed the western prairies, and the Indian still dared assume his rights against the white man. The primitive prairies of Illinois have not been entirely destroyed. Here and there has been left something of the primitive that the plow had not turned under. It seems a pity, rather a stupidity, that some section of this marvelous landscape has not been set aside for future generations to study and to love—a sea of flowers in all colors of the rainbow.

Along our railroad rights-of-way one meets the last stand of these prairie flowers. What a wealth we would have if our prairie roads could be lined with this rich carpet of colors, miles of flowers reflecting their colors in the sky above, or millions of sungods (sunflowers) in the strong prairie breeze nodding their heads to the sun that had given them their golden hue. But perhaps this is too much to hope for, as man seems unappreciative of these gifts.

One evening at the opera in Ravinia, an outdoor theater, when my thoughts were not much in sympathy with the foreign opera, as I did not understand the language, my eyes shot across the audience into an open glade in the woodlands where the brilliancy of the goldenrod in the path of the sun's afterglow gave to me an illumination worth a million operas, and I silently wished that those about me might have seen and felt this spiritual message from their native soil.

It goes without saying that many of these flowers which are so beautiful and fitting on the open prairies are often disappointing in a shady situation in the garden. Light and shadow and scale play a principal part in the art of making gardens. Some years ago I was honored with visitors from the University of Illinois. Amongst them was a famous photographer. He was looking for certain plants he wished to photograph, especially the tiger lily and the prairie phlox. The lily that once covered large areas in certain sections had practically disappeared. Only small groups or single plants were found.

<hr />

Jens Jensen, *Siftings* (1939; repr., Baltimore: Johns Hopkins University Press, 1990), 56–61.

The phlox that once was so numerous on the wet prairies nearby had entirely disappeared.

On a cultivated meadow we found one plant. The professor disagreed with me that the plant he saw several hundred feet away was a phlox. I insisted on investigating, as I was sure I knew the plant well. We found it to be a phlox, a plant the plow had missed. The professor then determined that it was a new variety. I took my hat and placed it between the sun and the phlox, and the color changed. So the high priest learned a lesson that day. The lack of a knowledge of this difference has caused many mistakes in gardening.

The early creeping phlox often covers a sandy slope so densely with its white flowers that it looks from a distance like snow, but placed in a crowded border it would mean nothing. Its color value and its real beauty would be entirely lost.

It is not always the great outburst of color, like the blazing light of the setting sun, that touches us most deeply and brings into growth beautiful thoughts and ideas. The little dune violet, that hides away its skyblue face from the gaze of man and the tramping feet of the many, speaks more profoundly and sings more deeply into the soul of man than the gay cactus plant with its large yellow roses covering the sunny slopes of the duneland within a stone's throw of the shy violet.

Never shall I forget a May day in the woods of northern Wisconsin when snowdrifts were everywhere. Along one of the snowdrifts, on a little sunny slope, the trailing arbutus was in full bloom. Years ago I had seen flowers alongside high snowdrifts in the Rockies and had marvelled at the close association of spring and winter. Now I had found the same lesson down on the plains of Mid-America. If you want to see the trailing arbutus at its best, where it sings of spring and sings of winter, you must see it in full bloom alongside a snowdrift.

I might go on indefinitely speaking about these many plants I have met throughout the years, each having a sermon for me. There are but few plants that do not love company. On the other hand, most of them are particular about their associates. The spiritual message or character of the individual plant is often enhanced by its association with other plants which are attune with it. Together they form a tonal quality expressed by an orchestra when certain instruments in chorus bring out a much higher and a much finer feeling than a combination of others. The different plants are then given a chance to speak their best.

I have often marvelled at the friendliness of certain plants for each other, which, through thousands of years of selection, have lived in harmonious relations. But I have also found through years of experience many plant combinations that are not friendly. Years ago I experimented with a great many varieties of spring bulbs. They were planted on a slope of our ravine, which I thought was ideal for them. Some soon disappeared; others have remained for more than twenty years but have not visibly increased. Others have run all the way to the bottom of the ravine, and on their march they have completely destroyed a stand of maiden-hair ferns. That I made a mistake was evident.

When travelling through Maryland and other eastern states, I have been amazed to see the disturbances caused by the introduction of the Japanese honeysuckle, not only to the forest floor but to small trees and shrubs. This shows the ultimate danger of transplanting plants to soil and climate foreign to their native habitat. The great destruction brought to our country through foreign importations must prove alarming to the future. Many of these importations will in time become the sparrows of the plant world and destructive to the beauty which is ours.

The motives and compositions in our native woodlands, our hills, our valleys, our river bluffs and our swamps, our cold and rugged north, and our sunny south are unlimited. Such wealth and such refinement speak well for the art of landscaping in our country. Examples have been given, but before a plant is used in the composition, it must be tested—whether it likes to be alone or in a group, whether it enjoys lowlands or highlands, whether its character and color sing with its surroundings.

My departed friend, our great western poet, Vachel Lindsay, had honored us with a visit. It was early morning when he called me to the open door where he was standing looking out over a clearing. There was a peculiar light over this little sun opening, caused by the reflection of the sunrise. The clearing was bordered by a simple composition of hardwoods with a few hawthorns, crab-apples, and gray dogwood scattered on the edge. The light had added an enchantment to this simple composition, and Lindsay, watching this, said to me, "Such poems as this I cannot write." Many years have gone by since then, many mornings and many evenings, and I have watched the clearing. I have seen it on cloudy days and in full sunlight, in the starry evenings and on dark nights and moonlight nights, but I have never seen it the same.

〰〰〰 JENS JENSEN (1860–1951) WAS BORN IN DENMARK BUT BECAME one of the greatest admirers and defenders of the midwestern landscape. As a child, he roamed his family's farm on the west coast of Slesvig, a duchy claimed by both Denmark and Germany. According to Robert Grece, little is known about Jensen's early years, but from Jensen's writings it is clear that he became imbued with a deep affection for nature during that period: "This great love for the out-of-doors, for its history and its beauty and its spiritual message was... woven into the lives of my people."

When Jensen finished his time in the German army in 1884 (it was his fate to come of age in a period when Germany controlled his ancestral home), he made up his mind to move to the United States. Grece notes that this decision was not based on the desire to seek political freedom or economic opportunity, but rather to escape the disapprobation of his family for marrying beneath their class. (He also had no interest in running the family's farm.)

A couple of years after arriving in this country, he procured employment as a street sweeper for the Chicago West Parks District (this was before the Chicago parks were consolidated under one district). Jensen's innovative use of native plants, and his skills in solving problems and working with people, propelled his rise to superintendent of Union Park in 1895. A year later he was promoted to head Humboldt Park, then slated for expansion. By 1900, however, he was let go because of his opposition to the corruption that had taken hold of the parks.

He stayed busy promoting park reform and the preservation of local natural areas. (He authored an important part of the 1904 *Report of the Special Park Commission*, which was a key element in the creation of the Forest Preserve District of Cook County.) In 1905, with reformers in control, he was made superintendent of the Chicago West Parks. Two of his greatest projects were the development of Humboldt Park and Garfield Park. After another five years, Jensen began devoting more time to his private practice. Over the course of his career, Jensen worked on well over six hundred projects including parks, private residences, hospitals, subdivisions, visitor facilities, golf courses, schools, and corporate campuses.

Of equal importance to Jensen's legacy was his advocacy on behalf of the preservation of natural areas. Early on he would take his family on excursions to the Chicago hinterlands. He helped found two of the region's first conservation organizations, the Prairie Club and Friends of Our Native Landscape. Through these efforts he became confidant of many of the area's most influential conservation supporters, including Henry Chandler Cowles, Dwight Perkins, Lorado Taft, Stephen Mather, and Mrs. Julius (Augusta Nusbaum) Rosenwald.

The first of his major conservation endeavors reached fruition in 1913 with the establishment of the Forest Preserve District of Cook County. Most of the areas he identified in his previously mentioned 1904 report eventually became part of the district's holdings. His second major project took longer to accomplish, and that was the protection of the Indiana Dunes. He asked Julius Rosenwald to purchase several thousand acres of the Dunes for the creation of a school of horticulture, forestry, and landscape design but was turned down. Later, Henry Ford showed more interest, but he, too, eventually backed off. (I have mused that Rosenwald was one of the country's greatest Jewish philanthropists and Ford was to become one of the country's strongest supporters of Adolf Hitler, but they seemed to have agreed on the Indiana Dunes—it was a beautiful area worth preserving but not by either of them single-handedly.)

With the death of his wife in 1934, Jensen moved to Door County, Wisconsin. There he created a school "where the wilderness and the cultivated meet," and students "have a deep feeling for the art [they] want to study" (Grece 140, 143). Named "The Clearing," Jensen worked tirelessly to keep it operating. Fortunately, after his death, loyal supporters of the school have continued to ensure that Jensen's visionary school continues to instill the values of its founder.

Source: Robert E. Grece, *Jens Jensen: Maker of Natural Parks and Gardens* (Baltimore: Johns Hopkins University Press, 1992).